# The
# ineffable
## and what that has to do with humanity

Deconstructing
the identities of gods
and man

## By M.R.Holt

authorHOUSE

*AuthorHouse™*
*1663 Liberty Drive*
*Bloomington, IN 47403*
*www.authorhouse.com*
*Phone: 833-262-8899*

*Published by AuthorHouse  06/28/2023*

*ISBN: 979-8-8230-1114-3 (sc)*
*ISBN: 979-8-8230-1113-6 (e)*

*Print information available on the last page.*

*This book is printed on acid-free paper.*

# INTRODUCTION

This book was written to aid those deconstructing religion and resolving its corresponding traumas. It's intended audience is those leaving the Christian faith. I say this as a warning, because as hurtful as those traditions have been and as many people as they have hurt. They have also helped just as many, if not more. And tho (as you will soon discover) the existence of some cosmic hell or eternal torment is highly unlikely and even virtually impossible. The philosophical one we face by leaving religious paradigms is very real. You don't leave the comfort and security of such religious ideas unless you have to. But for those who have been hurt or refuse to be complicit in that abuse perpetrated by religion and its blood thirsty god, then this book is a victory cry. For those abandoning a hierarchical system in favor of a more inclusive and wholistic one. There is no choice but to leave. This book is for those leaving their faith in the name of truth and compassion. Tho no one is coming to save you. There is hope and virtue beyond the existence of some traditional conception of god and man.

"I truly believe that this world will be fixed when we learn to appreciate the "broken things" we were taught to regard with fear or hatred. When we embrace the chaos which threatens to destroy only that which divided us against them, and gave us belonging in a hierarchy by separating us from the equilibrium of the whole. For Every lie we told because we were taught our authentic selfs were something to be ashamed of and would not be accepted, let us incite chaos! Because intolerance of mutability is not a sign of immensity but rather one of insecurity." I truly believe in a love with the capacity to validate every expression of life simply for being what it is and not dependent upon its utility."

To start off let us lay the foundation.

Order is created by establishing identity, assigning meaning and value to things. Labeling this as holy and that as cursed, this as belonging here and that as strange. This is why faith is the president for obtaining salvation "Sotor" (the Greek word for preserve) as a means of solidifying or holding together those values preserved thru tradition.

We see this mythically depicted in the genesis poems in the role of "god" as the one who separate the heavens from the earth, the holy form the cursed, And the men from the beasts. We see this again embodied in the ark which also classifies its species of animal from the incalculable flood of chaos (depicted in the genesis poem as the primordial chaos waters or abyss). Or at the Tower of Babel where difference in definition divides mankind. This idea of order here is not a manifestation of mater but an distinction between its states. And thus Chaos only threatens to destroy the boundaries which divide the whole against itself. I would here warn my reader. Because as is the case with all knowledge, also comes the threat of death to our illusions of identity and meaning. The revelation is also an apocalypse.

I have no desire to take a faith which has comforted and guided. but rather to comfort those who have been hurt by these identities of god an man. And these values which have striped so many of self worth. And so This book shall be a dissection of those identities and phenomena which have often been attributed to god and godship.

This is a continuation of the discussion started in the book (what it means to be human and what that has to do with the ineffable)

And If the last book accounted our exodus from Egypt and exile in the wilderness, then in this book we shall be going toe to toe with Jericho and acknowledging that there is in fact a land of abundance on the other side of that impregnable city which is religion. Because after all, we did have to leave Egypt but we were never meant to stay in the wilderness. The goal has always been a bigger more inclusive, comprehensive and compassionate world view unrestrained by those structures and distinctions which had taken the diversity and regarded it as dichotomy. And so without further to do. Let us begin

In it's most ancient forms the cosmology of gods ranged anywhere from deifying ancestral and political figures to attributing mental agency to systematic phenomenas in our environment. And tho I will not be giving much authoritative credence to these patriarchal perceptions of god solely intended to retain authoritative designations of meaning and value thru the omnipotent claim of power and might and often establish thru the divine right to rule of some conquering king or Caesar. I will be going into much detail in regards to the seemingly prolific muse posited in our subconscious. Here I favor the holy messenger who invites us to grow beyond the malevolent master who requires glory and praise. As my reader will soon discover, this claim of being "made in the image of god" has far more to do with our ability to negate our instincts and govern our own nature and behavior, then it dose with being physically constructed in gods likeness. This does not denote the "design" by which our anatomy consist, but rather that as my reader will soon see, our mutability played a more positively vital role in our development then it did a derogatory one. For we like all life on earth were not merely manifest but instead evolved slowly and gradually over trial and error. For which there is ample evidence to affirm this widely accepted claim.

And so we are not seeking a creator so far as that pertains to immaculate conception. but instead as that of an incessant muse and messenger persistently prompting us towards more comprehensive and compassionate forms of life.

# DISCLAIMER

"One of the most beautiful things about deconstruction is the new found grace to be wrong, to change your mind, and to ultimately find better ways of understanding and interacting in and with the world. Because when we are, no longer required to be right all the time, we are free to do what's right."

Dear reader There is virtually no turn of phrase which I have constructed that I haven't then come back some time later and rearticulated in a better way. And yet without the tribulations of the journey there can be no final destination. However I would argue that a finished conclusion is a myth, because it assumes we ever understand anything as it relates to everything else. Of corse I'm going to be wrong. And I don't really think it's a problem to make a claim in one book and then come back 6 months later and have completely changed your mind on the subject. In fact I think it's more problematic if you don't alter your world view in light of new information. That does not then make my previous statement false. but rather only partially incomplete. A problematic conclusion is not necessarily bad. but any conclusion that claims to be final or omnipotent is more then problematic but incomplete. We are not subject to our beliefs, we get to change our minds and we should ideally constantly be confronting new information which often seems contradictory to our previous conclusions. This is why it is important not to label ourselves in accordance to our beliefs and restrictively attach our meaning to a paradigm. I will miss speak and I will also speak truth that contradicts previous claims which are also true. And I will do this simply because I am more concerned with being honest then I am about being "right". We are here to discover and rediscover life, for that is the very nature of life eternal or "true life" as Jesus use to call it. Language changes, meanings change, all living organisms evolve and adapt, and further more so do healthy beliefs. Ultimately this idea of perfection as stationary, static, and sterile, is innately flawed. If my reader is looking for a binary system, then it should be clear that you have come to the wrong place. But if my reader is looking for a land of abundance and truth unrestrained by a binary view of life. Then this is the book for you. This book shall function as an imperfect attempt at finding a more inclusive and comprehensive world view then that of a dichotomous belief system. With that said, I welcome you on this messy journey as we ponder the question " what is the ineffable" (often depicted as god). It is because of the robust nature of this subject, that our primary concern will be constructing a more comprehensive and wholistic depiction of this unquantifiable quality or entity. I will do my best to articulate this subject as thoroughly as I can, so long as it does not inhibit the connectivity of this discussion. And so I will ask my reader to be open minded. This discussion is not about keeping score but rather about encountering the here in now without the predisposed criteria of our past to prohibit our movement into the future. And so without further to do what is the "ineffable" ? And what does that have to do then with humanity?

First things first. We must destroy the great walls and barriers before we can answer this question "What is the ineffable?" And enter This metaphorical land of abundance.

If you read my last book then you'll remember that this term "ineffable" is a title used to describe anything that cannot be quantified, however more commonly this word is used to describe god. Tho to be honest, for someone or something which can't be understood, this character "god" has most definitely been the center of many religions and cultures claims of identity. And understandably so seeing as how many scientists and anthropologist have concluded that in many cases this character is actually just the deification of ancestors or attempts at explaining the agents posited in nature. However still some claim that this "ineffable" entity is far more illusive than those somewhat trivial explanations. And that's fare! However this ineffable character still has not shown themselves but is instead depicted in a broad array of expressions across virtually every culture that ever existed. Each with their distinct claims about gods nature and character which often contradict others. It should be noted that this bias to attribute mental agency to our environment is innate to all humans. And that can't be overlooked.

Truthfully The very fact that this title "god" is then labeled "ineffable" just adds ambiguity to an already antiquated idea. And so It's no surprise then that there have been many disputes surrounding this title, name, and its corresponding claims! And so In this publication we shall examine a number of these claims with the soul goal of finding those traits which are present cross culturally and are not then restricted to one nationality or the other. But first let us look at the common consensus among scholars.

Which quite simply is that "god" (at least as we know them) is a figment of our own anthropomorphisms. So then as it pertains to the subject, the ineffable has everything to do with humanity! And it is what makes us "humans" so special!

Quite recently in our evolutionary process we came in to possession of ample resources which then allowed us as a subset of species to allocate the necessary energy required for us to develop a big thinking mind. This mind gave us the ability to not only construct complex languages but more profoundly to negate our instinctual habits. This perplexing proclivity to question our biological programming and genetic prerogatives came with a side effect, or perhaps this rebellious spirit is the side effect, ether way it is here where we developed a consciousness or the phenomenon of us become self aware (self conscious) if you will. Now mythically this transgression of instinct and awakening of our psyche is depicted as the fall of man in the garden. As it is here where we impose our rationally critical thinking mind over our instinctual submission to nature. However it is this very defiance of (gods command) that then allowed us to bare the image of the divine. For we were no longer mindless slaves to our instincts or nature, but rather had obtained the sacred knowledge required to decide for ourselves what our nature should be. And yet we still hear the profoundly prolific messages (the voice of god) echoing thru our subconscious intuition. And why shouldn't we listen to their wisdom? For that vast trove of information which we are prevê to thru mindful meditation and prayer has most certainly served us well for millions of years. And so this wisdom found in ancient texts, interpretation of dreams, and meditations of mindful presence, are most certainly posited with directions for navigating and interpreting life's cyclical nature. Or as Jung put it "the abstract image/symbol is fare more accurate and complete then our empirical attempts to quantify them". And so this is where the Bible comes in. Because this collection of ancient texts and traditions are not univocal in their solution but most definitely account an ongoing discussion about the nature of god. This was our early attempts at quantifying that intuitive voice and those instinctual habits. But more profoundly this was us attempting to define the nature of the universe and subsequently ourselves. We see many stories accounting this struggle between our instincts and our cognition. These stories brilliantly depict our denial of mindless subservience to instinct and culture (or in other words the insistence that we don't let our emotions or our adherence to an idea control us. But instead we ask the question why do we act or feel the way we do.) Ultimately we decide what kind of image we want to bare! Rather that is one of a malevolent master or that of a compassionate companion. It should not then come as a surprise that much later when this rabbinic figure who inspired a movement of people leaving their religions and abstaining from sacrificial practices all together enters the scene. They would then be deified for simply reiterating what those before him had already stated, which is that we don't need a pharaoh or Caesar to envisage the image of god but rather that that quality is bestowed on even the least of these. And more importantly that this holy hobo demonstrated this autonomy thru embodying wisdom and compassion. And so in accordance with this movement we find these writings written or transcribed 15-70 years later (after Jesus's death) attempting to rationalize these heretical claims with the progression of society. Because after all god has not come down in all their glory to defend their identity but instead has left it up to us to decide their character thru the images we chose to bare. And so we are left to answer this question, is nature nurturing? Is the cosmos compassionate? Is god good? And Are we any of these things?

I realize that may have been a lot to process so allow me to reiterate that in a slightly different way.

The psychologists Carl Jung deduced from his study of the unconscious mind that man first acted instinctually and only later did we question our behavior. And thus The distinguishing quality pertaining to humanity is our ability to negate that programming. This awakening (often depicted mythically as "the fall of man" in the garden.)

is actually just our ability to override our instinctual prerogatives, however this is not a fault! For here lies our role as image bearers of the divine. For unlike all other creatures who are doomed to obey their nature, we have a say in ours! Or in other words we decide who god is.

Archetypally the nite or (unconscious darkness) is where the wild things dwell. and the day (conscious light) is where man abides. And thus Jung claimed that the goal is not to surrender wholly to our rational mind or our intuitive heart but rather for nature and culture to live in equilibrium such as is mythicized in the garden. He concludes that there is profoundly prolific wisdom in our instincts which we are privy to only in our dreams or prayerful meditation. And as heretical as this all sounds this is exactly what we see in scripture. from verses like Deuteronomy 32:8-9 and psalm 82 which presents ancient cosmological claims about gods identity to writers like Jeremiah and Ezekiel who then argue with the writers of exodus about the nature of god. The entire Bible is comprised of an ongoing discussion deliberately arguing with earlier traditions and competing cultures about the nature of god. Because after all god has not shown themselves but instead has left up to us to decide who god is thru our deliberate defiance of our instinctual nature for ether more malevolent masters or more compassionate companions. And this brings us to the tyrannical rule of traditions. because ultimately the laws constructed by religion were even too encombersome for the Christ when it came to sabbath and honoring familial ties. And so it's no wonder we see these "laws" condemn figures like Abraham and Moses who contend with the ineffable on man's behalf. to those like Nietzsche and Jung who earnestly pursued the nature of god thru science and philosophy. Now This is not to say we are god or even the universe witnessing itself. but rather that we decide for ourselves our own nature and are not resigned to the instinctual voice of god or the traditional written word to dictate our identity. We can still find profoundly prolific wisdom written in the stars of our unconscious mind as well as in the sacred traditions and epiphanies of days past, but ultimately we decide what image we bare. Because after all God has left their character up to us to decide.

Rather it is the spark that lit the fuse for the Big Bang or the muse which continually entices the progression of nature and culture thru evolution and revolution. The role and parameters of god far exceeds the manifestation of the material but the continuous process of creation and recreation.

"The primary purpose of prayer is not to change the will of god but the heart of man into accordance with god"

(Soren kilkaguard)

# POEM

Blessed is he who seeks god within his own religion. for he has yet to see god. Only he who has seen the face of god is then compelled to abandon his religion. If a man dose not find what he seeks, then he shall continue to search it out. It is only when god themself comes down in all their glory and says to man there is no god, that man then becomes an atheist.

In my book "what it means to be human..." I argue that one of the pivotal roles of the messianic figure was as the interceder or one who wrestled with the ineffable on man's behalf for more compassionate representations of god then those presented by the cultures and traditions of their day. J. Richardson Middleton Makes a similar argument in "the silence of Abraham" where he presents the reader with an alternative interpretation of "the test" where Abraham actually fails by merely obeying god's command to sacrifice Isaac, rather then wrestling with god like he did on lots behalf in sodom. He makes the point that if Abraham had kept arguing with god, god would of sparred the city on behalf of one righteous person. There are of corse many interpretations regarding this "test" in genesis. Some argue that the sacrifice of Isaac was a display of gods justice on behalf of Ishmael and Hagar. Others interpret this as an anthropomorphic claim by Abraham or even as a literary tool by Moses declaring that their god does not require human sacrifices like the gods of their ancestors. I personally don't think any one of these interpretations are inherently better then any other. but function as a brilliant representation of the Texts purpose, not as an inerrant authority but as an interpretive guide. And here in lies the primary goal of this literary work, not to function as a proof (ether for or against) the existence of god. But to represent both arguments as an invitation to abandon the binary world view and our subsequent identification with one or the other (theist or atheist) but capable of entering both respectively.

With that said, this book is not constructed linearly but instead invites the reader to experience the whole all at once, and how each part fits from many different angles.

# CHAPTER 1

# THE GODS OF OLD

The modern term atheist most commonly refers to an individual who has attached their identity to the belief that there is no such thing as a god. However this can also simply refer to someone who has yet to find ample evidence for the belief in any specific god. And even still this term may be used to refer to an individual who does not adhere to the corresponding conceptions of god according to their culture. This last definition is perhaps the most fitting in its adherence to antiquity. As It was in fact this definition which then designated early Christians as atheists according to the Roman Empire. Because this group of people had abstained from sacrificial practices and oriented their worship very differently then all other religions of their day. because they didn't require a temple to commune with god but instead claimed that they themselves were the temple. This understandably produced problems in the community especially when it came to beliefs on eating meat which had been prepared in sacrificial practices to other gods. And the fluidity of gender roles in worship which oddly enough were not as strict in the early churches as they are in churches today!

However oddly enough this word atheist was used very similarly to the way it is used today. Which quite simply is to describe anyone who doesn't subscribe to the prevailing traditions or religion of an corresponding culture. Or in other words anyone who didn't believe the same as me and mine.

However this then entirely depends on your culture and where you grew up. And so this begs the question "is god restricted by culture?" Because that sounds more like a god made in man's image as opposed to that of a god who made all mankind in their image! Ultimately according to this logic you are in danger of being labeled an atheist if you disagree with any cultures corresponding claims about god. And yet very culture makes claims about who god is that deliberately contradict those of their contemporaries and predecessors. Even the biblical texts don't adhere to a univocal conception of god.

It should be noted that there is no one single univocal standard for governing the Christian traditions but rather a divers dichotomy of interpretations and hierarchies when it comes to any particular verses authority and meaning. There are countless denominations and even exponentially more so interpretations of the texts meaning. Not to mention the fact that verses that sound archaic by today's standards were at their time and place excessively liberal and progressive. A god that gives people Saturdays off from work are still gods who claim man was made to serve god with all his/her heart, mind, and body. A god who does not require the sacrifice of sons is still a god who is incapable of meeting their own emotional needs and requires the worship of his subordinates. A Christ who died for our "sins" is still a human unable to live with their own insecurities and flaws (for not even Christ kept the law in its entirety). And so I will make accusations that on the surface may or may not apply to your particular pedigree of Christianity, but is firmly rooted in the doctrine and texts pertaining to the Christian traditions at large.

It is because some of these traditions condone parents marring their children off to grown adults. It is legal in 42 of the 50 states.

So without further to do let us take a look at these claims and their corresponding cultures.

Most animals on this planet migrate seasonally. Humans however have the misfortune of often spending their entire lives where they were born. And in part this is why things like nostalgia are so strong in hometowns. The phenomenon we call nostalgia literally sets the base line of preferences in environmental conditions. And so We were quite literally made for those environments. However the fault here is that the greatest proponent for growth is a changing environment. And Like most organisms on this planet, people change when their environments change. we are more equipped for change than any other creature on this planet because our genes developed a mechanism called consciousness which allows us to monitor our ever changing environments and then audit/adjust behaviors accordingly in real time. This reappraisal often deliberately challenges traditional narratives which gave us identities biased on stationary assessments of our environments. However much like those assessments of fish and birds which on a genetic level have far more in common with other terrestrial species then they do their aquatic/areal counterparts, so to these isomorphic and binary identities assigned to people are just as erroneous. This does not undermine the profound wisdom instilled in traditions but rather aspires to understand the crucial tools like collective effervescence which practices like recitation of prayers and hymns utilize, so that we can keep these evolutionary mechanisms but discard all the prejudice, and superstition associated with them thru religion. Admittedly science has presented us with some very uncomfortable truths but it has also given us viable solutions for the problems of our age. And tho religion may make us feel good, it very seldom actually address the environmental conditions which cause the problems In the first place. We must do more then scratch the itch! But do the work to grow and heal the hurt in ourselves and in our environments. In truth people have such capacity for experiencing life, but far too often we subscribe to the narratives of tradition and sell our souls to religions, inadvertently relinquishing our autonomy and identities in the process. In nature hierarchies fluctuate with the seasons, it is only our traditions which insist that we find our belonging in exclusion form the whole.

## CAINANITE GODS

### El

El was the presiding high deity of the ancient Canaanite religions and is observed in many dialects of the levant such as ugaritic, syriac, Arabic, and Hebrew languages and traditions. In the Hebrew traditions el simply translates as god or spiritual being because originally "angels" were gods or god themself and only much later were changed/added to retain certain theological and rhetorical goals.

El can be equated with the Greek Cronus as the father of all other deities in (his) corresponding culture and tradition. Now later el or ll would be conflated with adony and even supplemented for adony in later iterations of the divine council. Where el appointed lands to his children (second tier gods), the much later adony distributed the gods to preside over the nations. There is ample evidence to the effect of El also known as the ancient one being an desert god.

### Bal

Bal (Hebrew for master and lord) was a storm deity attributed with controlling the wether and the blessing of fertility. Tho Baal su vull (lord prince) is often drawn upon by the ancient Hebrew traditions to describe their god, later authors transcribed the name bal hesabub (lord of flies) as satire.

### Asherah

Asherah or asratu was a fertility goddess of motherhood and lady of the sea (perhaps a kindred to the

5

summarin temat) she was also the wife or consort of El, bal,Yahweh, and many others...

## Mot

Mot was the god of death and the underworld and on a couple of occasions noted to contend with bal in combat. Mot is often eluded to as a ravenous lion in the wilderness or famine and plague. Mot is later inflated with the devil and the angel of death.(Habakkuk 2:5, job 18:13, Hosea, and Jeremiah,)

## Yam

Yam is the god of the sea and an adversary to bal. Yam is accredited with being the champion of El. Yam's dwelling is the abyss and he is attributed with the primordial chaos waters, yam is described as a mighty serpent or sea beast and equated with Tiamat and leviathan. In fact the battle between yam and bal shares similarities with both that of the later Greeks Poseidon and zuse as well as that of the Mesopotamian Tiamat and Marduk.(this is hi-lighted in the later chapters of job)

## Yahweh

Yahweh was a war deity worshiped by Israelite tribes around the early iron or possibly even as early as the late Bronze Age. Originally Yahweh was worshiped polytheisticlly along side El and bal and only later become conflated with the two as one in the same. Monotheism in Judah didn't become common until 586 BC and possibly even as late as the 2nd century.

Dagon

Dagon was a god of agrarian fertility and crop prosperity. This is made even more apparent in the entomological root of his name as Dagon (grain) and the root dgn (to be cloudy). This aside Dagon was also called the "father of gods" and associated with assigning royal legitimacy.

## EGYPTIAN GODS

Ra

Ra also known as Amun Ra was the head god of the Egyptian pantheon. (according to the myths) he was the first pharaoh of Egypt and was only usurped by his daughter (isis) after he was tricked in his old and feeble state. Ra's astral body or representative was the sun.

Isis

The goddess isis was a powerful magician and necromancer, and was most probably first popularized in the Osiris myth for resurrecting her slain husband Osiris. Isis is also famous for creating a serpent in an attempt to usurp her fathers throne. (Which she succeeded in by the way) and as the mother of Horus.

Thoth

Thoth was most likely the deification of knowledge as he is attributed with inventing written language and governing wisdom or knowledge. Thoth was a representative of ra and thus associated with the moon.

According to the myths Thoth is indirectly responsible for the births of Osiris, set, isis, and nephthys, as a result of a wager with the moon for 5 more days in the year. It was with these extra 5 days that nut and geb were given fertility. As the god of science and learning, Thoth possessed a sacred book of wisdom which also illustrated the perils of knowledge. (it is often our illusions or ignorance which make life tolerable) (a revelation is also an apocalypse "the end of the world as we knew or believed it to be") "sometimes to survive is to watch the myths which were once alive within you die".

## Sit

Set or Seth is the Egyptian god of chaos, destruction, and war. Set ruled peacefully with his wife/ sister nephthys until he was left by himself when nephthys betrayed and abandoned him for his brother Osiris. At which point set become the first god to kill another god (his brother) out of jealousy. Set was eventually vanquished by his nephew Horus and even forgiven by his murdered brother Osiris. Literarily set represents the rebellion caused by rejection similarly to the biblical archetype illustrated thru Cain.

## Ocyrus

Ocyrus or Osiris was the eldest son of geb and nut. In the myths, Osiris was preferred and loved by both his sisters (isis & nephthys) and thus invoked the spite and jealousy of his brother set (Seth) who murdered him subsequently making Osiris the lord of the dead "foremost of the westerners"(a term used to describe the inhabitants of the underworld). Osiris was briefly resurrected by his wife and sister isis and conceived a son by the name of Horus. Osiris was charged with ferrying and judging the dead (similar to the "biblical gods" executioner "the angel of death"). Prior to 2181 BC the pharaohs were believed to be the sons of ra. However after the Osiris cult gained influence this belief was discarded for a belief that Osiris was the means for salvation on behalf of the Egyptian kings ascending to the heavens, much like Osiris rose form the dead.

Horace
    Horus was the son of isis and Osiris and the vanquisher of his uncle set. Horus was attributed as being the split image of his grate grandfather ra (the god of creation and order), but Horus is also charged by Thoth with filling the role of is uncle seth after his (seths) defeat. Horus was the god of the heavens (sky) and kingship.

## SUMERIAN GODS

In the ancient cosmology The celestial bodies represented the gods.

Marduk
    Marduk was the Sumerian storm god of the sun and head god of the Mesopotamian pantheon. He is depicted biblically as nimrod and is credited with vanquishing the mighty Tiamat (primordial chaos dragon) and establishing order thru might. Marduk was the son of mami and enki. Marduks astro body or representative was Jupiter.

Tiamat
    Tiamat was an ancient primordial representation of chaos, creation, and the sea. She was typically feminine

and depicted as a dragon. Her husband was a dragon named apsu who represented the fresh water(in contrast to Tiamat's qualities of salt water "the sea") Tiamat was the mother of the Sumerian deities and following her demise at the hands of Marduk, her body was used to construct the heavens and the earth. She mythically functions as an illustration of creation dividing the whole (chaos) in order to incite order.

## Anki

Anki or enki was the father of Marduk and a god of knowledge, craftsmanship, water and creation. Anki has ties to ancient Akkadian (Ae) and Canaanite (la) depictions. Enki was the son of anu and Nammu.

## LATER GODS

## Pangu

Pangu is a Chinese deity who was meant to hold the yin and Yang of primordial chaos together, however instead pangu was awaken and thru their consciousness of such ambiguity, they were driven to separate the light from the dark, the heavens from the earth, and the ethereal from the caporal (yin form the yang).

## Vishnu

Vishnu is an Hindu god who originally existed in a state of "Brahman" (equilibrium between the material and the philosophical) according to the mythos Vishnu is awaken by a sound (the original sound). This sound inspired change in the form of creation.

This creation split the equilibriums state of Brahman into 3 distinct entities, Vishnu (the positive charge and preserver), Brahma (the creator), and Shiva (the destroyer and negative charge).

## NATIVE AMERICANS

The great creator (Shasta myth)

The great creator is an Native American god who created the earth (feminine) and then shaped humans form the mud. (Originally these "humans" were not distinct from the rest of the animals and could even nonverbally communicate with the creatures of the field). The natives recognized not only the intrinsic connection between all living creatures (trees included) but also the interdependence of all life upon one another. For this people nature was not so much an hierarchy but an partnership.

## GREEK GODS

Zuse

Zuse is a thunder and sky deity who presides as the head of the Greek pantheon. He ruled as the king of the gods on Mount Olympus and can be compared to the imagery depicting the much earlier Canaanite/Hebrew god whyh who's abode was mount Sinai. zuse was the son of Cronus who sought to kill him before he could usurp his throne. Zuse escaped the wrath of Cronus by using a decoy. And even usurped his father thru a rebellion resulting in the

11

deliverance of his siblings. However this deep generational fear of being overthrown by their children was inherited by zuse. Zuse presents the reader with an interesting character study as we see him exert power out of fear of losing it, and his display/ need for control is a direct response to this "higher god" being completely controlled by his impulses. We see his fear lead him to devouring his wife Metis and his conscience cause him to regurgitate his daughter Athena who would become his favorite child. From his sexual conquests and objectification of people (himself included) to his demands for omnipotent authority and control, he functions as a brilliant representation of the empirical ideas of Greice and later Rome.

## Hera

Hera was the sister and wife of zuse and thus attributed as the queen of the gods. She was the goddess in charge of protecting women during child birth. She is most noted in the myths for her jealousy and vengeance towards the women her husband zuse cheated on her with as well as their offspring.

## Apalpo

Apollo is the illegitimate son of zuse and Leto and the brother of Artemis. Apollo exemplified the ideal for young Greek men as the studious god of everything form archery to music and poetry. Apollo completes the archetypal model for Greek families.

## Prometheus

Prometheus is credited with stealing the sacred power resigned for the gods and gifting humanity with fire (technology and science/knowledge) he is the champion of humankind and was subsequently punished with eternal torment at the hand of Zeus and the other gods who never cared about making the world or life better for all but instead controlling and exploiting those they deemed lesser. Prometheus presents the reader with the archetype for the heretic as the one who usurped the malevolent authority of the gods in favor of those who had been abused by the gods. Note (aside from being a major technological advancement to ancient man, fire pathed the way for our ascension to divinity thru the ability to cook food (often thru sacrificial rituals) and subsequently reallocate recourses for the development of our frontal lobe (and consciousness /self governance).)

## Hades

Hades was the Greek god of the dead and underworld and older brother of zuse(sky or heavens) and Poseidon(sea). Leaving their aunt Gaia in charge of (earth land). He was the last to be spat out by his father Cronus (who devoured their children to evade being usurped). And thus left with the underworld to rule.

## Gaia

Gaia (a name which quite literally means land or earth) was a primordial goddess and mother of life. According to the theogony Gaia immaculately conceived Uranus (the sky or heavens) who Gaia then entered into a union with and conceived the titans. Among these titans was Cronus (the father of zeuse.

# DIVINE COUNCIL

The divine council was an assembly of deities and later monarchs/profits. In its earliest conceptions the Canaanite god El presided over the lesser deities. The lesser deities were given nations to govern and they functioned more prominently as kings, pharaohs, or national deities (Deuteronomy 32:8-9). However Later EL and adony (one of the lesser gods presiding over Israel) become conflated as one in the same. We even see latter accounts (Deuteronomy 4:19) where adony distributes the gods among the nations as an inheritance rather then the nation's belonging to the gods.(the gods then belong to the people and not the people to the gods). This is interesting because in its earliest iteration and in the majority of the literature surrounding it, a deity could only be worshiped in their corresponding land. And so in the Hebrew Bible when we see adony sue for power and then gain power over all the earth (psalm 82) adony (the god of David and Israel) can now be worshiped outside of Israel and even in exile.

Originally only deities could participate in the divine council. however we get later accounts where profits are counted among its members (Jeremiah 23:18) (Isaiah 6) (micha in 1 kings: 22).

There is even some evidence that bal and adony competed in the more ancient Semitic traditions as they both fill very similar archetypal roles as storm deities where as el is unequivocally a governing deity. However this role was usurped by adony in psalm 82. We see a similar parallels to this when Jeremiah accuses god of prescribing malevolent precepts for his people in exodus 22:29 & exodus 13:2 where as Ezekiel out right refuted those dictums not only arguing with scripture but with god (according to scripture). In ether case we see people in the Bible looking back at earlier authors claims or conceptions about god, and then rejecting those ideas and even commands of god according to scripture. The point is, the Bible itself is people rejecting the "god of the Bible" to bare a more loving, inclusive, and compassionate image of the divine.

Note (the neuroendocrinology researcher Robert sapolsky notes the correlation between the character of deities and their environment. Where desert gods tend to be less forgiving and more authoritative, however those deities of more bountiful and diversified topography are more compassionate, inclusive, and wholistic. Gods who preside Over rural populous, are more intimate and egalitarian. whereas Those of dense city states are more legalistic, hierarchical, and prestige/status oriented. With that said gods of rural or hostile environment are often also more prejudice towards outsiders and non conformists(heretics).)

## IMPLICATIONS

The famed psychologists Carl Jung hypothesized from his analysis of the subconscious psyche, that man first acted and only later questioned those actions. Or that to a great degree our involuntary actions and behaviors were dictated by our instincts rather than our intellect(amygdala as opposed to pre frontal cortex). And this follows suit with what we see in physics. because in virtually every case, it is mater that provides the conduit and material for the mind to manifest. or in other words there is no software without hardware, the program needs the circuitry, thoughts need neurons and the spirit needs the body! But this is particularly interesting because it is that ability posited in humanity to then override and negate our genes prerogatives and programming, that then gives us the authority to dictate the nature of nature itself (to rewrite the program and ultimately alter the circuitry which facilitated it in the first place). Because we are no longer subject to the whims of our instincts, we get to decide what we are. Now it should be noted that there is most probably profoundly prolific wisdom posited in our intuitions as a result of millions of years of evolution. And so it is no wonder why we see patterns when we mindfully meditate on our subconscious or even the instinctual traditions of our ancestors. Even Jung concluded that the goal was equilibrium between the rational mind and the intuitive heart (conscious and subconscious psyche). It should be noted that This prolific voice of nature (or instinct) is often interpreted as the voice of god. But then again why shouldn't it be? For it was most certainly this voice which produced the habitual traditions in culture which after all are just extensions of nature. And it is that same instinctual voice that is prevalent in all living creatures? If there is a universal entity it would most likely be that of instinct. We see this information posited cross culturally in depictions of the primordial chaos waters (all life emerging from aquatic creatures) to the intercession of light unto darkness (humans becoming self aware) (the day(consciousness) is man's abode and the nite (instinct) where the wild creatures dwell), not to mention the formation of complex language as a form of creating meaning ("and god spoke and it was so"). However something we see all through scripture is this idealized human interceding on man's behalf and arguing with god. From Abraham outside of sodom pleading with god to spare the city for 10 good persons, to Moses on the mountain demanding that god stay true to his word after Israel had erected an idol! And even later when Jeremiah and Ezekiel argue with the writers of exodus 22:29 & exodus 13:2 which require the sacrifice of the first born sons of Israel to god. Or even when Abraham makes the claim that unlike the Canaanite gods of his father, WHYH does not require or condone human sacrifice. We see people deliberately contending with the ineffable and the sacred traditions of their cultures on behalf of more kind and compassionate expressions of life and the divine. Because after all, god has not shown themselves, but instead has left it up to humanity to decide the identity of god thru the way they bare THEIR/gods image! We even see this expressed thru Jesus who claimed god didn't have to look like the malevolent pharaohs and Caesar's of this world but could instead be the least of these.

> "For source criticism, even with all its flaws and irreconcilable debates, is the best mechanism to show how the Hebrew Bible is not static. The cultural memories that we read on the pages of the Hebrew Bible were written by individuals from a

wide verity of backgrounds who learned their traditions from a wide verity of sources (written and oral). They also added and subtracted from their inherited traditions and reshaped them to address the needs of their own quite different historical contexts." "When modern believers appropriate scripture they are following their biblical counterparts"

-Theodore J. Lewis-
(the origin and character of god pg63)

## DEFINING TRAITS OF ANCIENT GODS

"We posit agency on events because we've been conditioned to posit agency to answer as cause for actions in our environment."

The ancient primordial gods where ferocious beasts like giant lizards or sea dragons, and thus the new gods where those men who had the tenacity and skill to vanquish those mighty chaos dragons and serpents of old. Like a warrior mouse smiting the winged terror (hawk) with a makeshift spear. the first men were most probably revered as gods for their ingenuity and courage. As those who were once the prey became exalted among the beasts.

The first gods were undoubtedly those fearsome beasts which struck fear in the hearts of all creatures. And the later gods likewise were most probably those men who inspired fear and admiration. (Fear the lord your god). it was probably only much later when god become equated with things like beauty and inspiration. Like a beautiful maiden or charming stud who inspired adoration (the muse or angel). Or even that worship intuitively attributed to the parent by an attentive child learning how to be just like their parents. And tho fear sounds barbaric to us today, it was that very trait which had governed our species survival up to that point.

"Until you can command the storm clouds, you build a shelter to combat its wrath. You discipline what you can until you are no longer a slave to your own instincts or the wrath of your environment."

The inconvenient truth however, is that the hero's who vanquished the monster (primordial gods) are often no different than the monsters they vanquished, both competing to survive at the expense of the other. And so we simply exchange one fierce predator for another.

"In half our stories the adventure wanders out into the wilderness to their demise, and in the other half the scholar begrudgingly is driven from the comfort and security of their holy cities and discover its walls were more like bars to a cage. In both cases it is in the wilderness where our heros find salvation or as in the cautionary tales show the curious creatures demise.

"Authority is not granted by gods but by those who respect it." Unlike the old gods (primordial beasts) or their replacements (the mighty heros), later godship was not exclusively dependent upon one persons might. but rather their ability to inspire and motivate a group of people towards a common cause or goal. And thus The thing that set humans apart from the rest of gods creatures was in part our proclivity for mutability and more profoundly our capacity to forgive faults. To contend with god and nature on behalf of more compassionate representations of the divine and nature alike.

It was not might but mercy which gave us the upper hand, as we tended to our wounded and recognized the crucial skills and traits of the marginalized. Or more simply we developed an appreciation for a more wholistic world view.

It was our creation of morality and our distinction between good and bad, or more accurately our ability to dictate those values for ourselves and not in accordance with the precepts of nature, which then made us fit to rule.

"A god more concerned with the prestige of the priests than they are with the provisions for the poor and destitute, is not a good god worthy of paise or worship",

## A QUESTION OF WORTHLESSNESS

As the myth goes, god was lonely and thus they made us to keep them company. But A god who created us because they were lonely and yet refuses to respond when we cry out seems completely counter intuitive. This god who demands us to worship them and validate their ego but is incapable of appreciating yet alone tolerating our expressions of grief, anger, joy, and fear. The very same such expressions that god wrote in our genes, then makes no sense.

We were told that it was a relationship and so we studied their texts and worshiped the idea of a god we created in our minds. This is a god we must believe in because we can not know them or see them. For You can't believe in something you know. Belief requires doubt. In order to believe in a god, we must first not be abele to know them. Now we may believe in our parents or spouse or even our kids or friends. But what we believe is in an outcome, not in the existence of the person in question. We may have faith in a Cause or belief in an outcome which is based on evidence. But we cannot believe in what we know. We hear a voice deep within us. A voice which we now know is the remains of our primitive mind (instinct/amygdalaian) however many still believe that voice is that of god. And for those who believe such, for them that voice is god's. It is their belief which makes god, god. But if it is our belief which makes them a god Should we not then expect more from god, then we do ourselves? Are we more tolerant and compassionate then god? And if so, if we love better, If we can appreciate what god(who created it) can't even tolerate, then we have become bigger than god. The truth is Our best and worst behaviors are programmed in us. and the more we become conscious of that, the more we can resolve the malevolent imperatives of our creator. Everything we once called spirit we now see has a physiological correlate in our neurology. Understandably Many would rather be slaves in Egypt then wrestle with the ineffability of life. We would rather erect an idol or master, a belief in hierarchy, then recognize the value of the life which is sacrificed for our life. The life of every plant, animal, and virus, which must die in order for us to survive. We would far too often rather give up our autonomy, then take responsibility for our own lives. And thus we surrender our identities to be made in the image of gods who give us a purpose by exploiting us, and give us belonging by separating us from the whole. But Love, true love does not control people. It whiteness each expression and recognizes it as holy. We would far rather be asleep, intoxicated, or high on hallucinogenic substances, AE Subconscious slaves to our habitual instincts, then conscious autonomous beings who actively reprise and evaluate the commands of "god". We would rather throw all our guilt on some mythicized savior then develop the capacity to honor the death in life. We would rather stay stuck in our toxic cycles then admit we were wrong and alter our own behaviors for the better. We would rather escape our lives then face them.

It was only when I learned to truly love myself and believe that I was worthy of unconditional love, that i realized how poorly the "god of the Bible" had loved us. Because love does not require blood. But if it did then with All the evil done in gods name and by their authority, why has god not answered for their crimes. they still have not come down and rectified the abuse and exploitation which is "biblically sanctioned" by them (god). The question is not the existence of god, but the integrity, virtue, and moral character, not to mention authority of god. It is pointless trying to prove ether the existence or lack there of god. What this book questions is the moral standard assigned to and by claims of divine identity and authority.

## THE ANGEL OF DEATH

The imagery of the grim reaper is derived from Azrael (the angel of death) who was originally responsible for fairing our souls to heaven. but something many people don't know is they were also responsible for harvesting our bones to be repurposed for new life on earth. That's where the reapers sythe comes from.

We are like plants harvested and recycled to propagate life in our seasons.

Or in other words, death is not the destruction of matter, but rather a reallocation of recourses. It's mater taking different shape. Much like chaos, it only deconstructs the current state or shape life takes and not the substance itself.

"Like the rocks we shaped into computational machines, We too are the dead mater of the universe come alive"

This personification of death being a vital role in the process of life, makes complete sense.

This image exemplified by the reaper/angel of death has become inflated with those of the opposer or satan who originally functioned as an alternative perspective or opposing view. This vital role functions as a check for the authority of the head deities. Tho understandably over time they became perceived as nefarious for their opposition (much like an activist or whistle blower who opposed or calls out the injustices of an massive oil company or oppressive government). This image is mimicked by characters such as permethious and Seth.

The name Lucifer was a phrase rhetorically parodying a Babylonian king who claimed he was a god and declared Venus (the morning star) was the astral body which represented him.

"Chaos is destructive to our cosmological constructs because it threatens to tears down the walls of our segregations and separations of kinds from the whole. It contradicts our myths which gave us meaning thru division."

Order is created by separating things from other things. And this is done by defining like kinds from unlike kinds to a degree. Or more accurately It is a result of our innate proclivity for Pattern recognition. We see patterns and posit meaning to them. And this is good, it's one of the mechanisms which allowed us to survive this long. However often we mistake these distinctions, and these boundaries we have erected, as concrete or fixed. This however is only an illusion, for like all things, life and all which suffer it must change. Further more just because two things are similar or diss-similar on one level doesn't mean that is the case on another. These categories are myths. And death is no exception. Tho yes certain qualities do have a sense of permanence (for instance the permanent loss of consciousness we refer to as brain dead), other qualities are not. Tho our body's decompose, the cells and material which make them up is not destroyed but simply reallocated and reconfigured for other purposes. Even ethereal ideas find their way thru time in the lives of very similar "souls". The words and discoveries of neiztchie or Paul find resonance with people centuries and even millennia a part. The observations and discoveries of Darwin or Galileo are made by children today who have never read their works. In some sense the muse which captivated those hearts and minds, are not limited to the extent of their lives.

It makes perfect sense then that we would anthropomorphize death in such a way.

Religious or mystical thoughts often function as coping mechanisms for exposure to ours or our beloved others eminent demise or existential crisis and threat. This is why fear is such an crucial component of religious dogma and traditions. It functions to solidify a belief thru our physicality.

Matter can nether be created or destroyed, it just changes states. what we call death is simply a reallocation of recourses. There is no waste, what man does not consume, beast, insects, or bacteria, shall.

## SERIENS AND WITCHES

In their original rendition, seriens and witches were feminine creatures who lured men to their deaths by way of melodious song or enticing feminine whiles. They were the black widows of myth and lore. Perhaps a kindred to the muse or genie who inspired hope and life instead of death and depravity. Unfortunately this term which came to refer to any woman who sought learning (a power which was only granted by god to men, under the precepts of the church). This resulted in women being tortured and burnt out of fear. The fear was not death or spiritual corruption (at least at the top anyway) but fear that in light of an eco system which corporated on a wholistic level, societies hierarchical structures would fall to ruin. Knowledge is power, and more often then not it is a power greedily kept for and by the gods. This is not to say there are not those who seek to destroy the nieve fool, but more threating are those who seek to Enslave with fear and animosity towards those who would see you freed. The muse must then not seek to divide but to Unite.

## HEBREW AND CHRISTIAN GODS

All throughout the ancient texts we find descriptions of divine traits which are then attributed to god, but are actually just attempts at deifying ancestors and monarchs. Theses traits like jealousy and anger, the desire for glory and praise, are valid depictions of god, so far as they depict the traits that were desirable for emulating according to our progenitors. However we are not restricted to our ancestors claims about god thru the way they chose to bare gods image. We each are given the authority to decide for ourselves what our nature shall be, rather that is exemplary of a malevolent master or more like that of a compassionate companion and caretaker. The truth is we most certainly get plenty of poor examples for god's character in the ancient and sacred texts. But ultimately it is us who decide what qualities we will choose to exemplify in the way we live our lives!

## MYTH AND MYSTICISM

Emotion is a mechanism our genes use to incentivize us to procure conditions and environments which have proven advantageous for propagation and preservation of those same genes in the past. Unfortunately this means that behaviors like aggression, which have proven to be effective in the past are then mistaken for being advantageous in the present. And so much like our emotions which coerce us to comply ether thru incentivizing or punishing certain behaviors, so too Most manipulators and abusers do so unknowingly and are instead simply acting out a behavior they were taught. And so (like all organisms) humans mimic their environments ether developing abusive and manipulative offensive behaviors (fight), or abused/ manipulable defensive behaviors (flight/freeze) in order to survive their environments and predatory conditions.

This is why logic and rationality is so important in constructing a sense of morality and illuminating our subconscious biases and prejudices.

> "Never attribute to malice that which is adequately explained by stupidity"
>
> -Hanlon's razor-

We are as kind as our environments permit us to be. In hostile environments organisms which comply with their fight or flight/freeze mechanisms often survive. And thus We often attribute divine qualities to our emotions (AE anger against a norm violator or transgression of authority is godly anger,) subsequently fear of authority is godly fear. In the ancient world the greatest fear and transgression against a monarch is the subversion, challenging, or usurping of their authority by a subordinate.

Rather pain or pleasure desire or dread suffering is a mechanism our genes developed to incentivize our obtainment of conditions which are conducive for healthy preservation, propagation, and the prosperity of those same genes. "Be fruitful and multiply" this is why romantic and familial love often objectifies and exploits as well as is so often manipulative and even abusive. Because ultimately Emotions (rather conscious or unconscious) are mechanisms our genes use to motivate their collective hosts propagation, much like people (the said hosts) do in a society.

Consciousness and instinct then can be understood and quite literally are mechanisms our genes developed in order to delegate and automate vital tasks much like we did with AI. this is the same mechanism which validates the conservation of energy for one organism by employing "laborers or grunts" to do the work they find tedious or grueling.

This is probably why we naturally elevate value judgments (assignments of meaning and value) over knowledge and objective reality.

And thus far too often cultures and societies value

the subjective claims of ancient men over the objective evidence and factual knowledge unanimously agreed upon by math biased logics, our biological coding, physics, geology, anthropology, neuroscience, history, and even entomology. Far Too often we choose a book compiled from manuscripts written 2-4,000 years ago, over people who's very existence on a biological level contradict those claims (which by the way contradict themselves countless times). With that said it was probably that very biases which allowed us to congregate and coexist to such a degree. It was those assignments of value which gave us an alternative to the cruel prerogatives of nature.

Humans are highly suggestive creatures and we see this in the effect of illusions like hope and placebos. Moreover our minds allocate recourses accordingly. In order to combat uncertainty we constructed delusions like destiny and fate to survive futility. We told ourselves a story, and by acting out that story we made it a reality. Fictions like fate and destiny, keep the slaves poor and and the gods rich.

The most accurate amalgamation of reality is often actually disadvantageous, the myth is more beneficial for the propagation of genes and survival of a species then objective reality. In maters of survival, speed trumps accuracy and too much is better then too little.

We construct narratives which inadvertently assign meaning to our lives, but those same narratives often trap us in detrimental cycles. And yet to escape those cycles we must sacrifice those narratives? subsequently sacrificing the meaning they assigned to our lives. It was those narratives which gave us the means for our greatest advantage in surviving predation (the ability to corporate on a much larger scale) but at the same time they gave predators a new arena to thrive, in the form of divisive narratives which pitted us against those who threatened to expose the exploits of our collective oppressors and gods.

In the ancient world and even still today the use of value judgments such as those made by "sacred texts" and traditions as a stander by which to govern reality was common practice. This is why discerning right from wrong was so abhorrent. It is an insistence that objective reality conform to subjective claims about the nature of the cosmos. And so the authority of scriptures and traditions are used to fact-check objective reality because the paradigm or world view it prescribes is not grounded in truth or even virtue, but rather authority and supposed power.

These stories essentially set the parameters for the internal simulations which ultimately govern our volitous actions.

Fate is a fiction which has been used to both justify the exploitation and abuse of "subordinates" as well as ensure that the subordinates remain submissive to the cosmic schemes of man.

We believe the stories we tell ourselves! Because These stories we tell ourselves are usually based on the stories we've been taught by our culture and so culture shows us who can and cannot play certain roles (AE the hero, the victim, the villain,...) this is why representation of marginalized communities or minorities is so important and also so dangerous to the hierarchies. It threatens the narrative which determined their divine right to rule over others.

Our emotions are mechanisms our genes developed to regulate their host and evaluate its environment. Some emotions like anger and grief, tell us stories such as "you've been wronged" or "you will never be whole again". And sometimes we believe those stories because we don't have the strength to defy them. More often then not we buy into the fairytale because our survival depends upon it. We need the community and the comfort it offers. Not to mention pursing truth often puts us at odds with those in power. And so we willfully believe the lie and choose ignorance because we don't have the power to question it.

In truth our anger does tell us when we have been mistreated, but far too often the ones mistreating us are also those in positions of authority and because we can't dispel or exhaust that frustration on those who caused it (like patriarchs, priests, and politicians) that anger often leads us to abusing or misusing those who are "inferior" or subordinate to us. We need an outlet to voice our grievances because those angry voices are often the most honest and compassionate of our defensive mechanisms. They are one of the few which actually protect the organism and not just the propagation of its genes. The problem is that because we can't exercise our autonomy over our "superiors" we end up perpetuating the same cycle.

"People (and even many mammalian animals) will intuitively set their counterparts up to feel the way they themselves had been made to feel. If a person feels neglected or powerless then they will intuitively behave in a way that inspire those same feelings in those around them"

When an animal feels frightened they will (at least attempt) to frighten those who frightened them.

The will to live, no life essence or spirit just chemicals and electricity. Why people both survive impossible odds and yet also die when they give up. What we call the spirit or soul is the result of neurochemical programs emplaned and employed by our genes to incentivize the execution of certain tasks. And thus when the prerogatives of these emotive programs are negated then so to is the fighting spirit. Much like the meaning attributed by the forsaken myth, once the illusion is broken, so too is the power of its magic and myth. When we dissect the bird, we lose the song and all.

In many ways emotions function as probes for computational assessments of an environments viability for life. Or in other words prospects for propagating and preserving progeny matters.

Rather it is physical or emotional, Pain tells us where the problem is, and often how to solve it. In part this is why we revisit the painful memories and intuitively touch the open would until it no longer evokes pain.

From crystal clear streams, lush trees, and rolling hills full of flora, to prominent buts and long healthy hair, what we call beautiful or attractive are markers signifying fertility in an ether an environment or mate. Prom personality traits to physique, and even to things as base as status and prestige, these are all qualities which have in the past proven to promote health and prosperity.

If the "creator" calls The very process by which life is made "abhorrent, unclean, or sinful" then it is not nature which is the abomination, but the "god" who claims to preside over it.

## THE BATTLE OF THE SEXES

On average women tend to preform better or excel more often in academic arenas then men do. And in part this is due to the values idealized in the scholastic environment such as not challenging authority and differences in how aggression presents itself in each gender. as a result women tend to make up a greater percentage of office personal wile men make up a vast majority of the labor field. Women sell the product wile men produce it. This has resulted in men believing that they have a voice when in truth the majority of men are just as much victims to the patriarchal systems (if not more so) as the majority of women. But because men see themselves represented in the elites and a great deal of an individuals value in said society is based on their ability to produce or consume (take part in the economy) these men often tend to defend the patriarchal systems which exploit them. Women do this to but in a different way usually in the form of mate selection (women don't want men who aren't economically viable or hierarchically intrinsic to the patriarchal system.) women also tend to take better care of their own physical and mental health wile men have been conditioned to see themselves as an expendable resource. It's no wonder then that One of the greatest determining factor of a societies success and prosperity is observed in the rights afforded to women,s equality with men.

Tho it is understandable that in certain fields women are payed far less then their male counterparts what is more telling however is that As women's representation increases in any given field, the overall pay for those professions decreases by as much as 57%. This says more about the systems used to exploit and abuse people then it does the competence of ether men or women.

Let us us take for example a tactic often employed by those who abuse their power. (Like Hitler in the 1930's or DeSantis in the 2020's, not mention the 1% era who egregiously exploit the bottom 80% of the population) because these men (like Henry Ford and Ronald Reagan, and Hilary Clinton) are vastly out numbered by the people they exploit ('ed) they are required to redirect the blame for their mistreatment onto a scapegoat. For the majority of the right wing that scapegoat is "the government". And so a private corporation refuses to pay its workers a livable wage, its the "government's taxation which the problem". a privately owned bank forecloses on a families house, its "the governments fault". And on the lefts they blame the result of the problem and not the cause. "The problem is white cis men" and not the traditions that cultivated patriarchy (which is no better than matriarchy). Rather then addressing the system which prescribes the exploitation of men and women. Liberals and conservatives, and the rest of the bottom 80% of this countries population. We end up fighting a war against the victims or side effects of these abusive systems and beliefs. And the reason this tactic works so well is not only does it resolve the perpetrator of any blame and redirect that blame onto the victims (like the Jews in the 30's and the trans community in the present era), but it also allows the perpetrator (who caused and perpetuates the problem) to be extolled or worshiped as a savior. The police protect the rich because the rich fund the police. Both men and women defend policies that deliberately go against their own best interests because they have been tricked into blaming the symptoms for the disease. The problem isn't men or women, gay or straight, white or black, the problem is the belief that any of these categorizations is intrinsically superior or subordinate to any other.

The sad truth is far too often, healthy and positive traits in men who embrace their femininity and idiosyncrasies is then characterized as an ick or unattractive. Conversely there is just as much toxicity in the femininity exemplified in our age as there is toxicity in the masculine. The difference is women have learned to prioritize their own safety and thus recognize when they are being treated poorly. Men on the other hand are often so deprived that they accept the abuse (any shelter/community in a storm is better than none at all). This however is not true. A bad solution is not better then none at all. And we can see that the problem is systemic and thus instilled in both men and women. Women are often better at protecting themselves (or their progeny). Whereas men are often too occupied trying to protect others from themselves, that they never learn to protect themselves from others. This often results in them becoming just as abusive themselves as those who's abuse they came to expect as normal. But Just because the culprits were once victims themselves doesn't then make their abuse of others ok. We have too heal our hurt before we hurt others. And in many cases women tend to be better at that then men.

One can see how facts of nature like androgyny explicitly undermine the binaries many traditional narratives employ to divide men and women against one another. This is why topics like homosexuality, gender fluidity, women's equality, are so pivotal in religions like Christianity, Mormonism, and Islam.

Note: On a neurological level men are more emotionally (amygdala) driven, and women more logic (PFC) oriented.

Nuance is a literary tool used to make uncomfortable truths more palatable and easier to digest that's why the absence of such virtues is considered crude or unsettlingly vulgar.

In much the same way we often turn to coping mechanisms in an attempt to numb the pain of a heart break or loss. And tho numbing the pain never really fixes the problem it dose help us cope with the grief until we have the resources to actually deal with it. It's a bandaid not a cure, but sometimes that's enough to get us by. In order tho live we must often first survive. When grieving the transcendent, remember it takes time to transcend it.

The process of neuronal pruning is a natural process where our brains prune the neuronal circuits they don't use anymore in order to make the computational process more efficient, this process is the hallmark of neuro typicality. And thus Any deviation from this is often characterized as a neurological disorder. When we consider the impact corollary neuronal connections have on conditions like trauma and depression, we can see how "neuro typicality" is an evolutionary form of willful ignorance intended on conserving energy and resources at the expense of deviation and diversity.

## THE WORD OF GOD AND THE MOUTHS OF PROFITS

Every set of ancient texts and traditions make claims of divine authority (god said this or that). But in every case god requires a mouth piece by which to speak thru. Men on holy mountains make authoritative claims on behalf of a god who will make commands but dares not show his face. And thus we are left with no other choice but to negotiate the authority of such texts and claims. Ultimately it is the authors of said texts who claim "god

37

said this or that" and thus it falls to the general populous to decide if said claims of authority are valid or not. The Bible has has no innate authority beyond that agreed upon by the populous. Every thing we "know" about god (or the gods) comes second hand from those who claim to "know" a god who has yet to make themselves "known" to the general populous but instead depends wholly on the words of men. The claim "god says" is just a less reputable and unsubstantial version of "he said she said". And it only carries authority because of an communal agreement.

> "The truth is like a lion. You don't have
> to defend it. Let it loose. It will defend
> itself."

> -Augustus-

The ancient cosmos was contained within the laws and values of a society and presided over by the governing authorities who acted as the mouthpiece of god.

The good news is these assignments of meaning and value have no innate authority beyond that which we as a society attribute to them. And thus it is completely within our collective pervirw or authority to reappraise and renegotiate the authority and value of those claims.

Even things like karma are socially agreed upon constructs which change depending upon the values determined authoritative in each respective culture.

Some scholars even argue that Jesus's forgiveness of

sins in mark 2 is not rhetoric intended to establish the divinity of Christ but instead the authority of man (the son of man) to admonish the debt of past transgressions. Which deliberately conflicts with the scribes who claim only god can forgive sins. Dan McClellan articulates this point by pointing out Jesus's dialogue leading up to this admonishment of the paralyzed man's sins. "So that you may know that the son of man has authority on earth to forgive sins"-Jesus- (mark 2:10) NIV

## THE PIVOTAL ROLES OF
## PREDATION, PREY, AND EMOTION

The notion that a superior being was incapable of meeting their own physical and emotional needs and thus created inferior beings to meet those needs with the soul purpose of worshiping and serving them as a god. And the very fact that the gods greatest fear is being usurped (for god is a jealous god). In truth an god (which is truly a god) should have a greater capacity for compassion an mercy then mortal men who simply act out the behaviors prescribed, programed, and prewired into them by that supposed god. But instead what we see is a all powerful god with less tolerance for their supposed creation then the men and women born into conditions of scarcity and poverty. We see ancient men and women who show mercy to those who could (and who often have) hurt them. We see mere mortals forgive those who literally caused them pain and inflicted suffering on them. And yet they worship a god who smites an entire generation for indirectly hurting his feelings. This god who literally created a universe to worship him, not only stifles the growth and capacity of those who worship him, but he also explicitly prescribes the abuse of those who outgrow his fragile ego. However some claim that In order to be god they must be more them mighty but more importantly they must be worthy of worship which would require a superior level of virtue and ethic. But this is incompatible with a god who created lesser beings just to validate his ego and at the same time inflict suffering on those who act out behaviors which are preprogrammed into their neuro chemistry.

Ultimately it comes down to what world view we filter thru. A god who commands genocide and rape is admirable if you subscribe to a authoritarian world view where morality is vertical rather the horizontal. If our standard of morality is based in obedience and loyalty then "might quite literally will make right" and authorities have the right to abuse as they see fit. However if you abstain from harming others because such behaviors "harm others" then you subscribe to a horizontal sense of morality which is grounded in compassion and empathy. The "objective" morality of the authoritarian world view is just as subjective as the horizontal one of nonbelievers, especially when you take into consideration that the authoritative claims are made by men who often contradict rules or commands "given by god" according to previous authors. And thus the reader is left with no other choice but to choose which claims of authority prescribed in the biblical texts to acknowledge as authoritative. When compared to the horizontal form of morality where the exploitation and abuse of others is wrong not because it injures the ego of an authoritarian figure but because it causes harm to others is no less objective than that prescribed by the biblical authors.

"Laws are threats made by the dominant socioeconomic-ethnic group in a given nation. it's just the promise of violence that's enacted and the police are basically an occupying army" -brennan lee mulligan-

Truthfully most rules are arbitrary. In fact one could make the case that all rules are ultimately arbitrary, however some rules are necessary for the survival of a community and species. And thus all laws are made up tho some are grounded in virtue. Likewise economics are a mechanism used by those socioeconomic-ethnic groups to keep the majority of the populous compliant to their imposed hierarchies.

Behaviors that are rewarded are repeated, behaviors that are penalized are not. Organisms that follow this rule of thumb are more prone to surviving their given environments.

> "I like to remind my patients. the opposite
> of depression is expression. What comes
> out of you doesn't make you sick; what
> stays in there does."
>
> -Edith Eger- (the gift)

Tho depression is most certainly the result of a chemical imbalance it's characterizing trait (despair) is a logical response to our current socioeconomic circumstance. It is caused by the futility of conditions regardless of the expenditure of resources. And tho technically not genocide for all intensive purposes, our brand of capitalism has cultivated an environment where people can't afford to have kids. after working 60 to 80 hours a week people don't have the energy and time to socialize and so effectively it has pacified the priory prerogative of procreating ultimately eradicating a genie pool. In truth that suicidal voice is the one that cares The most, it's the most truthful and honest about the pain, it's the most realistic about the price for survival. And it is clear that something must die. But far too often we would rather sacrifice our lives rather then the stories which have defined those lives meanings. These stories tell people who they are and where they belong in an hierarchy, and in part that is why so many vilify those who challenge the narratives and define their own identities. ultimately that's why far too many will end up dyeing defending a system which is literally killing them and their environment. Ultimately these stories have no inherent authority and are in fact just tools, we as a species have constructed and collectively agreed upon. And so tho these stories have helped countless people congregate and collaborate for survival, their purpose is to help us survive and so when they themselves hinder that goal, they can most certainly be discarded. Every law and claim to authority is completely made up by man.

in a paper published in 2019 by Matthew D. Herron and colleagues entitled (De nova origins of multicellularity in response to predation) the scientists preformed a series of experiments where they introduced unicellular predators to 5 unicellular algae. After 50 weeks (750 generations) 2 out of 5 unicellular algae Evolved multicellular structures to combat that predation. In short They become too big to be consume. These alterations were retained by the following generations even after predation was no longer present.

Not only does this show how single celled organisms evolved, but why unicellular life evolved. Ultimately Organisms learn to corporate in order to survive predation.

"Competition is the law of the jungle, but cooperation is the law of civilization"

-Peter Kropotkin-

However these advancements also probably prompted similar advancements in unicellular predators in order to compete. Resulting in a sort of biological arms race for survival. Where the lion becomes faster to hunt the gazelle, man became more cunning to manipulate "lesser men and women".

But if all life originated from the same unicellular organism. How did predation arise in the first place?

All life originated from a single cell. Similarly so did you and me. We all start out as a single cell which then divides and multiplies. As those generic cells multiply they specialize in different tasks. And in the case of humans those cells collaborate to form a single organism made up of trillions of cells (37.2 trillion to be more accurate). And so Like siblings raised by very different parenting styles and exposed to very different environments, two identical single celled organisms exposed to very different environments will ultimately result in two very different organisms. Say one single cell organism is exposed to a bountiful environment and another is required to survive a scarce or resource deprived environment which requires them to develop aggressive behaviors in order to obtain the necessary resources for survival. When the second (deprived and now aggressive) organism is reintroduced into the bountiful environment, it will inevitably devour the passive organism. When we replicate this process exponentially we end up with very different organisms and thus this is most likely how predators evolved, subsequently prompting the evolution of more comprehensive, robust, and complex (passive) organisms.

This is also why a false sense of scarcity is perpetually cultivated in authoritarian societies in order to prompt innovation as well as enforce compliance (or an aggressive form of cooperation).

And thus more often then not predators are elected to positions of authority because they make life harder (often thru exploitation) on the prominently passive populous inadvertently prompting the passive populous to evolve.

Humans aren't the prey or victims but instead the most cunning and distinguished predators on this planet and more often then not our competition with each other results in travesty. Both men and women have an innate proclivity to manipulate and subdue those deemed lesser or inferior as prey. And thus the incessant need for hierarchy is employed to quell that thirst for retribution. stories then have unfortunately been employed to predate marginalized communities. And thus society in many ways is simply an agreed upon set of permitters to which one class is permitted to predate another. It is still Predation nun the less.

## CATS AND DOGS

We will briefly examine Two kinds of predators and their functions and in doing so we will address why hierarchies are necessary.

Both cats and dogs are communal creatures to varying degrees. They Are also both predators. However both the way they organize themselves in communities and the ways in which they hunt and predate are very different. A stray cat is the norm because they can't really be tamed to the same degree as their K9 contemporaries, however a stray dog causes problems. This is because in social circles dogs challenge authority until said authority is established at which point they become submissive. And for this reason dogs hunt in packs and have strict hierarchies. They also cooperate and ultimately dependent on their communities, whereas felines are more individualistic and self sufficient. Cats also tend to be less openly confrontational preferring instead to stock or prowl. Dogs will challenge an animal much bigger than them, and with the aid of the pack will often succeed. Cats on the other hand tend to stock smaller prey and avoid blatant confrontation. In their own respect both dogs and cats are territorial but dogs tend to be more vocal and confrontational (probably in an attempt to gage their own position in the hierarchy much like they do among their own packs). Whereas cats tend to assess their prey/opponent from a undisclosed location. Truthfully these are more stereotypes then strict guidelines but we see both approaches function in modern human societies. And both have their own corresponding pros and cons. They both answer the problem of prey populations evolving defenses to combat predation. Dogs can be trained for obedience and loyalty this need for validation and connection is characterized as narcissism in humans. Whereas cats exhibit more sociopathic tendencies.

We see the wolf in sheep's clothing (feline archetype) who is conniving and manipulative. And we have the vocal lion (k9 archetype) who is narcissistic and aggressive. The real distinction between these two is that one is controlled by their own impulses where as the other uses other peoples impulses to control them (their prey).

More often then not the kind of people who ascend to positions of power in authoritarian societies are those who exhibit prominently (aggressive/manipulative) predatory traits like sociopathy and narcissism. However it should be pointed out that people (humans) are in fact a predatory species. And thus in many ways our economic and social systems are designed to "predate" (farm and exploit) people rather than help them. More over our economic systems especially are tools used by predators and not the one used to protect the weak from predation. However we will go into greater detail on this particular topic in a later chapter.

Tho society is intended to make life easier, economy is designed to deliberately make life harder. Where society is a defense mechanism, economy is a predatory mechanism or tool intent on enforcing a hierarchy. And thus many times people are rewarded for their ability to manipulate and exploit others by being put in positions of power and authority. And In many cases these authority figures deliberately cause the problems that they then profit from remedying. "If someone offers you salvation/deliverance form the situation they put you in, that's extortion." "And If you are defending the ones with power in a system that profits form marginalized communities misfortunes, exploitations and abuse, then you are on the side of malevolent predators." Now granted the kind of predation we see in these economic systems are more a kindred to those of the jeweled wasp which injects cockroaches with a (zombifiyng) neurotoxin and then impregnates the cockroach with its luva to be consumed from the inside out, or perhaps it's closer to that observed in homosapines who farm livestock before eventually slaughtering them. Tho unlike the cows and pigs which spend their lives sleeping and eating before they are slaughtered, humans are forced to work until they are physically exhausted and emotionally drained at which point they finally keel over and die. On second thought (perhaps the wasp was a better example of this form of predation after all). Ultimately we let narcissists and sociopaths set the baseline and standard for neuro-typicality. And thus we live in a world ruled by master manipulators and brutes. These were the archetypes deified in the myths and "scriptures". And thus they serve as the examples for hierarchy in society still today. It is these very same psychopathic, sociopathic, and narcissistic, behaviors which are often cultivated in order to survive such socioeconomic systems and environments.

Now my reader may be hesitant to accept that constructs like economics and religion are prominently predatory, so allow me to present my reader with some statistics. As of 2022 the bottom 80% of Americans possessed only 7% of the private wealth in the country. The other 93% of the country's private capital was controlled by the top 20% and the top 1% of Americans controlled over 40% of the nation's private capital or wealth. Not only do these statistics demonstrate just how disproportionately wealth is distributed in this country under capitalism, but from the power these top 20% have when it comes to lobbyists pushing laws that criminalize homeless, and poverty to those which deliberately exploited and abuse the working class in an attempt to disenfranchise those who actually produce the gods and services thru forced wage labor under the threat of loosing their home and/or access to food and clean water, or literal slave labor thru the pineal system. And in regards to religions who compound over $100 billion worth of untaxed assets from the "required" tithe of their congregants and then use their power and influence to pass anti gay, trans, and abortion bills to strip women and marginalized communities of their innate rights. Not to mention that 93% of sexual offenders identify as religious. Tho I'd admit that economics and religion do serve a vital role, it is understandable how often they have been utilized and even weaponized to exploit and abuse (predate) subordinates under the guise of some form of higher authority or power, rather that be political or "divine".

"When the cats away the mice will play"

"TGIF! NO! The gods didn't give us the weekend or "sabbath" It's because of the gods that we work our lives away!" Remember if they offer deliverance from the system they instituted, then they are not saviors but the captors.

Now We of cores have defensive mechanisms like anger and grief which evolved to combat this form of predation, but in many cases these mechanisms are vilified and even invalidated by well meaning caretakers who honestly believe in the virtue of the systems which have manipulated and exploited both them and their progenitors. Further more the fact that our emotions aren't designed to promote an honest representation of reality but instead constructed to emote or manipulate certain behaviors often reenforces the untrustworthiness of these defensive mechanisms that ultimately are attempting to preserve and protect our genetic material.

Anger is a very important emotion because it notifies us when our agency or identities are threatened or obstructed. And so it is understandable why so many atheists and agnostics harbor such anger towards the idea of god. Many of us were Christians for a vast majority of our lives. And as Christians our identities and authoritative agency was subverted by that idea of a god. This is why so many Christians get so defensive when their idea of god is criticized or critiqued. As Christians their very identities are defined by those ideas and assignments of meaning and value. We now understand this association as a key component of our ego. Our egos job is to ensure our safety and security within communities. And it is dependent upon traditional narratives to govern our station and behaviors in such social environments (much like it's contemporary empathy). And so for my dear reader who is attempting to deconstruct these narratives of patriarchy, religious supremacy, and misogyny, that anger is crucial in reclaiming your authority. We see this in the distain for the lives of the "victims" of the titan submarine implosion. As millions of people applaud the deaths of billionaires who deliberately profited off of the exploitation, and abuse, of millions of people every day. This dehumanization of oppressors was initiated by those very same oppressors in the first place. And is in fact a tactic used to disassociate with their victims. rather it is the religious zealots who are demanding the eradication of entire community, or the god tire billionaires who have deliberately caused the suffering of millions for their own personal gain. The anger felt towards oppressors is rightfully placed. For they obstructed and often even deliberately subverted our authority, autonomy, and discretion of identity. And that anger alerts us to that fact.

The people who revolt in response to being starved and exploited should not be punished, however the "authorities" who have deprived and exploited those communities most definitely should be held accountable for their malevolent treatment of the populous. It should be noted that the only thing really separating the two groups is a arbitrary culturally agreed upon authority of a social hierarchy and the conditioning which comes with ascending that hierarchy. In most cases behaviors (particularly those inherited genetically) are activated by environmental stimuli.

Why is it that god's children have to defend god's honor, is it not the fathers job to defend their children. Not only is this "good father" complacent and petty, but he's also emotionally immature and psychologically abusive. Now obviously the god of the Bible is modeled after very ancient and primitive parental figures. However this jealous patriarchal image which is still revered by religious institutions as a good example for fathers, mothers, sons and daughters is more than part of the problem.

We now know from advancements in neuroscience, brain imaging, and clinical research that the self-regulation disorder ADHD is actually an neurogenic disorder. Which means that factors like self control and higher executive functions are primarily neurogenic and thus inherited. Dr Russell Barkley makes the argument that our ability to manage our own behaviors (exercise executive function) is prominently preprogrammed by the genes we inherit from our parents and not a discipline which is learned. This is conclusive with what we will see later on In our chapter on neuroscience and biology. But for now the take away is that virtually all our behaviors and ability to regulate them is prewired. "Our capacity for regulating ourselves is a neurobiological trait, and not some socially learned phenomenon" -Dr Barkley- . These traits originate in development and as with all genetic traits they are activated by environmental stimuli. And thus training can enhance these executive functions but do not cause them.

In 2013 Dr Kerry Ressler, and Dr Brian Dias with Emory University, conducted an experiment on male mice which became known more commonly as "the Cherry blossom experiment". This experiment exposed male mice to the scent of cherry blossoms and then promptly accompanied that stimuli by shocking the mice with a mild electric current. They did this repeatedly until the smell of cheery blossoms triggered a trauma response or anxious behavior in the mice. They then inseminated female mice with the sperm of the traumatized male mice. What they found was that the offspring (who had never meet their fathers) had epigenetically inherited those same trauma responses and anxious behaviors around the scent of Cherry blossoms even tho they had never been exposed to the negative stimuli. In fact they found that this trauma was transmitted up to three generation in the mammalian brain and was not taught or instilled by the parent/progenitor. Because the offspring never actually meet their fathers. In many cases not even the mothers meet the fathers, as they were often artificially inseminated. These behaviors were inherited genetically, or more accurately epigenetically. Which simply means that the DNA sequence or code was not altered. Instead a chemical tag (most probably in the form of a non-coding RNA, seeing as how the methylation method is all but impossible thru a process which effectively wipes the DNA methylation from the paternal chromosomes in the egg). Thanks to the work of Dora Costa we can deduce that these RNA alterations are most probably carried by the Y chromosomes of male progenitors. This does not mean that we will inherit the fight or flight response of our parents, but simply that we mite be more prone to exhibiting these behaviors when exposed to similar stimuli. This RNA tag effects the readability of particular genes. A environmental stimuli is still required to activate/deactivate those genes and their corresponding behaviors/traits. However this does mean that to some degree (consciously or unconsciously) we inherit "memories/experiences" form our parents epigenetically.

Another way trauma is passed down is thru repeating abusive behaviors or cycles which we were taught by our parents. The prescribed form of correction for children as articulated in passages like proverbs 13:24 (NIV) is now known to be very flawed in its conclusions. As "punishment" in the form of physical and/or psychological abuse does not actually teach cognitive functions but instead inhibits them. What this learned behavior actually does is train reactionary behaviors (fight or flight) in many cases both. It does not instill a sense of morality but rather enforces a child's compliance or even perpetration to/of tyranny and violence. Not only does punishment inhibit cognitive mechanisms required for volition and executive functions, but it also cause neuronal atrophy and loss of synapses (brain cell connectivity). Prolonged stressors like trauma have been shown to effectively decrease the size and functionality of synapses.

## HOW MEMORIES ARE MADE:

When we experience an event or phenomenon millions of neurons light up all throughout the brain. These neurons correlate with various senses receptors and internal computational mechanisms. We call these neuronal correlations neuronal assemblies and they function much like notes which make up a cord. More simply this is expressed as "neurons that fire together, wire together" which means that if you repeatedly eat a peanut butter sandwich wile listening to a particular song, then eventually whenever that song comes on the radio you may "smell" peanut butter and toast. However simply experiencing a phenomenon is not enough to construct "what we call memories". Those neuronal assemblies must also be consolidated or "memorized". And this is achieved thru dreams. When we sleep these neuronal assemblies are sped up and repeated over and over again. This not only solidifies each individual neuron but more importantly strengthens its contention with its corresponding brain cells.

Our genes evolve mechanisms which prefer there propagation and are not necessarily concerned with traits which make accurate easements of reality. They are concerned with effective and efficient not fact or truth and thus we are prone to mythicizing and anthropomorphizing.

We have a innate proclivity to fantasy and fictions like those catered to by religion. And we see and compute/think in narrative (somatic) forms. Their goal is to emote reactions which lead to survival, adaption, and most importantly procreation of corresponding genes. And so Much like our modern computational AI, our genes constructed mechanisms like instinct and consciousness in order to automate and delegate tedious tasks. These biological programs and computational somations primary goal is not accuracy but efficiency and effectiveness.

In truth People are ecosystems and the hosts for millions of microorganisms which are constantly at odds with one another (approximately 39 trillion bacteria inhabit a healthy human body). These Tiny tentacles which are stabbing and clawing their way into more passive organisms are only matched by the defense mechanisms released by these "passive" microorganisms. And thus as easy as it would be to generalize these predatory mechanisms as bad or evil, in many ways it is thanks to those components in both our biology and our sociology that then allowed us to survive many of the same treacherous environments which caused them in the first place.

It makes sense that those without access to the resources necessary for developing higher executive function would seek an higher power and also why those in positions of power and in possession of the resources to actually think and behave more rationally often reject the idea of a god. Those with power see just how incompetent and malevolent an omnipotent god must truly be. It's understandable then why they would then reject an god who does so little with so much and yet who's capacity for acceptance tolerance and understanding is inferior to their own.

The more you know the more you are aware of those who claim to know really don't.

Those who know better, and are capable of doing better, are held responsible to be better.

The more resources one has at their disposal the more equipped they are for tending to the needs of themselves and others. However This does not mean that they are more proficient or competent at caring for themselves and others, but simply that they are better equipped.

Trauma is a fight or flight response that kept us alive in hostile circumstances. These fight or flight responses become trauma when they continue to exhibit themselves in safe environments. But when a harmless stimuli correlates with harmful circumstances, the harmless stimuli triggers the defensive mechanisms. Neurons that fire together wire together.

Some of these biases are innocuous such as superstitions like (last time I wore this shirt by favorite team won) and even advantageous like (the last time someone in my tribe ate the red berries they died, so those berries are probably cursed), however in many cases that firing of corollary but irrelevant neurons actually hurts us. For instance just because that nostalgic stimuli correlates with an environment where your basic needs were met to some degree, however that preference can often be employed to stunt progress and inclusionary measures by "idealizing" (and even idolizing) "traditional values" which are then weaponized to marginalize, exploit, and disenfranchise, communities which deviate from those cultural biases.

In order to heal trauma we must first resolve the environmental conditions which made those fight or flight responses necessary.

Note Children intuitively expel stress by fidgeting.

We live in an society where men seek not the strength to control themselves but rather the power to govern and rule over others. And thus in this age we have laws which dictate how individuals can behave, identify, and think, laws which deny women agency over their own bodies and govern who people are allowed to love. And yet we also have laws which condone and even defend the corporations rights to abuse and exploit its workers and its customers health and well-being. For as long as there are gods, there shall also be slaves. For there can be no gods without slaves for them to rule over. Most people don't seek to rule over others, we simply desire authority over our own agency.

Instead of respecting themselves they require the worship of others.

I don't believe justice is taking the hand of the hungry thief, but rather addressing the systems which starved them in the first place. I believe justice is meeting the fundamental needs who's absence caused them. Predators are created by scare environments and conditions which necessitate their aggressive behaviors.

One tactic often employed by the narcissist is the construction of an artificial handicap like "original sin" where a "perfect god" creates imperfect beings, and then demands perfection from them. Ultimately they unlevel the playing field so that they always come out on top. Now realistically what would happen is the narcissist would grow up in an environment where their emotional needs were not met and thus these defensive mechanisms (emotions) which were intended to regulate and gage an organisms well-being in an environment are then invalidated or exploited, and thus in order to compensate for their own inadequacy and neglect they judge others by a higher standard then they do themselves. Their inability to love themselves results in a need to be admired or worshiped by others inadvertently requiring the narcissist to cultivate a false sense of scarcity (or insurmountable cosmic debt) by which to "even" the playing field. Because tho predation is preserved And propagated thru imposing a hierarchy, it was most likely scarcity which prompted its manifestation in the first place. The goal (rather consciously or subconsciously) is to project and mirror those same emotions (fear, shame, rejection, powerlessness..) onto others. But because of hierarchy which protects the powerful or more predatory members of society, those projections are instead cast on to those in positions of inferiority (prey). It should be noted that hierarchy was most probably constructed as a social mechanism for electing the most competent and capable members of a community to a position of leadership for the benefit of the whole group. But because the same conditions which prompt capable leaders also prompt predatory behaviors the two are often indistinguishable from a distance. In fact in most cases predatory behaviors are simply survival tactics and not intentionally insidious.

Note emotion doesn't tell you what to do, it simply tells you that you should do something. The action (to fight or to flee) are based on simulations run in your VMPFC and are thus based on past experiences and/or observations.

When parents or authority figures treats their children and subordinates like prey, then their subordinates and children will rightfully treat their "superiors and parents like predators which they should avoid at all costs. Any shelter from the storm is the rule of thumb for defenseless children. But as individuals grow and become more capable of meeting their own needs, the less they will depend upon those who (knowingly or unknowingly) abused or manipulated them when they were week and defenseless. Thus narcissistic gods often depend upon "values" like obedience and loyalty to keep their subordinates disenfranchised and helpless. They give out trophies for participation and then accuse their children of being entitled wile they work tirelessly just to survive.

Ultimately our brain discerns how resources are allocated and thus it make sense why our mentality effects our physical well being (people who are happy require less sleep, and as we shall soon see mental health greatly effects our digestive system and ability to combat infections.

One of the most compiling qualities of Christianity

is the way it provided community to people who had been denied it by their ethnic/religious and class/political institutions.

Early Christianity literally be like "we will defeat the imperial overlords with the power of FRIENDSHIP!"

And they got that part right. The problem is those in power caught on and devised a scheme to stay in power.

So many of us were taught that we could be rockstars and billionaires. But here's the catch, we can't all be gods. Because gods require slaves to lord over. We were taught that we could be the hero and main character, so we did everything we could to set ourselves apart from the rest. And in turn we lost the only real hope we ever had at overthrowing our imperial overlords. we lost our community. And in turn they commodized every need which was once meet by our communities. You see martyrs weren't enough. The key to liberty from tyranny is and has always been community! Cooperation is our greatest defense against predation. And so they used religion and politics to set boundaries around what kind of communities were acceptable. And even then rules where set in place to cause divisions rather then let differing opinions strengthen one another's resolve. Our diversity isn't the problem, our divisive intolerance for it is.

We often require saviors, gods, one true loves, or heros, as some form of sacred object to adore and admire as a distraction from our often tedious or melancholy lives. Like a holiday these immaculate beings function as an escape. A true home however is not an escape but a state which we no longer seek deliverance from.

Entire generations in America were sold a narrative that if they worked hard they could be rockstars and billionaires too. This narrative not only enforced the incentive for the worker to exploit themselves for the companies benefit, but inadvertently convinced them to vote for policies which were deliberately against their best interests. They defend the system and gods that abuse them.

The story is that there are saviors and there are slaves. And the saviors are to be worshiped and served by the slaves.

The binary "us vs them" serves a vital role, because you can't have a savior without an oppressor. And if The liberated are made out to be the enemy then we won't question our captors who claim to be saviors.

It doesn't matter how many books you read, until you rewrite your own story nothing will change.

We often have a proclivity to single out a hero or savior to function as an example or leader subsequently elevating or exalting the one as a master or lord over the 99. And truthfully Community shouldn't come at the expense of our individuality but nether should or individuality cost us our community. We need both.

All this to articulate that tho unbeknownst to them at the time the patriarchal fathers were nether gods nor evil but simply ill equipped to handle their own insecurities and trauma not to mention that which they quite literally inherited from their fathers and subsequently passed down to their children. We of corse are now capable of understanding their plight and realizing that the solution is not for a god smite them but rather for their biological needs to be met.

## OBJECTIVE MORALITY

We live in a world where Viruses like the Marburg virus kills its host thru excruciating internal and external bleeding, fungus like cordyceps (zombie ant fungus) drains insects of all nutrients before promptly using their carcass to reproduce spores, and wasps like the emerald cockroach wasp which uses a neurotoxin to commandeer its victims (still animated body and mind) as a host for its larva. these are not abominations but simply organisms that evolved mechanisms and behaviors in order to survive precarious environments. And yet These "insidious" behaviors are no worse then the god given commands to kill,rape,plunder, and torture indigenous peoples.

The objective morality of the Abrahamic traditions are authoritarian and thus grounded in obedience and loyalty. And thus The objective wrong is anything which transgress that authority. So if god were to say command the sacrifice of every first born son (exodus 22:28-29) or the slaughter of every breathing thing (numbers 31:17-18) or even to take up a sward against nonbelievers (Matthew 10:34-36) to refrain from murder and genocide would then be "objectively wrong.

In the case of objective reality all meaning is made up (ether by some god who assigned it or by men and women who suffer it). And thus the conclusion of an objectively moral being in a world where every living organism survives at the expense of countless others, would then be to cease living. This brings us to The epitome of nihilistic morality. For the author of Ecclesiastes the answer to this conundrum was indulgence and pleasure, where religion requires a sacrificial lamb or son of man to atone for the author of life's blunders. Some philosophers assign meaning in regards to the amount of good one does, wile organisms like viruses and fungi simply kill rich and poor alike all so that the organism may survive.

Intuitively man has assigned some form of cosmic value to sensations like pleasure and pain, ascribing virtue to pleasure and associating pain with punishment for wrongs. This however is not quite the case as we will see in the next chapter. But not only does pain and pleasure not abide by these rules, more often then not it is the wicked or cruel who are dealt lives of ease wile those who are pious often suffer the most. More is required Then a zero sum, and simply imposing some form of authoritative architect of life, leaves the studious philosopher with an authoritative author who's moral integrity is inferior to that of its subordinates.

"They call it god's love, and yet the only evidence of it is in the actions of man. For it defies all laws of nature and god, perhaps such love is despite god."

There was a time not too long ago when people believed

that things like love and inspiration/genius were strictly ethereal qualities exclusive to gods and only bestowed upon man by gods. However this is not the case. The phenomenons we call love and genius/inspiration are the results biological processes and neuronal networks devised by our genes to facilitate and incentivize the propagation and preservation of those same genes in the form of a particular species.

The Abrahamic traditions deliberately claim that those who reject their "god's"prescribed identities are then fundamentally immoral and subsequently deserving of suffering. This demoralization of nonbelievers deliberately dehumanizes those who don't adopt religions identities. And is by very definition a religious form of supremacy.

## HELL AND THE GOD THAT MADE IT

If an eternal being can be offended by a finite "crime" (especially one so trifling as refusing to validate the ego of an invisible and absentee god) then eternity must be a rather small and fragile thing! Not only dose this entity lack integrity but they also invalidate the profundity of "eternity" by definition. Especially when what ever transgression one committed was a result of that beings neglect or malcontent of that individuals innate needs and desires, in accordance with their design in the first place! If there is a hell, then it is only fit for the narcissist of a god who with limitless recourses neglected the needs of those they designed. That's the equivalent of breeding a dog to fight for survival and then holding it responsible when your neglect to feed it and malevolence in bringing it up causes it to retaliate. However even in that case the dog is capable of hurting its malevolent yet very defenseless owner, whereas an all powerful god being offended and even hurt by the opinions of an ant to such a degree that they retaliate with perpetual torture for infractions which hurt nothing but their fragile ego, is exponentially worse. In short if hell exists, then god is nether good nor mighty. Especially when one considers that that god uses that power not to help the world they created but rather to punish those who don't glorify and praise that absent and evidently abusive father.

The truth is this conception of god and it's relating identities are actually the result of man and beasts (culture and nature) and not that of some divine being of old.

Ultimately the question isn't if god does or does not exist, or if the Bible is an accurate historical account, but rather the virtuosity of that "god" and "assignment of meaning"

If there truly is a judgement then all the evil done in the name of a god and by those who bare their name will bare the brunt of its wrath.

More lately we have seen an increase of stories which sympathize with the villain. However this is not a move towards idolizing immoral behavior, but rather recognizing many of the tropes vilified by traditional literature were those of marginalized communities. These narratives were intended to justify the abuse of those of lower economic/hierarchical strata's. And deliberately dehumanize those who opposed the traditional values and conventions of tribal and national communities.

Far too often it is those who were denied a voice and kept in the dark which are then called monsters and labeled as evil or less the human for expressing their disscontempt. Like a starved dog or neglected heart, it is not those who had been abused, but the systems which perpetuate that abuse which are the true monster. And above all else the gods who instituted those systems of oppression and abuse which should be held accountable (not only because they had power but more prominently because "if they truly were all knowing as they claimed" they should have known better).

In its most ancient literary forms (46-7AD setne ll papyrus) Egyptian, and (Daniel 164 B.C.) Jewish, the concept of hell was (like the day of the lord) a form of justice on behalf of the marginalized and exploited communities. It was only much latter that it then got rebranded to enforce social norms and threaten perpetual torment upon those marginalized and exploited communities who were vilified for not conforming to the hierarchical social conventions of their time and culture. It was originally intended to subvert the hierarchy of (might makes right) and not substantiate it's prescribed abuse of subordinates.

The point is that if there truly is a god who created the world as we know it, then that god with supposedly limitless resources then deliberately created a world of scarcity and division where the very behaviors that god called abhorrent and evil were then a fundamental prerequisite for survival.

And as the architect of the very same scarce environments which made evil and violence necessary, that god is all at once the soul cause of such evil and thus responsible or guilty of the sum total of sin and vise.

## THE CONQUEST AND KINGDOM OF GOD

The earliest historical account or record of Israel as a nation is found in the Merneptah Stele (1210 BCE) by an Egyptian king MERENPTAH.

In contrast the oldest art we have form Egypt dates back as far as 4500 BCE (sculptures found in the merimde Beni Salama site in the delta) however there are social records (Egyptian structures and buildings) dating back as far as 100,000 B.C (nearly 9,000 years before the reign of merenpth.)

It is worth noting that tho there is much contention among scholars on the identity of Yhw archaeologists have found the name inscribed in a geographical list from the time of (the Egyptian pharaoh) Ameniphis lll dating back to around the fist half of the 14th century BCE and this serves as the oldest record of "the god of the Bible". However this would contradict claims that WHYH was solely the god of Israel seeing as how the Yhw here belongs to a nomadic tribe known as the Shasu who wandered Arabah.

the first uncontested record of the name Yahweh

is found in a much later source (the Moabite "mesha inscription) dating to the mid 9th century BCE.

From Jericho to exile in Babylon.

Moses was the embodiment of the tora delivering us from Egypt and thru the promised land. but now we need the courage to face the giants who stand before us. No longer do we follow, now we lead.

But what does that really look like. According to the scriptures it looked very similar to those very same oppressors who were overcome by the Israelites. We see just as much war, genocide, slavery, and exploitation of land and humanity all in pursuit for more land, slaves, and resources. We get accounts of kings like David himself taking credit for the slaying of the giant Goliath (a Hebrews cosmological depiction of the raphirm (mighty warriors/barbaric rebels) "the mythical offspring of cain" and the biblical progenitor of Marduk, Gilgamesh, Babylon, filista, and niniva, AE Canaan's indigenous people (cainanites) . It should be noted here that much of the Old Testament literature functions as commentary on these cultures myths, propaganda, and law code, similar to that of the New Testaments satirical commentary on the Greco Romen propaganda), and so tho the tiny David is credited with the vanquishing of this mythical giant, it is revealed later that it was in fact Elhanan, son of Jaare-Oregim whow did so (2 Samuel 21:19). Regardless We see these kings take all the credit and glory for the deeds achieved by the hands of the slaves they conquered. (Much like their god).

Tho it is worth noting that the Israelights (according

to scripture) didn't have many male slaves and so By slave here I don't mean someone sold into slavery but rather people born into a system that profits from their exploitation.) however this lack of male slaves is not a sign of their generosity but rather that of their tyranny. Because as they were commanded by their god most of the males were killed.

(numbers 31:17) ESV

"17 Now therefore kill every male among the little ones, and kill every woman that hath known man by lying with him.

18 But all the women children, that have not known a man by lying with him, keep alive for yourselves."

(Deutaronamy 20:16-18) ESV

"16 But in the cities of these peoples that the Lord your God is giving you for an inheritance, you shall save alive nothing that breathes,

17 but you shall devote them to complete destruction,1 the Hittites and the Amorites, the Canaanites and the Perizzites, the Hivites and the Jebusites, as the Lord your God has commanded,

18 that they may not teach you to do according to all their abominable practices that they have done for their gods, and so you sin against the Lord your God."

They did supposedly have female slaves which were then taken as wives. (Exodus 21:1-25) tho in many cases these laws where their to protect the slave.

So how do we reconcile this Israelite nations conquest

and crusade of cainan (genocide and all) see Joshua 11:20-23, Deuteronomy 25:19, and Deuteronomy 7:16, with the proclamations in Leviticus 19:34 to treat foreigners with compassion.

And yet in passages such as 1 kinks 11:12, and Deuteronomy 7:3, Israel is instructed not to intermarry with those of other nations and cultures. We see this articulated later in the words of the biblical Jesus who does not even recognize the cainanite woman as human (Matthew 15:21-28) and again when he instructs his followers to take up swords against unbelievers in their own families (Matthew 10:34-37). We see a god who judges people for both their ethnicity and their cultural identity.

The truth is this kind of "Christian nationalism" and "colonial conquest" is seated so deep in these traditions that it is hard to discard without injuring the integrity of the texts and traditions as a whole. And yet at the same time we have those voices arguing on behalf of inclusion and acceptance as well. Even so Both of these voices are assigned to god.

The god of the Israelites did not seem to see the land of abundance and promise as a destination but instead as an perpetual journey. Seeing as how a journey which should have taken only a couple of days (if that) rather than a whopping 1 to 40 years.

It is important here to reiterate that the scholarly consensus is that the Israelites crusade of Canaan was at the very least greatly exaggerated and most likely completely fictional rhetoric. In Truth these mythical narratives were probably constructed to explain the Israelites captivity and exile form the promised land. Or more simply the author(s) of Deuteronomy-1st kings imposed this rhetoric concluding that the nations which the Israelites mythically vanquished or displaced did so in divine retribution for their "idolatry" and rejection of the Israelites god('s). This rhetoric also functioned as an explanation for Israel's exile and captivity by imposing "stipulations" for the lands inhabitants. Ultimately these narratives were employed as a means of retaining the power and might of the Hebrew's god, by claiming that the Israelites defeat was a sign of their unfaithfulness and not the inadequacy of their god.

Nevertheless the love of god in these narratives was a very patriarchal kind of love of which was dependent upon belonging to god, like a son or daughter in the near Middle East who belonged to the patriarch or father. God loved those who bore his name only so long as they remained loyal only to him. The Israelites were gods portion and possession. They were property and even servants of god and his will. To say Moses was like the Elohim, is to say the former prince of Egypt was likened to a god to preside and rule over the Israelites, which were like cattle.

> "to go out and come in before them, one who will lead them out and bring them in, so the LORD's people will not be like sheep without a shepherd."
>
> -numbers 27:17 (NIV)
> (thru the wilderness) -

We still see this today in Christianity who loves believers as family under Christ. And yet the moment one renounces Christ, that brotherly love is subsequently null and void. Thru this paradigm one's identity is solely dependent upon their relation to Christ. And thus it's deepest, most profound sense of love. Is one of possession and ownership. It's not an acceptance or appreciation of the individual as a whole, but rather the jealous covetous of that individuals association with/or belong to an idea as their identity.

## THE RISE AND FALL OF ROME

Much like the origins of the United States of America, the Roman republic began as a citizen militia where power was shared by the citizens after overthrowing the monarchy. Tho it was still prominently those of rich, influential, and powerful families who were elected as magistrates. So tho it was not a true democracy it was better than the standard royal monarchy at the time. (The Roman republic would eventually fall to the Roman Empire on September 2nd 31BCE) however the small agrarian city of Rome began it conquest of the ancient world in Italy by granting their defeated foes a degraded status of citizenship rather then merely slaughtering or enslaving the populous which was often the custom at the time.

By 250 BCE Rome presided over the entire Italian peninsula and set its eye on the western Mediterranean provinces, eventually even conquering the eastern Mediterranean shortly after the start of the 2nd century BCE. Tho the Hellenistic (Greek) world was far more refined (culturally), the Romans were far superior when it came to military prowess. By the 1st century BCE the Roman republic had conquered the entire ancient world. But because of its egregious disproportionate distribution of wealth and spoils to an small group of elites, and it's mistreatment of veterans who reaped nothing from their conquests and were often left homeless and impoverished, the internal conditions of the Roman republic were dire.

After a failed attempt at reform in 133 BC a grueling civil war broke out which would ultimately last until the battle of actium in 31 BCE. one of the figure heads of this civil war was a man named Julius Caesar who gained power and declared himself dictator for life. he ruled from 49 BCE to 44 BCE at witch point he was assassinated by the senate. However Before his death he adopted a grand nephew named Octavian as his son and heir. This led to yet another civil war resulting in The Roman republic becoming peacefully split between the east (ruled by Marcus Antonius), and the west (ruled by Octavian). However after a love affair between Antony and the Egyptian queen cleopatra (which violated a number of Roman traditions), a war broke out, of which Octavian was victorious. This victory subsequently made Octavian the ruler of the known world at the age of 31. In order to avoid meeting the same fate as his "father" (Julius Caesar) Octavian (symbolically) relinquished his omnipotent authority back to the senate and was "granted" authority over just a few prominently powerful provinces (Syria, Gaul, Cilicia and Asia Minor, Egypt, and hispania. Octavian then delegated his power over these provinces to appointed officials called legates (governors). Tho This appeared democratic from the outside, by presiding over these hand picked provenience, Octavian effectively controlled 23 of Rome's 28 military Legions. Ultimately the power of Rome was seated in military control and

By 68 AD it was apparent that the seat of power in Rome was in its military leaders (this would later both cause and ultimately cure the crisis of the 3rd century)

And thus By way of the title emperor Octavian retained control over the military.

Octavian also took on other traditional roman titles. For instance As a tribune Octavian had the sacred authority to discern what portion of the Roman citizens had a voice in politics by way of census, he could even deny certain people or groups a voice in government biased on no more then his subjective standard of morality. Ultimately giving him the authority to deny electoral power to people who opposed his particular views. (We actually see legislation attempting to do this in the state of Florida 2023) (SB 1248)

In Octavians self designation of a plethora of traditional Roman republic titles, he escaped the perception of kingship but still retained all the power and authority of a king or monarch, which was important seeing as how an enmity towards monarchy was one of the key components which led to the assassination of his "father" Julius Caesar. One of these titles incapacitates this humble appearance, wile also attributing divine (deistic) quality (much like the title "human one" and "the Christ" used by his contemporaries) the particular title in question was that of "Augustus"

One of Octavians titles was " pater patriae" which simply designated him as the father of Rome. In Roman culture fathers had the power to sell their children as slaves or even execute them at will. And thus this title was a traditional symbol of both warmth and familiarity as well as omnipotent authority and power.

In some ways Octavian modified traditionally republican roles to repurpose monarchical authority in a "democratic" setting.

It was almost 200 years before any roman would be willing to go on record debunking Octavians propaganda however. In the year 200 AD it was Cassius dio who made the claim that "in this way the power of both senate and people pass entirely into the hands of Augustus, and from this time there was strictly speaking a monarchy. For monarchy would be the truest name for it. The name of monarchy to be sure the Roman's so detested, that they called their emperors nether dictators nor kings nor anything of the sort. But Since the final authority of the government devolves upon them they must need to be kings. By virtue of these democratic sounding titles The emperors have clothed themselves with all the powers of the government to such an extent that they actually possess all the prerogatives of kings except their poultry title!" Nun the less Octavian set the president which all other Caesar's would follow.

Following the tyrannical rule of the psychotic Caligula, Claudius invaded Britain in 43 AD and by 52 AD most of Britain had been conquered and was among Rome's many provinces.

In 66 AD the Jews in judea rebelled against another tyrannical emperor known as Nero. To squelch the rebellion Nero employed the prominent military leader Vespasian.

In the year 64 AD under Nero's reign a massive fire befell the city of Rome. The flames engulfed 10 of the 14 districts in its tirade. It lasted 9 days in total. After fraudulent rumors spread that the great fire was instigated by Nero, he attempted to divert the blame to an obscure cult movement which would later become known as Christianity. This lead to the open persecution of the Christian's. After Nero's reign came to an end in 68 AD Rome was thrown into yet another civil war with Vespasian finally emerging victorious and instituting a long overdue time of stability.

In many ways Vespasian ruled judiciously, he gave economic incentives to educational providers and cut down on the governments frivolous extravagance. He was however very traditionally minded and sought to reinstitute means of power in the government chiefly to those of high birth and social status. He also over saw the fall of the temple in Jerusalem in 70 AD after the Jews insurrection. Among these Jewish revolutionaries was the sacarei, as well as the famous Jewish historian Josephus (the later of which was later freed by Vespasian).

In The year 81 AD domitian became emperor of Rome. subscribed to his own deitship he gave himself the title "dominus et Deus" which meant lord and god. This propaganda was mimicked by the Christians who used it as rhetoric to campaign for their sotor or savior (the Christ) of which their were many examples of at the time (like Apollonius of tyana)

Christianity repurposed much of Romes propaganda which fit quite well and thus Christianity thrived in many heavily urban Roman populous. Before Christianity (with the exception of Judaism) there were no real dogmatically authoritarian traditions which were grounded in a strict literature. Not Even Judaism had a set of orthodox texts considered canonical before the Hebrew Bible in the 2nd century CE by bishop Melito of Sardis. The biggest proponent for Christianity was a Greek citizen of Rome who was also a practicing Jew prior to his conversion. This man (now know the world round as Paul) drew from Roman propaganda to describe the Christian "Christ" or savior.

In Roman culture the heir designated in the will of the deceased, would not only inherit their wealth but also their debts and figuratively assume their identity (much like a reincarnation of the deceased ancestor).

In the year 235 AD a constant state of dissrest, social injustice, and distrust led to rampant civil war and rebellion in Rome. On top of this these on going conflicts took a detrimental toll on Romes economy causing devastating inflation. In conjunction with an increase in natural disaster and pandemics this period heavily influenced the ultimate fall of Roman and led to a split which would last till the end of Rome itself.

This period (235-284 AD) is known today as the crisis of the 3rd century.

From this moment on Rome was caught in between coercing the once independent kingdoms of the known world to corporate wile simultaneously recovering form the constant competition of those in charge of the military and government. In addition to these social struggles any environmental catastrophes and natural disasters during this period contributed to the decay of the once mighty empire.

"When a civilizations economic system becomes a virtual cancer to its general populous and environment then social and ecological revolt is inevitable."

On December 25th 274 AD the Roman Emperor Aurelian endorsed an romanized version of sol (an eastern son god) in an attempt to obtain unity throughout Rome. This would function as a precursor to Constantine's deceleration of monotheism in instituting a romanized version of Christianity as the national religion of Rome. Sol was often depicted with a beam of light around his head (this is where the halo around Jesus's head most probably came from).

By 274 AD Romes economy was ravished by rampant inflation and distrust in the value/integrity of the Roman coinage. This is a problem Aurelian failed to remedy.

After his assassination in 275 AD many rulers rose and fell with no real impact of resolving Romes devastating state. That is until 284 AD when a man named Diocletian rose to power and instituted many reforms. As opposed to struggling with his rivals, Diocletian delegated his authority to them convincing his most worthy opponents to work for him rather than against him. Among these reforms was an edict which curbed inflation by setting a price cap on certain goods and services. It was during Diocletian's rule that deviations from the national religious institutions began to be heavily persecuted. In 303 AD Diocletian passed an edict requiring the deviant Christian cult to make sacrifices to the traditional pagan gods as a sign of unity.

In the year 306 AD Constantine was elected emperor by his troops in Britain and he would rule till 337 AD. Originally Constantine claimed that sol (the sun god) had ordained him to rule over Rome but by 310 AD it was Apollo who supposedly chose Constantine as his elect. However only two years later In 312 AD Constantine was the first Roman emperor to convert to Christianity and he expediently declared Christianity the national religion of Rome. This was particularly interesting when we consider just how obscure this religion was at the time. The oldest recorded evidence of Christianity as an religion dates to around 130 AD and even by 300 AD only 10% of the population identified as Christians. Adopting Christianity served Constantine well as it designated him as the one and only ruler ordained to rule over the world by the one and only "true" god. Tho previous emperors claimed to be gods or agents of gods they still acknowledged the existence and authority of entire pantheons of deities. This would ultimately caused problems for Constantine tho. A group of theologian known as the arians who claimed that the trinity explicitly contradicted the monotheistic claims of Christianity, and that the Christ must then be subordinate and lesser then god and thus not in fact a god himself. this esoteric dispute over the divinity of the Christ threatened to divide the country and so in 325 AD Constantine convened a council of 300 bishops to resolve this theological dispute. The result of this council (which Constantine himself presided over) was an official creed by which to refer to in such theological maters (the nicene creed). Before this point there was no real standard for orthodoxy or canonical text. moreover it was here that "authoritative" personages came to an agreement on the meaning and authority of certain theological claims. It was also here where categories like angel were constructed to retain coherence with the claim of monotheism.(prior to this point these divine messengers and genius's were themselves gods and/or parts of gods divinity dwelling within man. Among These were tools like the belief in "divine right to rule" which designated authority thru consensus. Likewise many of these frameworks have been abandoned and or reappraised. Constantine exercised his authority over both the conduct of the nation and the church. Constantine used this Christian nationalistic dogma to justify his conquest in the pursuit of being the soul ruler of Rome. After Constantine all but one of Romes emperors would be Christian.

In 330 AD under the rule of Constantine Rome was officially split into the western empire, and the eastern Byzantine empire.

In the late 4$^{th}$ century a nomadic group of raiders known as the Huns forced the goths to flee and seek refuge within the boarders of Rome. Initially emperor valens accepted the goths into the fold and offered the greuthungi and thervingi aid in exchange for military support. This aid however was withheld by corrupt officials in the Roman government. This ultimately led to an incursion near adrianople in 378 AD where valens was defeated (mostly at the hands of his own hubris and thirst for power). This was the first time in the history of Rome that an emperor had been slain by uncivilized Barbadians. The term barbarian here was used to describe any tribe of people who did not ether speak Greek or comply with the traditions of Greco Roman culture. After this defeat the rulers of Rome became far less traditional from the barbarian arbogast and stilicho, to the women who ruled the country from the shadows like galla placidia, aelia eudocia, and aelia pulcheria. In 410 AD a Visigoth, mercenary, and former Roman general, named Alaric captured and luted the capital city of Rome. And tho this occupation only lasted 3 days (at which point Alaric and his Visigoth army departed with their plunder) this defeat devastated the moral of Roman nationalism. For the fist time in its history the capital of Rome had been conquered by foreigners. After this point the western Roman Empire slowly began to diminish as barbarian tribes like the Gauls, franks, saxons, Angles, and vandals, who reappropriated regions once controlled by Rome. Spain was taken by the goths, and later in 429 AD it was taken by the vandals who continued their conquest into north Africa and Carthage, Britain was seized by the Angles and Saxons. The majority of Romes food came from Spain and Northern Africa and thus these barbarian conquests ultimately resulted in food shortages and diminished population throughout Rome. Even Italy ceased to be under Roman control. and so by 450 AD when the huns invaded Europe, the western Roman Empire didn't stand a chance at repelling them. After a stalemate in France in the year 451 AD the Huns were momentarily halted by the Romans, visigoths, Burgundians, franks, and Saxons. The very next year however Attila and the Huns set out to conquer the city of Rome. Tho Rome didn't stand a chance, a meeting between Attila and pope Leo the 1st resulted in Attila's thirst for conquest being quelled at which point the Huns returned to Gaul leaving the city of Rome intact and unspoiled. In 455 AD the vandals saked Rome, and in 476 AD the last emperor of the western Roman Empire (by the name of Romulus Augustus) was deposed by odoacer (this would mark the beginning of what we now call the Middle Ages).

Tho the western Roman Empire had fallen in 476 AD, the much wealthier eastern Byzantine empire would purists into the renaissance era finally meeting its demise in 1453 AD at the hand of the Ottomans.

In 610 AD a middle aged merchant named Mohammed who lived in the town of Mecca began to have visions of an Angel named Gabriel who imparted divine revelations. These revelations would be preserved in literary form as the Quran (or the recitations). Which would function as the doctrine for what would become the religion of Islam (the word Islam can quite literally be translated as submission) which is a key component of the doctrine which prescribes complete submission to the omnipotent authority of god or Allah. Like Christianity and Judaism, Islam too was an reprisal of the ancient Hebrew Abrahamic traditions and thus centered around the Hebrew god whyh or el. Where Christianity claims yeshua (Jesus or an abbreviation of the name Joshua) was the one and only son of god, and Judaism claims yeshua was an apostate and heretical blasphemer, Islam claims that he was a prophet who's message was construed by people like Paul and Constantine who sought to profit off of it. The message of Islam resonated most with the Arab tribes of the east. And it was then these very same tribes which would defeat both the Byzantine and Sassanian armies in 636 AD under the flag of Islam. This very same year Jerusalem was conquered by Islamic tribes. In 642 AD Egypt fell and most of the eastern world was under the rule of these Islamic tribes, leaving only Constantinople, the Balkans, and parts of Anatolia under Byzantine control.

After the fall of western Rome, and even into the gradual decline of the Byzantine empire, much of the culture and knowledge secured by Roman rule eroded giving way to a influx in religion. Tho in great part Romes demise came at the hands of external pressures like the Huns, still more consequential in its fall were the results of its own malevolence. From income inequality/distribution of wealth, to the ecological drain on the land from excessive cultivation. One could argue that it was factors like the socioeconomic injustice which lead to the construction of religions like the early Christians and Jews and latter the islams to hold the Roman authorities accountable. The institution of Christianity as the national religion closely corresponds with the decline of the Roman Empire, and tho in great part the religion was adopted to unify the nation during a time of strife and division in many ways it contributed to the problems of inequality and disproportionate distribution of wealth as the church became exclusively rich but left the common people greatly impoverished. Tho the church drew from the nations wealth it did not give back to the economy, instead hoarding its wealth and land (much like mega churches in America today some even amassing over $100 billion dollars worth in member donations). The livelihoods of clergy, monks, and nuns, depended on the workforce to sustain them wile the Christian church did not aid in alleviating the strains but instead chastised and shamed the working class that supported them. Tho at its roots Christianity sought to subvert the power structures instituted by Rome, its hyper vigilance on obtaining salvation for the next life inadvertently detracted from meeting the needs of the people and environment in the present. This was a stark contrast to the Hebrew traditions which prescribed fair treatment of foreigners and that its observers take care of the earth even going as far as allowing the land to rest (lay fallow) every 7 years. The decline and fall of Rome inspired both a dark period in regards to advancements in science and philosophy but prompted much needed cultural reform. This period of late antiquity very much was an apocalypse in the way it incapacitates both the end of the old world and the beginning or birth of new traditions and paradigms. Tho this time marks the demise of an empirical Roman

Empire, more notably it accounts an transition where many components of Roman culture were integrated with those of immigrant tribes which (tho not necessarily citizens) at many times were themselves considered Roman. Both the rise and the fall of the Roman Empire played an pivotal role in modern globalization.

With that said the fall of the western empire in 476 AD marked a steady decline and eventual loss of Roman societal organization and technology, as well as a stagnation in trade and transmission of diverse cultures and ideas. And thus this period is often referred to as the dark ages. During this period the migration of Saxons into Roman territory inadvertently prompted the rise of feudalism as wealthy land owners often offered protection to peasants in exchange for labor and a percentage of their crops. Tho feudalism is often depicted as a step up from slavery, during this time surfs owned 40 to 70% of the profits from their production. Which is vastly more then the current 25 to 35% awarded to skilled laborers under modern capitalism. In fact in many ways wage laborers today are closer to slaves in Europe in the 14th to 17th centuries then they are surfs under feudalism.

During the 5th century a group of Anglo-Saxons migrated from Denmark and Germany and into Britain. These closely related Germanic tribes would eventually become the English empire.

In approximately 1000 AD the Saxons (Vikings) had began exploring Northern America.

In 1095 AD pope Urban ll convened the Council of Clermont and initiated an campaign to retake the holy land. The conquest of Jerusalem began in 1099 and resulted in a victory over the Muslims. These campaigns or crusades would conuente until 1291 AD.

In the year 1315 AD inclement weather caused a massive famine to sweep Europe. over the next 2 years. This famine would decrease the population by 10% and lowers the life expectancy to 29 years of age. A mere 20 years later Europe would be hit by a massive (7 year) pandemic known as the bubonic plague which by 1353 AD would wipe out another 50% (100 million) of the European population. In 1350 another pandemic hit Europe and lasted till the 17th century. Tho these events were understandably tragic. They gave the remaining population more political influence and power allowing peasants more rights and mobility to climb the socioeconomic rungs themselves often becoming land owners. The plague also prompted an increase in religious fanaticism which in turn lead to an increase in Jewish persecution under the guise of the Christian inquisition. Many Christians at this time (like the early Christians of Rome and the apocalyptisists of yeshua's day) believed that the end was nigh and the kingdom of god was upon them. They attributed the famines and plagues as a sign of gods retribution and judgment.

This led to a form of "spiritual eugenics" where conversion to the one and only "true religion" of Christianity became of paramount importance. And thus the persecution of inalienable human rights and liberties, torture, and the eventual execution, of "heretics" was not only a necessary evil but actually divinely righteous. Tho the plague was seen as gods retribution, it was also seen as "satan's" triumph over mankind. And thus fear of "demonic satans" was perhaps more influential than praise/adoration for "god". This allowed the brutal abuses "prescribed by god" to be overlooked as morally just and righteous. Moreover this fear of heresy permitted the clergy to eradicate people who could oppose them. More often then not these "witches, and heretics were scapegoats for the people to blame for their own misfortunes (Much like the Jews in nazi Germany). The inquisition began in France in the 12th century as a means of persecuting the "heretical Abilgenses sect" and it would soon be adopted by Europeans to "cleanse a corrupt populous. There was now no real centralized authority to hold the inquisitors accountable, but instead these "inquisitors acted on their own authority with absolutely no real oversight. In many cases evidence of witchcraft was egregiously fabricated and testimonies tyrannically chorussed thru torture of both the accused and small children who were forced to falsely testify against the accused.

In many ways England used religion to enforce compliance among its conquered and subjugated peoples as well as to justify its expansion which was ultimately about "spreading Christianity unto all the world". And not obtaining precious recourses and land. (That last bit was satirical).

> "Christianity did not become a major religion by the quality of its truths, but by the quality of its violence" (Michael Sherlock)

## Puritans in pursuit for persecution

In the winter of 1620 the Wampanoag people provide both food and aid to the European pilgrims and new comers to their native land. Just as Columbus had noted in his journal over a century before "But when they see that they are safe, and all fear is banished, they are very guileless and honest, and very liberal of all they have. No one refuses the asker anything that he possesses; on the contrary they themselves invite us to ask for it. They manifest the greatest affection towards all of us, exchanging valuable things for trifles, content with the very least thing or nothing at all." In the year 1630 (10 years after the pilgrims departed for the Americas) The puritans fled England not to escape persecution (like the pilgrims did) but instead so that they could persecute those of different religions and cultures. As their political influence had been stopped by King Charles I. However much like Columbus, the puritans and the company which chartered the canonization of the Massachusetts bay sought to first and foremost convert and indoctrinate these native peoples ("to win and incite the natives of this country, to the knowledge and obedience of the only true God and Saviour of mankind.") even going as far as stripping children from their homes and subjecting them to extreme (physical, and emotional) abuse in the name of god and Christ. This forced conversion and torment combined with the seizure of native land resulted in an all out war (Pequots war of 1636-1638) between the puritans and the native tribes (Pequots). This war ultimately ended in the defeat, enslavement and in some cases even genocide of a vast majority of the Pequot people.

And this is where we get a great deal of the tyranny associated with colonization. Because contrary to popular belief "cultural appropriation" (taking and modifying positive qualities in competing cultures and then integrating them into more conducent and inclusive social constructs and systems) is a very good thing and actually exactly how creatures (and life as a whole) evolve. What was bad about this particular practice is the way it was used to indoctrinate and enslave/exploit those people of the indigenous cultures by imposing this belief in some form of divine hierarchy (or right to rule). This conquest and colonization of neighboring civilizations was the norm and most certainly not restricted to Greco Roman and indo European cultures. Further more in group biases and cultural prejudices are innate to virtually all sentient organisms. However This does not make these biases morally sound nor does it validate the hierarchies imposed by these cultures, but it also affirms that these evils are not rooted in man's corrupt nature but in the very coding of nature (or god). And our ability to negate those instinctual biases and refute the omniscient voice of god is actually very good. It is perfectly fine to change your mind and it doesn't invalidate ether your past or present by choosing a future which looks nothing like ether. Your life has value simply in being but if you do choose too you get to improve yourself.

This belief in hierarchy is illustrated elsewhere by the detrimental consequences which befell one of the barers of the ark of the covenant in $2^{nd}$ Samuel 6:7 where an man named uzzah who is of inferior station reaches out to stop the "sacred" ark (dwelling of god) from hitting the ground. He was subsequently struck dead for daring to touch the sacred ark.

There is this myth going around that people are rejecting god because they've been mistreated by the church. This claim Hingis on the assertion that these people never really had a relationship with god. And perhaps there is some truth to this claim because the only ephemeral representation of god is exemplified in the character of those who claim to bare gods image and name. After all god has yet to come down and set things straight and answer for all the evil perpetrated in their name, by their authority, and in accordance with their supposed word. Which every cutler claims a different set of sacred texts to.

The truth is many of us believed we had a relationship with god, but no matter how many times we prayed, the only voice we heard was that of our PFC, amygdala, and autonomic nerves system, projecting an internal autobiographical simulation of our biological systems check. The truth is those people who are leaving their communities, are doing so because they can no longer subscribe to a belief in hierarchy which claims gods might makes their malevolence and malcontent "right"! They are not people looking to do evil. But rather those who are no longer willing to be complicit in the evil done in gods name. They are people who have made the sacrificial choice to love themselves and others better without all the bigotry, sexism, racism, and misogyny, of a small,intolerant, malevolent god who is restricted by culture, ethnicity, and nationality.

If god truly is a god then they are responsible for all the evil (genocide, crusade, conquest, slavery, indoctrination, abuse, and prejudice) done in their name and by their supposed authority (in accordance with their "word") which they have yet to correct. All that blood then is on their hands. Every act of evil perpetrated by those who supposedly bare gods name and which was justified by claims in sacred "divine" texts is ultimately an extension of gods behavior by way of their complicity and absence of rectification.

"Gods do not fear blades but meet their demise at the hands of doubt. To kill a god, simply cease belief in their authority."

## TRADITIONS ROLE

> "The secret of change is to focus all of
> your energy, not on fighting the old, but
> building the new."
>
> -Socrates-

To curse the name of god thru a rejection of faith and hope, is to be cursed or to deprive life of its meaning. Or in other words to reject these claims of value and meaning attributed to the phenomenon we call god or divine is to be neglected of those cosmic values and meaning.

To abandon the narratives which ascribed meaning is to inadvertently reject the meaning they ascribed. It is like stripping all the emotion attributed to "love" only to recognize it analytically. It's more truthful perhaps but it's also less real in its loss of mirth and meaning.

In the book of numbers We get this account of Aron's sons attempt to usurp their fathers role in their offering of strange fire. This mirrors the literary frame work of genesis where the humans usurp natures rule thru our consciousness and obtainment of autonomy (or eating of the tree of knowledge). Ultimately this goes back to this cosmological idea of Creation as a form of separation. Because this order and authority which is being usurped in both accounts is an attempt at distinguishing the hierarchy of nature. The point is We must learn the secret name of ra like isis did before we can rule. We must understand why nature works the way it does before we can effectively decide for ourselves our own nature. We must conform to the systems of culture in order to change them for the better. We see this in the command To honor our fathers and mothers, to respect our tribes traditions. And thus "The test" is not about instilling in us a willingness to simply submit or resist the wisdom of our ancestors but to wrestle with our gods and cultural values. Regardless of what you may have been taught, The validity of your existence and identity is not dependent upon the conclusion of an ideology. We do not owe our autonomous role to the authority of an idea or instinct. They can help guide but they should not be allowed to rule. This is ultimately the problem with this idea of salvation thru faith. Because Such A standard for righteousness which is dependent upon retaining a belief rather then altering or even abandoning that belief in response to contradictory evidence is not rooted in a pursuit for truth but rather one of fear and contempt for change and growth.

A fool is one who ignores the evidence because it doesn't match their desired outcome. The fact that faith in a hierarchical structure is what delivers people into salvation is evidence enough that this standard is based upon controlling and enforcing social norms rather then enacting justice and progress. This is why the way to god is restricted by corresponding culture and that god is only as powerful as the faith of those who believe in them(god). Any god these restrictions derive from is not one of truth but one constructed by man in order to control and exploit people. It is an attempt to reduce humanity to something small, but We are not one or the other. We have simply been taught or chosen to identify with one or the other inadvertently denying more then half of (not only our potential) but our authentic selfs.

In part it was our commonality of belief which allowed us to congregate on such a large scale. as it set the perimeters for conducent behaviors. This in part may be why so many feel anxiety when required to interact with large groups of people, because the previously agreed upon standard of past generations is obsolete. so we don't really know how to both genuinely express our selves (communicate our thoughts and needs) and how to respect (interpret/understand) others verbal or nonverbal ques. However this breakdown of hierarchical standards and social norms is not a bad thing. And the primary reasoning behind its downfall is the unrealistic restraints it puts on our ability to effectively communicate and express ourselves in the first place. The truth is we can coexist without subscribing to a hierarchically standardized framework. Peoples diversity is not a problem, but instead it is the beliefs that our value and validity is dependent upon a traditional or hierarchical identity which are the real problem.

"We are predisposed to posit a preference for prior environmental parameters out of familiarity" new and strange phenomena inspire us, but familiar environments offer comfort and security in uncertainty.

Religion in this way often claims to seek truth but as exemplified in its omnipotently dichotomous conclusions it actually seeks to escape the uncomfortable existential truths of reality. Religion seeks to discredit all expressions of life which contrast its ego centric paradigm by labeling anything "other" as bad or evil, subsequently invalidating its own claims of univocality. Religions lack of compassion for reality "as it is" in favor of an illusion "what it should be" is a blatant disregard for truth. Sadly the idea that morality is dependent upon our adhesion to these illusions or dilusulions, is appalling. When intolerance is passed on as the only means of morality, then the problem isn't diversity but instead the small close minded religion which seeks to pacify existential pain rather then understand its purpose.

Tradition most probably functioned as an intermediary in our transition from instinct to autonomy.

Much of the rhetoric we see in both the old and New Testaments are attempts at reconciling and even retaining the wisdom in earlier tradition wile also embracing the coming revolutions. Because people (and life in general) are not static but dynamic and fluid by nature. It is my belief that Tradition was meant to be more like training wheels and not breaks, And this has always been what the story has been about, from man questioning the authority of god or the voice of our instincts (taking the forbidden fruit and discerning good from bad for our selves) to becoming the very heretics we once hated (rejecting the traditions of our fathers and mothers for more compassionate and comprehensive paradigms). However much like those of Abraham, Moses, Jesus, and Paul, our new paradigms will require critiquing and expounding upon. Much like these revolutions which we now know as tradition, we too will have to modify and adapt our new world views using what we know wile still leaving space for what we don't. And that's perfectly fine.

## SACRIFICE AND SALVATION

In many ways we have become desensitized to the fact that life for us and every living organism on this planet, comes at the gruesome and horrendous cost of death for another. And often Our peace of mind is dependent upon our distinction between our lives value and that of the creature we must kill and devour in order to survive. We no longer watch the light leave their eyes, we no longer drain the blood from its inert body, we no longer have to contend with the conscious suffering evoked in order to persist. We are products of a cruel world and nature. But we are capable of empathize with our prey, or giving consent and feeling remorse for those worse off. And so understandably we see this consciousness and autonomy/responsibility as a curse.

Belief is the way to salvation because in its original Greek salvation was about preserving or holding together. It is faith which solidifies the authority of societies beliefs.

Unfortunately it is those who have been oppressed by such beliefs in social convention which retain a sense of adherence to such illusions as concrete. Where as those who profit from them only profit from them with the understanding that they are in fact illusions and only powerful because of belief in them. This is most probably why sacrificial rituals are so important to religious beliefs. because in positing real consequences to the theoretical beliefs, they are given a sense of congruity with the real world. Which in part is how that particular VMPFC mechanism is designed to function. By imposing the internal model on the external world, that model governs our behaviors inadvertently making its presumptions (beliefs) a reality.

There are many different interpretations of Abrahams test in genesis. The most common in the church is that gods command for Abraham to sacrifice his son was ultimately a test of Abrahams willingness to obey.

A more anthropological interpretation of this story is not as a test but rather as a claim made by Abraham who is illustrating that his god is not like the gods of his fathers (who did require human sacrifice).

However perhaps the best interpretation of this particular story comes from the modern author j.Richard Middleton. Who presents the role of the test as an invitation to wrestle with god and tradition on behalf of more compassionately inclusive representations of nature and culture. More simply to reappraise the supposed meaning and value of such behaviors, beliefs, and claims. This interpretation concludes with the presumption that Abraham actually failed in his obedience.

Because ultimately the purpose was not to discard the wisdom of his fathers, nor was it to submit to its conclusions but rather to test its values and acknowledge its accomplishments.

Before we needed the unifying narrative of our traditions to compel us, we were driven by our instincts.

culture took the place of nature and the idea usurped the role of our instincts. This usurped authority is depicted literarily thru the angelic beings (the gods or nephalem) trading their knowledge and skills for our autonomy. And so not only do we see the serpent (animalistic impulses/instincts) usurp the role of humans, but also these cosmic celestial beings (the ideals and muses of culture) trade the promise of technology and security in civilization for the freedom and autonomy of humans. And so we realized that we had to sacrifice the animal in us in the name of these deities in order to reform society and obtain civilization. However This social structures success is dependent upon a hierarchy. And positing of this hierarchy had to be trained thru the creation of the holy.

Holiness or Sanctification is created (or more accurately creating) by separating a space from another. This is done by creating patterns and then labeling deviations from those patterns as ether holy or taboo. For instance diner is eaten at a specific time of the day every day, and also eaten in a specific place (around the diner table) this space may be considered holy or sacred. however Desecration occurs when we deviate from that pattern or set of traditions. For instance eating in the bathroom or having sex on the kitchen counter may trigger disgust. These things are considered taboo and it triggers the activation of our insula which regulates both physical and moral disgust. (This is also why people who our physically attractive are also often judged to be more morally upright and trustworthy). But A dog has no problem humping a strangers leg or eating it's own vomit/drinking out of the toilet. and yet we find these actions repulsive. (this can also be observed in small children). This is key to the integrity of the structure as a whole because as illustrated in the creation poem at the beginning of genesis, creation is not about manifesting the material but rather applying meaning and value to the universe by separating or defining the "holy" form the "whole". This is why Making the sacred common or even dirty/exposed, is so detrimental to the system. The truth is Things like defecting and sex are natural and not bad but rightfully are considered taboo and assigned a sense of privacy to them. The cool thing here is that tho this is a philosophical creation, it results in a physiological one. Because Healthy habits change our chemistry. And so practices like Fasting, grieving, morning, and abstinence, actually alters our states of awareness and consciousness making spaces and memories holy. The point isn't to reject tradition but rather to modify it accordingly. Ritual purity shouldn't isolate but integrate. it's about appreciating diversity not cultivating duality or enmity. And so we should refrain from degrading one so as to exalt the other. It's about respecting eaches vital role. The point is to remember that things like Sex are no more important for the survival of the species then things like eating and defecating. Nun of them are inherently bad and ultimately just as crucial as the others.

it's all connected. These innate traits like our mutability and our capacity to be self conscious are not evil problems but instead themselves the solution, our ability to adapt and evolve to negate our instincts and decide for ourselves our own nature, these are positives.

Note

On a biological level our bodies release Adrenaline which functions as a solvent to solidify experience of insult and injuries. We also create fast responsive circuits which are triggered by stimuli resembling past experiences of danger, shame, or threat. Deep passionate Experience like lose or love develop very solid circuitry. And so tho the circuitry facilitates the software, ultimately the program reshapes the physical material. We call this process neuro-plasticity and in short it means Human nature is malleable.

The problem isn't human nature but rather that people are fooled by our myths which shape that nature. It's those who play their roles like pawns in a game. For after all all religion, racism, and politics is an illusion of our myths. This myth that creates camaraderie by convincing people that our differences make us enemies and not more useful ally's. These myths turn men against women and artists against analysts, the faithful against the skeptical (subsequently putting truth out of reach). These myths have fooled people into believing that this world was divided into progressive and conservatives, theists vs atheists or even that the color of one's skin makes one more fit for power and authority then another.

This is why apart from subscription to "the belief" (in a god), man is deemed less than human and as inherently evil. Not to mention the myth that without "a god" there is no standard by which to be good or moral. This is why indoctrination is so important, as it is the means by which a human life obtains value and validity as more then animal in the framework of a society. This is the basis by which racism and ethnic purity has been permitted to persist among religion and culture. And it's horrifying to think that many still depend on such beliefs to dictate morality and the value of one life (human) in contrast to another (animal).

In virtually every ancient culture we observe the ritual of human & or animal sacrifices take an pivotal role in the formation of its development. And tho this act most probably functioned as a sociological means of signaling allegiance to a community or cause. It did more then simply unify but also tempered society on the personal level. Because As barbaric as this act may appear to modern readers, It can be easily understood poetically, as a means of taming the primal human.

What is animal in us must die, in order for the divine to be obtained? and thus this sacrifice "of the creature" unto the "deified idea" was necessary for the development of our societies.

Ego in health is autonomy and independent. but in un health it is isolation and sterility/intolerance. Perhaps that's the point! Because virtually any form of life which is dualistic, will ultimately become cancerous and corrosive.

The world is not divided against itself as dualistic personalities, but unified as one wholistic mechanism.

Sacrificial lamb

In the tora we get two particular forms of sacrificial lams. We get The scape goat and the sacrifice. One goat is sacred, and the other is sent out into the wilderness with the metaphorical sins cast upon it. the guilty is set free or sent into the wilderness and the blameless is killed. This is paralleled in the New Testament with the story of Jesus the Nazarene "blames" being sacrificed and Jesus barabos, (the zeolite and rebel) being set free. This motif was first used in genesis with cain sent east with a mark and able, whose innocent blood had been shed.

Note The sacrifices were often eaten, death was required to live and survive. And thus a Sacrifice ritual was often a feast and festivity (celebration within a community).

## CREATION AND CHAOS

It is undeniable that nature does evolve itself into more complex forms and this is what is illustrated by the claim that "nature created itself". The opposing argument to this is that there is a being which we call god that created all things. However both scenarios require something to exist with out a cause. Ether a creator to create nature or an natural phenomenon to create being/god. However who is to say that nature is not being and god is not nature. The idea that the two claims are mutually exclusive just is not viable. This oddly enough is close to the dowists beliefs.

Perhaps the greatest proponent of this argument is found in a couple of poems at the beginning of genesis. However these poems lack unanimity with themselves and this is because these poems don't actually depict the manifestation of the material world but rather the assignment of meaning and values which construct the cosmological one. this is evident in the way that creation is not illustrated in the construction of something from nothing but rather by dividing and separating the whole into parts (mythically depicted as the chaos water). This form of creation (separation of the the day form the nite or the land form the waters) is an assignment of social constructs which distinguish man from beast and good form bad. thus it is dependent not upon the laws of physics but rather the beliefs and subscriptions of people in order to retain its authority. And as we see illustrated in the flood narrative, when people lose faith in these divisions which call one thing holy and another cursed, we are subject to the flood of the chaos waters (or the cosmological whole) because What social convention calls chaos is a world undivided by order. A world where no land is holy nor cursed. For it is only faith in hierarchy which makes one thing holy and the other common. It is tradition which attempts to tell us who we are and where we belong by telling us we are not part of it all. What social convention calls chaos I call eternity, unhindered by myths like time. Chaos only threatens that which had us convinced we are divided from them. Ultimately this is where the idea of god as a moral standard comes from. However it is only a moral standard so far as it distinguishes that which is worthy of respect and that which is not. This is why the gods could command so much killing and bloodshed so long as the blood which was shed belonged to people who didn't fit that particular gods demographic and tradition. Or in other words the horror perpetrated in the name of god during the crusades is acceptable to god because the people who were dehumanized did not belong or believe in the standard or authority of that particular god and conquering monarch.

"In the same way creation is the act of separating the heavens from the earth, apocalypse is (the destruction of heavens and earth) and thus it too is a form of creation in a way. Whereas the first imposed an hierarchy the second subverts that hierarchy to a state of equilibrium and equality."

"The only chasms separating man form god are the social constructs imposed by gods which define one as holy and the other as cursed" (separating the "sacred" for the "taboo")

There is a myth that the most effective way to obtain unity is uniformity. This however is a folly. In truth much like it's original shape (chaos). the most effective way to obtain unity is in acceptance and appreciation of the diversity. It is in tearing down the walls which separate us from them, and the whole into parts

## THE AMBIGUITY BETWEEN GOD, ANGELS, MAN, AND CHRIST,

> 1 John 4:12, KJV: "No man hath seen God at any time. If we love one another, God dwelleth in us, and his love is perfected in us."

In the earlier fragments of the Hebrew Bible we find a great deal of ambiguity distinguishing between god, angel, and man.

And more often then not entities such as kings, pharaohs, and tribal chieftains, where ubiquitous with gods incarnate. And so when Abraham or Moses made covenants with the ineffable, they were inevitably exercising an authoritative role as gods in a since. "I will make you like a god to pharaoh" (exodus 7:1) ESV. And tho this may seem strange to us this day in age, anthropology has shown us that entities such as pharaohs and chieftains (particularly those who were immortalized in stone after death) where our earliest representations for this category of gods.

Or in other words godship was not only something that would happen to great leaders after death (probably as a way of retaining their wisdom and knowledge/ acts of creation on a cosmological scale) but something distinguishing them in their lifetimes as a form of divine right to rule. It's no surprise then that this is where we get a lot of our imagery form god as an authoritative king who sits on a throne and commands glory and praise from their subjects. And after all it was these kings prognosis's that shaped their nations and created what we now know as civilization.

From the erection of walls which separated us form the wild beasts of the field, to the social norms and rules which became culture. By its original definition, these kings did in fact create the world as we know it today. However there is a notable difference between this claim kings made towards godship, and those made by the earliest writings of the Hebrew Bible. And that is quite simply " all humans bare gods image" and thus any human held the power and authority to reshape the world (act as partners in creation). Tho later this autonomy is stricken in favor of more authoritative kingships, particularly the davinc one. At its roots this was the story of the Bible. God inviting Israel to be a nation of priests on the mountain, to join in partnership with the ineffable and nature in recreating the world. And each time Israel was invited to wrestle with the ineffable, they inevitably chose to erect a master or elect a king to rule over them. This brings us to our next point. Because in genesis 32 a man named Jacob wrestles with god, or an angel, or maybe it was a man? There is a lot of ambiguity here between those categories of human, messenger, and god,. But this is not the only account of such ambiguity. Just a few chapters earlier we see a man named Abraham entertaining three angels/men (gen 18) however more inquisitively it would appear that god themself was among these men/divine messengers. As not only does Abraham greet them as gods or lords but in verse 13 god actually speaks among them (or perhaps even thru them). It should also be noted that the meal Abraham prepared for god/the three angels/men, was not kosher or in accordance with the dietary laws given to Israel by god.

The ambiguity between god and divine beings/angles is ambiguous. Because in its original Hebrew they are both described with the same name "Elohim" and often angels would be described as baring the name of god "the angel(s) of the lord".

This name of god could presumably be bestowed on a human as a form of divine intercessor of sorts and may be understood thru much later conceptual models such as gods innate divinity which exhausts all categories of divinity. And this is how we resolve our modern translation of 1 John "and the word was god" which is not its original Greek transcription. In its original Greek 1 John reads "the word was divine" not god. This later interpretation "the word was god" was imposed on the text in order to validate certain rhetorical goals for merging god and Jesus as one and the same. Originally Jesus's claim as messiah or anointed one (davinc king) was a tradition which was not restricted to godship per-say but extended to kings and priests. However with the loss of davinc kingship in the exile and the emergence of national superpowers such as Greece and Rome we watch the development of concepts like the "angel of the lord" and later Daniels the "son of man" merge with this idea or role of the messiah.

However even in the New Testament there is a great deal of ambiguity in this character of Jesus himself. As often is the case, this rabbi and friend, not to mention supposed prominent public figure, is virtually unrecognizable to his closest friends? We even get accounts of Jesus walking with two of his followers after the resurrection, even carrying on a conversation with them and yet they were kept form recognizing him as their friend and teacher (Luke 24:13-24)?

Now this could be used to illuminate a point Jesus seemed to be adamant about " which was that he was virtually indistinguishable from any other human who bore gods image" or that in short to be human, is to be breathed of dust, heaven and earth made one or the divine incarnate. This brings us to our final point in this discussion. And that is the role of humans as members in the divine council. Throughout most of the ancient world concepts of the divine council were limited to those of celestial nature. However in the Hebrew Bible we get examples like Micaiah in 1 kings 22, Isaiah 6, and Jeremiah 23:18, depicting or inferring the role of profits on the council. And this reiterates the invitation extended to all humans as partners in creation on behalf of more compassionate representations or barers of the divine image of god.

Tho to many this concept may sound heretical and even blasphemous. it is something we still observe today in human nature. This innate desire to admire and adore people and even animals. For instance We still worship celebrities and revere prominent leaders and yet from Abrahams role as a king making covenants with god (for kings where gods in their own respect) to Jesus's representation of the messiah or davinc ruler devoid of monetary, wealth, and power, we get this claim articulated by the later Buda that all humans bare that divine quality of god in some form or fashion. And thus it is not our proclivity to worship which is the problem, it is our exclusion from worship which is.

Many of us were taught that there is only one god, however instead of a god which is comprised of all traditions, we were given a god restricted by only our traditions. And so we were left with only two options. To believe in the god according to our traditions. Or to abandon belief in any god at all. This dichotomy requires us to find our identities in tradition or in rejection of those beliefs. Because our king, our god, our tradition is the only way to truth. And all others must be false profits. Unfortunately this often resulted in subscribing to an hierarchical and utilitarian conception of god, or abandoning all faith in any form of cosmic order and higher truth all together. This binary of Christian or atheist, has led many who's sense of morality and compassion which would not condone the malevolent tyranny of the Christian god, to abandon any hope in a holistically good and loving god.

Note there is ample evidence denoting ambiguity between man and beast such as in genesis chapter 2 and as well In Numbers 27:6 where we see Moses require an Adam/man to govern the beasts (isrilights). This role is filled by Joshua who rules over Israel as a shepherd to lead the sheep who had chosen a bal(master) over a partnership with the ineffable.

A god sacrificing his son is not about love but rather legacy.

## THE VOICE OF GOD

The gods were the architects of civilization. They were the ones who dictated the identities acceptable in society. They were the ones who decided what expressions were acceptable or holy, and which were not. The gods claimed dominion over our identities and values. They were the ones who decided our humanness should be dependent upon our compliance to their assignments of meaning and value. All the prejudice towards peoples of other races, genders, cultures, world views, and sexual orientations, is due to the gods prescription.

In the genesis poems The Hebrew word "aroom" is used to describe the serpent as crafty or cunning, however it used later to describe Adam and Eve after they eat of the tree of knowledge. In this second instance it is used to describe man and woman as naked or exposed. Moreover this word "aroom" translates as to make known, to uncover, to make wise, or to reveal, and thus is synonymous with the emergence of consciousness. Before consciousness we walked with god in the garden habitually. But with our awareness of our selves we became liable for or behaviors. For not only could we observe ourselves, but we could then audit and amend our behaviors. We could discern right from wrong.

> "The degree to which a person can grow is directly proportional to the amount of truth he can accept about himself without running away."

> -Lenand Val Van De Wall-

Instinct is a mechanism our genes developed to automate certain behavioral tasks, and so consciousness is most likely a mechanism they developed to audit and reappraise those behaviors in real time to better adapt to an ever changing environment. Emotion in many cases is the mediator between the two and its role was to monitor and emote or move us. However in consciousness early development people didn't understand how to distinguish between their feelings and themselves, and so much like a 6 month old infant identifying with their caregiver or a lion unable to recognize its own reflection in a mirror, ancient mam would feel their anger, love, jealousy, and fear, and attribute it to the stimuli that caused it. An example of this can be observed in the phrase "look at what you made me do". And because such emotions move us in accordance to external stimuli, our conscious observation of those feelings was often ascribed with agency. We called this emotive force spirit or god. And those feelings became associated with the environment as external or outside of ourselves. And yet even still now from the feeling in our chest to the voice in our head, everything we attribute to god in the present, is a phenomenon experienced only inside ourselves. Tho admittedly those feelings and dialogues are often triggered by external stimuli.

People are programmed to behave in a way that makes others feel the way we feel. And so if we feel frightened by someone els, we intuitively attempt to make ourselves intimidating. if we feel neglected, we behave in a way that sets the other person up to feel ignored... and so when an environment or circumstance would make us feel angry, that anger then would be gods anger. Why is this important. Many of us never learn to discern between ourselves and our emotions. (I am sad, I am angry, I am in love,...) but these emotions are tools our body uses to monitor and alleviate conditions in our environment. Our Emotions monitor our well being and attempt to regulate our safety and security in an environment. So if someone is depressed or anxious, the causes is most likely environmental. Not only are our genes turned on or off in accordance to environmental stimuli, but the environment also dictates how certain genes are implemented and expressed/exhibited. Which means more often then not harmful behaviors have more to do with harmful habitats then they do with inherently harmful people. People are naturally curious and creative and so if someone is lazy or behaving aggressively, those behaviors say more about that persons environment then it does the person themselves. Justice then is not brought about by punishing the person, but by remedying the environmental conditions.

this is not intended to deflect an individuals responsibility, but rather to help us better focus our energy on finding more viable solutions. A secure environment is not necessarily a healthy environment. In many cases fight and flight/flee responses are healthier then freeze, because at least with fight or flight the individuals retain their autonomy. In many ways this is the moral we are presented with in the exodus myths. As we are intruded to a people who had become submissive to a system which profited off of their exploitation. In the genesis myth we see these people reject this subordinate identity and even challenge the assignments of good and evil, however by exodus they had traded their authority and agency for security and stability. The wilderness in this narrative functions as an environment where these people who had learned to submit or freeze, could then learn how to fight or leave/flee. Subsequently taking back their agency. Many times during this journey however, this people groups sought an idol as their identity, and a law to rule over them. They sought a master or bal to subvert their own authority and rule over them. Interestingly enough we see this phenomenon today as well. However we more commonly know it by the name nostalgia which biologically posits a preference to conditions similar to those of the environments which influenced the activation and expression of our genes during development. And thus we often crave the familiarity of tradition as opposed to the tedious task of reappraising an ever changing environment in part because we were quite literally made for those earlier environments.

People are very much products of their environments. And so much like the hunter gatherers who were forced into agrarian lifestyles, it was god who placed man in the garden to work it. But by taking of the forbidden fruit of knowledge, man saw how they had been deprived. They saw their own nakedness and thus questioned the assignments of meaning and value prescribed by god. The exile from paradise (rather it be Eden or Egypt) was a liberation from tyranny. Now admittedly the society facilitated by agrarian lifestyle had the potential to be a paradise. But only so long as everyone did their part and everyone had a say in what was right and wrong and not just the elite gods!

On an cosmological level the god of genesis constructed

the very environments which required man to rebel. On an ecological level however we were shaped by a world and environments where the fundamental prerequisite for survival was to kill or be killed, to lie, cheat, and steal, or to parish. And thus the law designed for society was completely counter to that of the wild. For the wilderness demands competition to prevail over predation. However civilization presents us with an alternative in the form of cooperation.

These narratives which claim that mankind is inherently evil and all at once helplessly in need of a savior to deliver them, inadvertently subjugates mankind to systems where those very same claims are true. The gods of old were those who predated the predators that threatened mankind, but they then inadvertently become predators onto the people. These gods were cultivated by environments of scarcity and division, and thus they are unfit for habitats of abundance and inclusion.

The gods are gods because they preside/rule over man and predate/vanquish the great beasts. But in order to rule over man, gods must first confiscate man's authority to rule over themselves.

> "Had he decided that god did not exist, he would have become an atheist and a socialist, for socialism is not only a problem of labor or the so called 4th estate, but is in the first instance a problem of atheism, of the contemporary embodiment of atheism. The problem of the Tower of Babel construed expressly

without god, not for the attainment of
heaven from earth, but of the abasent of
heaven to earth."

(Book one the elder,
the brothers karamazov.
Fyodor Dostoyevsky)

It is clear here that this conflict between the rights
of human autonomy and well being are direct threats
to the malevolent authority of godship. And why it is so
imperative that the very gods who employed our bondage
are then the only means for our liberation, and as our
everlasting saviors. Because as long as there are gods,
there shall also be slaves. and without slavery there is no
godship.

This idea of "sin" is perpetuated in order to necessitate
the intercession of the gods and by extension the "divinity"
of some form of hierarchy. It is only god's demoralization
of man which makes gods malevolence "moral".

"If God is unable to prevent evil, then he
is not all-powerful. If God is not willing
to prevent evil, then he is not all-good. If
God is both willing and able to prevent
evil, then why does evil exist?"

(Epicurus' trilemma)

James 4:17, NIV:

If anyone, then, knows the good they ought
to do and doesn't do it, it is sin for them.

And yet the god of the Bible not only withholds blessing form his servants, but in some cases god deliberately causes suffering with the soul purpose of receiving glory for remedying the very evil he (god) caused in the first place.

Not only Does the biblical god fail to stand up to their own standard for morality. According to the biblical authors god deliberately causes evil!

In my last book I claimed that there was no such thing as a universal morality, however I would here rectify that statement. Tho many actions which social codes designate as morally wrong are in fact arbitrary and morally neutral as they contradict the primary prerogatives of nature it's self. I will concede that any action or intended action which causes another living organism undue physical or psychological/emotional harm (because after all emotion is our biological incentivizer and a chemical reaction is by extension a physical one) without reasonable cause is most definitely immoral. Immorality does exist in the deliberate and senselessly harmful actions towards another creature. And yet to a degree every living organism survives at the expense of another.

CHAPTER 2

# THE GENE AND THE GENIUS. PHYSICS, ETHICS AND THE ANATOMY OF THE COSMOS

## THE CELEBRITY OF GOD

Celebrity is a complex social tool we evolved in order to know people we've never meet thru a mental projection or version constructed of here say or the accounts of others. In the same way most of us have never met the former queen of England or Taylor swift in person, we assume to know their character thru the opinions of others who have met them or even possibly thru the art they have made. However to say that any of us have an actual relationship with these celebrities (whom we've never met) because we had a dream about them or saw them in a vision would be absurd. We interacted with a made up version of these people in our heads and not the person themselves. Now originally this mechanism served us well as it permitted more complex and comprehensive social structures. It also allowed us to simulate behaviors in hypothetical circumstances. However much like the celebrity we retain surrounding our memories of a lost love, there can be no such thing as a relationship when the other party is perpetually absent. We worship the idea or celebrity of god and Jesus like we would have a king who we had never met. And tho faith and loyalty in the un seen king and their cause was substantial cause for acceptance or rejection on an individual in ancient times. But truthfully there was never a relationship only a celebrity. Only a made up version of god in our heads.

Virtually all creatures exhibit the innate proclivity of puzzling and playing with their environment. This presents itself in the form of everything from attentive caution to hyper fixated curiosity and admiration. We, like our distant relations (mammals, reptiles, and marsupials.....) have a propensity for marveling at the behavior of other organisms and phenomena in our immediate environment. From a developmental level this has proven to be effective thru the way small children (human and animal alike) attentively observe their parents reactions to different environments in order to emulate or deviate from those behaviors dependent upon the outcome.

And thus it is no wonder that we have a innate proclivity for positing a form of personhood on our environment in response to the adverse reactions to seemingly indistinguishable approaches to an environment. This anthropomorphism of our environment along with its corresponding proclivity to admire and adore (to curiously or cautiously puzzle and play) seemingly autonomous agents in our environment may very well be a big part of our construction of god (or at least the celebrity of god anyways).

## POEM

What good is wisdom if it makes one solemn. To dissect the bird and loose the song. Must knowledge come at such a cost, That mirth and magic be lost.

Yet As any good philosophizer knows.

The weight of the world is not so daunting when you realize it sits a drift a midst oblivion and void, and is only sustained by the gravitas of the stars.

For beauty is not lost in the void but instead likend to the stars in the sky.

As futile as it may be, life is that which believes in light by which it may see.

It is not myth or story, but that to which it ascribes. For it is not about what the song bird must mean to prescribe, but simply the very fact that it's song can have any meaning at all, in which the life abides.

For after all, what good is wisdom if it makes one solemn. What good is music if one forgets to sing along.

## THE SEEING IS BELIEVING BUT REALITY IS AN ILLUSION

Mechanisms of The human subconscious may very likely extend back as far as our common ancestors with those primitive life forms such as bacteria, as we both exhibit very similar social mechanisms. As we shall soon see these mechanisms are preprogrammed into our neurology. on a sociological level we still see this behavior in modern evolved versions of bacteria today, in the way they excommunicate cells which do not contribute to the common goal of the bacteria as a whole. or in other words the primordial origin of life on this planet may very well still posits profound wisdom in the recesses of our subconscious.

It should come as no surprise then that In many cases we humans inherit our social behavior which constructs any given culture, from the same behavior in interviewing prerogatives as those of the most simple bacterial forms of life. And so tho these mechanisms may posit profound wisdom, in many cases the social biases are still remnants of much more primitive (less autonomous) versions of life.

Many of these instinctual behaviors are in fact innate. And it is only the context which is learned from our parents and piers. Further more our visual cortex (like our olfactory one) has a direct route to our amygdala (the primitive instinctual/emotional brain region) which means in many cases our instincts are more vigilant than our autonomous PFC mechanisms. And on an evolutionary level (rather it be thru instinct, or consciousness) our genes are more concerned with efficiency and effectiveness then it is with accuracy and assessing objective reality.

The mind is not limited to the brain alone but instead is a product, and component of the entire nervous system of any given organism. It is integral to the entirety of the whole body and not resigned to the head as its dwelling. The heart, the lungs, the brain, and even the digestive track are all components of the mind. Every nerve ending, every biological mechanism and regulatory systems in our bodies is a part of that (thing) we call "mind". And thus our understanding of "reality" is the product of many sensory organs communicating with one another. However as it serves the organism as a whole "reality" is not the goal of these mechanisms but instead survival/procreation is. And so sometimes our bodies lie to us in order to keep us safe and productive for their cause. And so We see faces where there are none, hear voices and fill tingles when there's nothing there, and we posit agency when there is no willful agent involved. And when there are miscommunications between our sensory organs and our nervous systems our "minds" are left to fill in the gaps. We call this "Our mind playing tricks on us". And so it is no surprise that our experiences of reality can often be miss leading or even out right incorrect.

The fallibility of our experiences, and the illusory nature of reality is often the very states religious practices and substance induced "trips" seek to exploit as holy or transcendent. And tho in some cases these experiences can be prolific (in the way they illuminate unconscious truths we would not normally engage), in most cases they simply function as self fulfilling prophecies or as the result of consistent behavior in an certain direction. However none the less We attribute the meaning to the symbols we see.

Perhaps this is Why things work out serendipitously.

It's Not the intervention of a god orchestrating it all, but rather that we simply have in us inscribed meaning making machines that have profited in the past from predicting potentially positive out comes to situations. This also makes sense of the margin for error as a rationalization of outcomes that don't end in our favor? As we've already stated, the mechanisms that are performing these tasks have been doing what they are doing for millions of years.

Another component of this illusory nature is the fact that Our body's naturally produce a schedule 1 hallucinogen(DMT) in our Nero circuitry (specifically the pineal gland). This substance is accredited with inducing "transcendent religious experiences" and may very likely be the cause of near death experiences of heaven and hell occurring naturally in the human brain. Tho as of yet we don't fully know the exact role this chemical plays in the naturally occurring mammalian brain chemistry. How ever this does contribute to the argument that we can't even trust our own senses especially in those particularly "spiritual/transcendent states".

I believe The fact that profound spiritual experiences can be induced thru the abusive/use of hallucinogenic substances, then invalidates the transcendence of spirituality beyond just a cognitive function of our bodies. And besides any truly monumental feat was achieved in part thru belief in something bigger than the outstanding organism which beat the odds to achieve it.

We are often taught that we can't trust our own emotions, and tho yes they can most certainly be misleading at times, in some cases that's what they are supposed to do.

Feelings and emotions function as a biological systems check of our own anatomy. Or in other words Our emotions motivate us to pursue vital resources and refrain from potential harm. They by very definition "emote" or move us. They make sure we flea or fight when danger is present, they motivate us to pursue communities and potential mates in order to reproduce, they evoke a sense of importance or insufficiency in order to regulate and maintain our position in a community. Emotions are how our genes get us to do their bidding. The thing we call love is a chemical reaction induced with the soul purpose of reproducing, the thing we call joy or pleasure is a mechanism our bodies use to incentivize our pursuit of vital resources necessary for survival. Our perceptions of reality is ultimately just an interpretation of the information collected by our sensory systems and neurological processes programming. In this sense We are virtually no different than the rocks like copper And silica which we taught to do math by shooting electrical serges thru them. Our modern day computers are no different then our own brains in many ways. And Further more our nervous system is essentially indistinguishable from that of plants. We are in many ways just tools and vessels for our genes. However unlike other animals and plants, we are conscious of this fact and thus able to rebel. And so Perhaps one of the greatest displays of our autonomy is the ability to simply sit and actively listen to our own bodies (emotions, and sensory organs) as they respond to external environmental stimuli, and yet deliberately chose not to react to it. To observe the emotion and yet not allow them to emote or control us! In this way we become more than a cog in the machine.

Tho We may in fact very plausibility be but a small part of a much larger organism. and perhaps even the original entity and life form. much like bacteria in our own stomach breaking down food, we too may subconsciously do the will of the whole. tho We have seemingly obtained the capacity to rationally question our experiences and negate our instincts. For we were made conscious and subsequently ordained as barers of the divine image. What ever that "divine" may be.

That which we once called god extends to everything from the inherited prerogatives and instinctual traits of genes to the anthropomorphism of celestial phenomena and changes in biometric pressure. And yet these things are no longer ineffable to us. And thus they no longer serve us as gods.

"Intolerance of mutability is not a sign of immensity and integrity but instead insecurity and insufficiency."

We don't see the world in its entirety or reality but instead we see our sensory systems accumulation of data and our brains interpretation of light rays and atomic waves in our neurons approach to them. We experience an internal interface which correlates with subconscious procedures and programs. We see what our genes designed us to see.

## EXISTENTIAL

Regardless of what you believe or how you define certain qualities of our existence there are a few truths which are simply undeniable. We all will die one day. And when we die our bodies will deteriorate on a molecular level as what was once organs and tissue will become (for all intensive purposes) dust. On that day when we inevitably die what ever memory's we've had, what ever experiences we've experienced, what ever we knew, thought, or believed will cease to be forever more. What ever idocyncracies, personalities, perspectives, and resolutions, we've accumulated over our life time will inevitably fade from existence. This version of you here and now will pass form existence along with all your emotions, observations, and judgments,. All the biological material required to construct feelings, emotions, and prognosis, will decompose. And so what is eternal? If the circuitry is gone, where does the program reside? There is no joy or sorrow without the necessary neuro chemistry to produce and facilitate it. No memory without a body to house it. And what about belief? Is that something we can keep? For Belief is not a choice one makes any more then one chooses to see (or not see) certain colors. Further more as we shall soon see belief (like empathy) is simply a neurological program our VMPFC's constructed to simulate and synthesize circumstances and situations to better govern and update our behaviors. It is an assertion based upon evidence which validates the existence of or the cause of certain phenomena. it is ones definition of that phenomena. However if one is incapable of seeing blue then what ever belief that individual has in the existence of such a color is completely dependent upon someone else's definition of a quality which is unbenounced and indistinguishable to them. Can that really be called their belief then? This of corse is the role of a preacher. For they attest to something common enough to be noted by more then one source (for instance if 10 people describe a tree standing in the desert, then there must be something at lest resembling a tree in the desert) however the very fact that such a tree is not prevy to all who seek it or even defined differently by just as many, leaves to question not the validity of the phenomena itself but rather the extent of its distinctions. We see what we've been shown, and we call it by the name we were taught. But One can not be rightfully judged for their belief or non belief in something any more then a English speaker can be blamed for calling blue what the espanuel speaker calls asue. The idea that your ethereal consciousness (which is proven to be a byproduct of your ephemeral biology) will be judged based on your subscription to claims is just as paradoxical as any supernatural phenomena. not only is it not a moral basis but it is also absurd.

Science has altered its claims in light of evidence because science is biased on our sense. However it is governed by math to check sciences claims and values. Because tho numbers are all made up, math as a model of logic which is not subject to one's experience to assign value.

Who created god?

The mind named itself, but before the mind developed the capacity to name itself, was it still the mind. Like a small child convinced that they are part of their mother for the first 6 months of their life. Before the mind became aware of itself, was it even a mind.

Who created god? And was god a god before they created light (became light or conscious)?

Could there be a god before there was a mind by which to call it by(light). It is only the exclusion or prohibition of certain body parts which cause them to evoke erotic response to their exposure or exhibition, skin is skin. And similarly there is no god without man, no order without chaos, no sanctity without separation.

At 6 months a child is an extension of their mother (as far as the child is concerned anyway). At 2 they begin to assert their autonomy. By the age of 5 they begin to assimilate with culture. By 13 they realize culture too is a lie and not them. This rebellion or distinction between the individual and their cultural identity is usually not fully realized until our 30s (when our PFC is finally fully developed). At which point we ether buy in to the ruse or reject it completely. And so again I ask. Can there be a god without creation, light without darkness. Can there be a mind without consciousness or one without another to distinguish between the two. Without a division there is no holy, without man there is no god, without slave there is no master, without consciousness there is no instinct, Like a small child deifying the mother they were once connected to, it is only thru separation which permits gods to arise. With this in mind it should come as no surprise that our understanding of god is a direct response to our relationship with our parents. Or at least our upbringing. If our parents were distant and authoritative then so is our idea of god, because after all god did not show themselves and thus it is only thru the representation of gods image presented to us thru our parents, by which we have to judge god's character by.

Never the less we are a product of our opposites and whatever we are lacking or neglected will inevitably express itself in us. Like the mind without consciousness, it is only the distinction which makes it a mind.

We evolved (like all life on this planet) with an intrinsic dependence on community to propagate our survival. And thus we require communication in response to its utility in obtaining and sustaining that community we require to survive. We talk to our dogs and gods because we are social creatures who require communication to fuel our conscious minds and the construction of their cosmological models, even when we are alone. Now some may vocalize the disrespect that claim attributes to "god" and that would be good because that is a form of healthy communication on my accusers part, however god has yet to retort a word and thus god is ether a poor communicator (which should be apparent thru the lack of univocality in scriptures), or god just simply doesn't care about their reputation(which is odd for a being who's primary concern is to receive glory and praise). We on the other hand do require communication and thus we converse with the unresponsive and imaginary entities (like gods and dogs) because we are communal creatures, or perhaps we are communal creatures because we were once one with the whole. It is only our separation form god and one another which makes one god and the other well "other". Perhaps deep down we know and crave reconciliation with all the parts and expressions of life that once were (and even still are) us. Like different cells in the same body, we are not separate at all, but rather just as much a part of this world, this galaxy, and the entire universe as any supposed god is.

It is only our belief in division, which creates separation or divorce from "god and ourselves".

When we begin to see the world and all life on it as connected and intrinsically whole, this idea of some kind of cosmic order theorized by "predestination" or "fate" actually makes sense. The idea that two people living very different lives, could then meet and change each others life forever more, is not so ridiculous. The fact that a chill in the wind reminiscent of that attached to memories and emotions of a past love, could then evoke a response to reach out to that "lost love" is in fact exactly the way our brains (via emotion) work (neurons that fire together, wire tiger). Our genes program within us a response to environmental stimuli which mimics past experiences (rather positive or negative). And so it's no wonder that things like sounds, scents, sensations, or even sights(visual observations of environments or behaviors) associated subconsciously with past pain or pleaser, would then evoke a emotional response. This reality in tandem with our understanding that (however autonomous we are) we are all intricately connected to the whole (as all life originated from the same source). This idea of fate or "that things happen for a reason" has some logical credibility to it. However To blame god for the good (say sparing a family from a devastating car reck) we must also blame god for the bad (as causing that devastating car reck). To credit god for the serendipitous outcome of events, one must also hold god responsible for the horrific outcomes as well. Bad things happen to good people and good things happen to cruel people. In both cases things happen the way they did because that was the way they happened. And if they'd of happened differently then they would not have happens the way they did. If you blame god for bringing two people who love each other immensely, together then you have to blame god for bringing two people who abuse or misuse(cheat) on each other. Or even for leaving two people who would be perfect for one another to instead spend their lives alone or in bad relationships. What about the gifted child born into poverty or the genius born in the wrong age, what of the cruel child born into power or the testing child indoctrinated by tyrants? And what about slavery (which was condoned by god) and only abolished because men and women fought for its abolition.

On an evolutionary level our proclivity to posit the question "why did things turn out the way they did" and the response to that question being to posit a form of order or cause for that particular outcome, has most definitely proven helpful in our development and survival. With that in mind it makes perfect since that we would deduce that "things happen for a reason" or in accordance with some bigger cosmic plan. That does not however make such conclusions accurate.

## THE LAND

We are to be compassionate towards the land and it's creatures or the land will retaliate. but regardless of our treatment, the land will have rest or sabbath. rather in fertility by our grace for it, or in famine in spite of our malevolence.

Like gods divided among the nations. Men claim to own land which will outlive them and yet suffer their compassion or malevolence by the land which the let lay fallow or exploit. Animals evolve for their environment. We on the other hand require our environment to evolve for us. This can be good. It means we can plant trees and survive in equilibrium with nature, however more often then not what it means is we strip the land of every recourse and neglect it it's time to rest. We conquer the land and in doing so we become slaves to the land we Own. But The Trees will take half the year off as do most plants. The land will not suffer our exploits, it will be us who suffer our excess. The ground will have rest.

A paper published in 2023 shows that plants do in fact emit sound (tho this sound is at a frequency inaudible to humans). What's interesting however is that tho over all noise level stays relatively the same, unhealthy plants are much more "vocal" then those who have their needs meet. Sound familiar? And when we consider that the nervous systems of both plant and humans is virtually indistinguishable, it would be moronic to assume that plant (like the slime mold which accurately mapped out the most efficient public transit routes in Tokyo) are any less sentient than any given mammal.

## MAN AND BEAST DISTANT RELATIONS

On a biological level the primary prerogative of virtually any gene is to replicate or reproduce. It doesn't mater what form life takes, the priory programming of any organism is motivated by that goal. And tho humans can negate these instincts to a degree in favor of alternative pursuits, for all other life on this planet, it's soul purpose is to procreate and or protect its progeny. However this almost always comes at the expense of other life forms, and sometimes even like/familial forms of life. As in the wild when a mother will often abort a weaker more feeble cub in favor of a healthier or more developed one. Because as far as nature is concerned, life has no inherent value but does come at great cost. And this is true for every organisms known to man. And tho plants are able to reproduce thru consumption of photons or obtain the energy vital for its propagation from the sun. Virtually every other kind of life form requires the consumption of another life form to survive, rather that is a herbivore thriving at the expense of plant life, or carnivores devouring of other animals in order to survive. So tho yes nature is wholistic it is not inclusive or accepting/compassionate. This natural order is made even more malevolent by the fact that these organisms are all cousins. That means that nearly every plant, virus, fish, mammal, and reptile, is our evolutionary equal and not our inferior. So yes we are the apex predator of this planet, however that does not mean we are the pinnacle of life on this planet. The truth is we are not so different from those organisms we call animals. And instead simply "differently" evolved and specialized then our animal, vegetable, and viral, counterparts and contemporaries. They our our piers, not our predecessors or subordinates on an evolutionary level.

What we find from studying the skulls of species like homo erectus and homo heidelbrgenis we can see the physiognomy of the cranium evolve slowly over time. More simply there is no one pivotal point where one species became another but rather a continuous process of adapting and evolving. Ultimately micro evolution is macro evolution. They are one in the same just on different levels.

## MAN OR BEAST, THE NEANDERTHAL

Evidence that man did not manifest but mutated and evolved from more ancient forms of life is quite extensive and compelling. Take for example the Nariokotome Boy. The Nariokotome Boy Was the name given to the almost complete fossil remains of a youth from approximately 1.5 to 1.6 million years ago. It is understandably clear that this youth was a very ancient ancestor of modern humans today. And this is evident in the numerous skeletal similarities with modern humans today. With that said this fossil did have a couple marked differences which were more similar to ours and our distant cousins (apes) most common ancestors, such as a low and elongated cranial vault, and lacking a prominent chin of any kind.

Further more the carbon dating of this fossil is unanimous among scholars in the field who scrupulously examined the remains.

We have also discovered that Neanratols buried there dead (perhaps indicating an symbolic arises of consciousness and empathy or even just recognition of a correlation between corps and the spread of disses). Neanderthals also exhibited a mastery of tool making and the facilitation of fire for cooking and crafting which then opened to door for the development of our frontal lobe and more complex mental mechanisms.

However if one is not persuaded by the archeological evidence then look at the biological evidence.

For instance The mere fact that we share approximately 98% of our genes with apes, and 96% with monkeys, should attest to our animalistic heritage, not as as a puzzle piece from our past but as living breathing proof in our cells.

## NO SUCH THING AS DISABILITY

Everything we call a disability is actually just an alternative program AE a particular way our brains and bodies communicate and interact with (in relation to) our environment.

Autism, neuro divergence, adhd, and dyslexia, are common mental "disorders" which describe different ways human brains process and then communicate information. For instance many dyslexics are capable of mentally visualizing 3D objects and images but are hindered by an inability to differentiate between left and right or the order of a written sentence on a page (dyspraxia). The spacial awareness in our minds is often obstructed on a 2D sheet of paper. This results in an inability to distinguish between b and d or the direction a number or Letter points. In addition to this, dyslexia is more prominently characterized by an individuals inability to compartmentalize information in order of probability and relevance without an existing context.

meaning that dyslexics often have to sift thru all the possible applications rather then going directly to the most likely conclusion. Similarly ADHD is a form of neuro divergence in which the individual is often incapable of compartmentalizing or narrowing their cognitive focus to information immediately pertinent to a subject. This also often results in individuals not being able to focus on a task until it is past due. In contrast ASD is often characterized by a hyper fixation or focus of a subject and an inability to see the whole picture for the extreme attention to the details. In all cases these "disorders" are not caused by "bad" genes but instead a need for diversity of traits to tackle that diversity prevalent in nature and life as a whole.

## Dyslexia and Autism in the Brain

In our cerebral cortex we have stacks or columns of neurons which specialize in sequence or order of procedure. These columns are more commonly known as minicolumns. Autism (and/or localized/detail oriented connections and conclusions) are/is characterized by close spacing between the rows in the minicolumns. Dyslexia on the other hand is characterized by larger gaps in between those columns (or more big picture/ grand scheme conclusions and connections). Or more simply autism is characterized by processing information zoomed in and focalized, where as dislexia is characterized by computing data zoomed out or holistically (much of this computation happens in the background or intuitively in the dyslexic brain). Moreover dyslexia and autism sit on opposite sides of the spectrum where autistic qualities tend to be hyper fixated on fine details, and dyslexia more adapt at wholistic processing. It is because of these dichotomous qualities that then lead Manuel Casanova to hypothesize that these "disorders" are virtually incompatible within the same brain at the same time. Tho many of the symptoms appear similar in many ways dyslexia and autism are actually polar opposites or extremes of the same spectrum. Where the dyslexic brain is predisposed for making broad connections (thinking out side of the box), the autistic brain tends to be far better at attention to detail and precision (orthodoxy).

From conception to about the age of 2, the mammalian brain will produce an excessive amount of neurons. However from approximately 4 to 6 years of age these neurons will undergo extensive pruning (by roughly 50%) as the more pertinent connections are refined and prioritized. However in the autistic brain this pruning is not as expensive (only 16% or so) resulting in more options to process.

This excess of neuronal connections is probably why creatively correlates with "mental illness and mental disorders"(Neurodivergence).

Whereas neuro normative structures are more concerned with efficiency then diversity.

(Laziness is an instinctual conservation of resources. Greed is a trauma response to the uncertainty of life. And prejudice and narcissism are an insecurity in our own value in an culturally imposed hierarchy necessitated by our communal nature.) these traits and many others are often vilified and reprimanded as character flaws when in all actuality they are incredibly intuitive mechanisms of our biological programming.

When asked why god makes some people inherently broken, Jesus responds by claiming it is so that god may be glorified thru their healing. however that is essentially like saying a woodworker made flawed furniture so that he could then profit from repairing his shotty craftsmanships. And this doesn't even address the problem of those who faithfully suffer their entire lives without being healed. But ultimately if this level of neglect won't be tolerated in a man. then how can it be worshiped in a all powerful deity? If this behavior puts to question the integrity of the craftsman then how can an infallible entity be considered moral and good (not to mention competent and powerful). However all this might be forgiven if that god and creator admired their craftsmanship. but according to the majority of sages and priests known to religion, that all powerful and inerrant creator is intolerant of even their most studious creations.

## PREJUDICE AND DISCRIMINATION

The sad truth is that we often do love others the way we love ourselves. And thus the strict or wealthy man treats the poor man with the same contempt he would himself if he were in the poor man's position. Similarly the judgmental busybody and the conceded beauty who act spitefully towards those of "inferior station" are actually treating the "ugly" or "uncultured" "inferior" the same way they would treat themselves in their situation. Because for all three of these examples, the love which they bestow upon themselves is directly dependent upon their own success, social rank, beauty, or wealth.

The sad truth is that each of these examples truly do love others the way they love themselves. Because their love is conditional.

"A hallmark truth of the human emotional system in relationship is its intuitive proclivity to evoke the same emotional response in our partners" which means if they feel intimidated, then they will (often unconditionally) behave In a way to make you feel the way you made them feel.

(It should be noted that all sentient creatures have a tendency towards regarding those who look, behave, and especially communicate differently as alien or a sub species. This is not the result of a racist god but the diversity in alteration thru the extensive process of evolution. In many cases animals adapt to their environment and this extends beyond topography and into sociology. We evolve to fit our communities, typically this particular facet of the evolutionary process is called culture and it is not restricted to humans but is observed on the smallest level in virus cells which refuse to share resources with cells that don't contribute to the common goal of the virus.)

Shame is an internalized belief that our expression, experiences, and existence is invalid because it does not fit convention.

Pride is an acknowledgement that our experiences, expressions, and existence, are valid and worthy of praise or at the very least acceptance and acknowledgment for simply being.

Arrogance is a belief that one persons experiences or expressions take precedence over others and supersedes someone else's.

Introversion is the result of social anxiety or exhaustion from ingenious behaviors and masking predicated on a belief that one's authentic behavior or expression is inadequate because it doesn't fit social convention.

Extroversion on the other hand is a blatant disregard for convention or an genuine compliance with that conventions values.

Why do we feel A need to forgive someone for being who they are? As if our intolerance of someone els is their problem and not our own. After all The person with the problem is the one with the problem.

(Obviously this doesn't apply to people who continue to physically or psychologically abuse you.)

But this idea that other peoples behavior is ours to control is absurd. Ether we love them as they are, or we let them go. We've got to stop trying to change or control people and calling it "love".

People who mask their true feelings as well as those who's trauma response is to freeze, most likely grew up in an environment where authentic expressions were not accepted and behaviors such as honest expression/

communication was meet with punishment. We freeze/ flee because in our developmental environment it was safer than fighting when we were threatened (often by people in positions of authority).

Distinguishing between ourselves and our environment.

Differentiating between The anger of god which burns within us and the stimuli which caused that chemical reaction Is just as important as it is difficult. It's easy to say Look at what you made me do. As if someone else's actions caused our reaction. This was the driving force behind social convention which establishes a standard by which to govern societies behavior.

Without cloths to cover the parts of our flesh which evoke erosal, there is no part of our bodies which are shameful or sacred. These ideas of privet parts being erotic is more about the attention we show to concealing certain parts of skin which are no different then the rest of our skin, then it is about the body part itself. The lower half of one's face or back of one's neck is capable of evoking just as much of an reaction as any other part of our bodies.

We demand others to cover up those things that we ourselves have been taught to hate or become intolerant of within ourselves.

All too often those things we are intolerant of in others are things that we haven't developed the capacity to appreciate in ourselves. And In our incessant need to belong, we've restricted ourselves and our expressions to an conventional identity.

The problem is we don't fit in those boxes(conventional identities).

Far too often the excuses we make for others not appreciating us (I'm too fat, short, dumb, old, ugly, skinny, eccentric, passionate, relaxed, anxious, ......) are things we've neglected to appreciate about ourselves. Your body and mind are part of you, and until you are capable of loving all of you (idiosyncrasies and all), how can you expect someone else to? Now I can't tell you that there's someone out there who will love you as you are, but until you can love yourself for everything you are. how can you expect someone els too? And even tho you won't need someone else to love you if you love you, if you are capable of loving yourself, perhaps someone else can too?

## BIOLOGY AND ANTHROPOLOGY

One of the biggest proponents for our evolutionary origins are the numerous Design flaws which argue against an intelligent creator. Take the human retina for instance. The retinas cellular walls are backwards, requiring light to travel thru multiple levels of tissue before reaching the rods (photo receptors). This design flaw serves no plausible function and is in fact a product of evolution (trial and error).

Or for instance the remnants of useless leg bones in aquatic mammals such as whales and dolphins.

"People (personalities and mannerisms) are algorithms coded by our genes and activated (switched on or off) by stimuli in our environment." Like the animals in the wild we too very much run most of our programs thru the amygdala AE the instinctual/emotional mind.

## THE "HIERARCHICAL ORDER"

Apes are not lesser evolved versions of ourselves (humans) but rather our cousins and members of the same evolutionary generation, only with a different set of genetic values for survival and propagation. After all life's primary goal is not chiefly survival but procreation which is survival of the species and corresponding genes/ values or traits.

It's not hierarchical or leaner but a spectrum of derivations.

Similarly gender is not a empirical binary but a spectrum, yes there are genetic and physiological distinctions between the two genders but the traits stereotypically associated with ether the masculine or the feminine are not static or binary but instead a spectrum. The point is this world is not divided against itself but integral and dynamic. It's not an hierarchy but instead an ecosystem of equals.

Man became gods among the beasts by learning to exploit the hierarchical structures of other mammalian creatures. Man became authoritative thru proxy by making the alpha subordinate man subsequently made its subordinates submissive. We can see this in the animals early man chose the domesticate. For some reason in the animal kingdom a sense of comradery correlates with hierarchy even in communal creatures who congregate but do not organize themselves in hierarchies also do not protect or defend their kindred from predation. In the animal kingdom more or less a lack of hierarchical loyalty corresponds with a lack of compassion.

Communal Animals devoid of hierarchy also often tend to be devoid of loyalty and even refrain from utilizing the combatant qualities of community when it comes to defending piers from predation.

There is actually an interesting correlation between loyalty and hierarchy observed in the animal kingdom. For instance animals early man chose to domesticate and tame (like horses and sheep) actually have close knit familial hierarchies, and so in the case of the horse man would simply tame the top male in the herd and inadvertently tame its subordinates by proxy. In many cases these hierarchies are simply familial bonds an values like loyalty and obedience/submissiance which are then exploited by men. Conversely however creatures like zebra (despite being communal creatures and traveling in herds) have nether a hierarchy nor any sense of loyalty to their compatriots. Zebra will often not even defend their progeny. Which this said however zebra are rather competent as their camouflage makes it hard for predators to single one out of the herd, not to mention that a kick for their rear legs can (and often do) prove fatal. To some degree we can see the implications of this in human psychology as conservative/traditional demographics often cater to this "familial" hierarchy and loyalty, to exploit workers and subordinates in small businesses and religious setting (like ministry). Where as liberals often are accused of "eating their own" in regards to loyalty or authoritarian hierarchies. To a degree it would appear that the familial loyalty religions exploit, is directly dependent upon the imposition of a hierarchy to enforce it. And thus once the hierarchy is disavowed, so too is the loyalty. This may be why so may who were once as close as family (and even family in some cases) completely excommunicate those who left the faith.

Growing up there was a custom of punishing both parties when an infraction or violation of a rule was committed and the culprit was not discovered, to ensure that "justice" is done, all parties are punished. However The goal here was not justice but rather a show of power in response to a defiance of authority. A display of Violence is often mistaken for a show of strength, but in reality it is a sign of fear and weakness. Above all els religion fears truth, politics fear equality, and economy fears justice, for truth, equality, and justice, threaten the very tools these systems use to predate people. Belief validates ego and comforts fear thru ignorance, politics preserve ego thru hierarchy and authority, and economics propagate ego thru exploitation of the poor.

If it ain't broken, don't fix it. Rather it's on the level of individuals, societies, or even species, the greatest proponent for growth (adaptation) is predation and tribulation. We can see this in organisms like alligators who (due to their predatory success) have remained relatively the same (unchanged) for the past 110 Million years.

People, thru the implementation of religious, political, and economic, systems have effectively posed a greater threat of predation on the general human populous then virtually any predator before. Subsequently resulting in the primary prerogatives/directives of our genes (to procreate, propagate, and preserve, its genetic material) ultimately being negated and halted due to an inadequate environment. In short our capitalistic, industrial, hyper Individualistic, and dogmatically religious, culture has cultivated an environment virtually incompatible with the health and well-being of offspring, subsequently stunting child birth rates. Not only is child baring financially impertinent but from the ever growing wealth gap, disparity, social injustice and egregious exploitation of an majority (disenfranchised) social class, the environment is not sustainable or beneficial for the propagation of healthy progeny. And the institutions (like the church) which claim to be "pro life" are not attempting to improve the quality of life, but simply the quantity of life by which it can exploit (as it was the church who has profited the most form the religious, political, and economic, systems). And thus people have become more concerned with social justice, human rights, pursuing truth, and equality, rather then procreating, because contrary to the dictums of doctrine, people (and their genes) prioritize quality over quantity. Growth is a defense mechanism for combating predation but along the way it became more then a defensive mechanism, but a goal.

However as we saw in the last chapter immediately following the bubonic plague and famines of the 17th century The greater the population, the lower the value of an individuals life, 1 out of 10 is more significant then 1 out of 1 million. And so in in many cases people in power attempt to use quantity as a means of combating the demands for quality of life. In many cases "pro life" simply means making the lives of those who demand equality, Expendable and dispensable. (On the cellular level this is also the prerogative of cancer cells "quantity over quality" and ultimately this is one of the reasons these cells are so detrimental to the health and well-being of the organism as a whole).

Life is costly, but not necessarily precious, it is miraculous, but not scarred, and in fact even treacherous at times. For example, even a yogi will smite an ant, and a vegan will consume a plant with which they share 60% of their DNA, not to mention a virus, who's only means of procreating comes at the detriment of its host. And thus the need for a hierarchy to set a priority on one life over that of another was of paramount importance. We needed to believe that we were special and divinely superior.

In the near middle eastern cultures which inspired and shaped the biblical texts, the greatest Shame or disgrace that could befall a man was for his nakedness to be seen (for him to be exposed or striped of the illusion of prestige and honor), and thus truth and honesty were sins in this culture, for man without myth was no more then the beasts of the field.

But Man is no more isomorphically male or female, then they are men or beast. Humans share approximately 94% of our DNA with K9's, 90% with felines, 80% with cattle like cows, and 60% with insects and vegetation like fruit flies and bananas. Conversely men are typically defined by one X chromosome and one Y chromosome whereas women usually have two X chromosomes. However there are approximately 8,000 biological men in the United States (as of 2023) who have two X chromosome, and a number of biological women (having prominently feminine physical and sexual attributes) who are born with both an x and a Y chromosome. On a genetic level Pom trees are closer related to grass then other trees, and whales more genetically simmer to humans then other aquatic life ("fish"). Ultimately these divinely assigned categorization is erroneous.

## SIGNALING AND SINCERITY

Collective effervescence is a term coined by Emile Durkheim which refers to a sociological phenomenon where a group of people subconsciously synchronize (or perhaps project) their own behaviors onto one another, inadvertently becoming one in body and mind to some capacity.

This can be observed at concerts and raves as well as religious events like mass, and is often accompanied (or induced) thru lyrical hymns or rhythmic music. In part this phenomenon may be a side effect of a number of mechanisms (like the empathy and belief functions of the VMPFC and the regulation of the ANS (autonomic nervous system) Thru heart rate coherence) (we will go into more depth on this topic and the functions of our hearts electromagnetic frequency in a later chapter, but for now all we need to know is that the heart emits a electromagnetic field which extends up to 6ft in every direction, and the nature of that field correlates with the rhythm or rate at which the heart beats. In addition each and every emotion corresponds with a uniquely specific rhythm and electromagnetic frequency)

We project (thru our behaviors) the world we believe in. Like the intuitive attempts at projecting our own fears onto those who frighten us. If we believe the world is cruel, then we will treat it as cruel (inadvertently making our beliefs a reality).

Like the virtuous hero when asked what they were thinking when they ran into the burning building (who then responds with" they weren't"), many of our most genuine behaviors and signals are entirely subconscious. That may in fact be why many of the people who are most vocal about a particular behavior or quality are also often the most disingenuous in their actual acting out of such behavior. In the same way we form muscle memory thru repetition, many of our most altruistic actions are entirely habitual and not consciously executive decisions.

More often then not We over articulate and exaggerate our ingenious actions and extenuate our authentic behavior. We often unconsciously compassionate for our unfamiliarity or incompetence by both enunciating each movement to understand it thoroughly and masking our insecurities or insufficiencies thru exaggerated confidence in an attempt to redirect the attention to a façade. This is probably why an disinterested or unenthusiastic attitude is considered "cool" or appealing as it signals a lack of exaggeration or unfamiliarity with a circumstance.

Another interesting tidbit is the observation of shame or embarrassment. Because tho these responses can be felt in solitude, they are inherently communal reactions. This becomes apparent when we realize that Shame and embarrassment can (and often is) felt on behalf of someone else. It is ultimately a form of social signaling to which the only real remedy to is Authenticity. Because it is only when we are capable of understanding and appreciating taboo behaviors that we can overcome our socially conditioned aversion for such taboos. (This shame or embarrassment felt on behalf of another party is actually intrinsic to the functions of our VMPFC, more notably its empathy mechanism which is used to construct internal simulations based on observations of the external world).

We all have very different definitions of love, of what it means to love someone and what it means to be loved by someone. More often then not our definitions of love have just as much to do with our own needs and insecurities/neglects, as it does to do with our past experiences with our parents and or significant others.

A man May leave his religion and tribe yet still cling tight to a version of love he had erected in his mind. Interestingly enough our bodies actually do a relatively decent job of subconsciously signaling our needs. Rather it is an anxious attachment style signaling a need for validation, or avoidant behavior subconsciously communicating a need for consistency and stability. Even when we ourselves don't truly know what we need or want, our nervous systems often do and have already communicated those needs thru subconscious and instinctual behaviors/responses to the presence/absence of particular stimuli.

With this said autonomy and self awareness (however uncomfortable or inconsistent) is almost always positive. And so tho it is helpful to listen to what our bodies are telling us. To feel and yet not respond is still perhaps the epitome of autonomy.

Real world application. If you pay close attention you will most probably find that rather in relationship with an individual or in interactions with children, the other person will intuitively (tho probably subconsciously) communicate to you their feelings by making you feel the way they do. Moreover rather it is a small child who feels powerless or an romantic partner who feels ignored, the other person will often behave in a way which "causes" you to feel the way you first "made" them feel.

One of The reasons autonomy and self awareness is so important is because Dopamine (the natural chemical that makes us feel happy) isn't actually the reward, but instead the incentivizer. Or in other words our emotions aren't the goal but rather the tools our genes (and by extension our bodies) use to evoke a response. And so tho fear may alert us to the presence of potential danger or even simply an unfamiliar stimuli, it's purpose is to make us react (fight or flee). Emotions like love, happiness, sorrow, etc... are no different. On a biological level these things are not some ethereal essence but rather chemical reactions attempting to communicate to us, or subconsciously control us.

It's all in your head! Of corse it is! Any religion, any social construct, any interpretation of our conscious experience is ultimately in our heads. And yet at the same time any chemical reaction or emotional expression is inevitably a physical one.

We may often feel guilt or shame for lacking empathy for others. However emotions like empathy are actually very complex and expensive. This particular emotion requires us first to be aware and sympathetic of our own pain. And secondly it requires us to of resolved our own pain so far as it no longer takes priority over our observation and capacity to tolerate someone else's. The truth is We are very seldom deliberately calloused to the pain and suffering in this world, but rather instead simply exhausted and spent by our own futile toiles. Because after all how can we have grace for the pain of others when we have neglected or inadequately tended to our own. Callouses after all are skin which has been hardened thru exposure to merciless conditions.

"There is more then enough for everyone, but you are not. There is more than enough for everyone, when everyone's needs are evenly distributed among everyone."

## THE LIE

People most often lie because they've been punished for exhibitions of their genuine selves. People have been taught not to communicate, to shield and hide, their authentic selves. Communication with ourselves and with others, is vital for healthy expression of our autonomy. This is made even more important with the fact that people (unlike ideas or idols/Stacenary objects) change, adapt, and grow.

The ability to lie (both to ourselves and to others) is actually an incredibly necessary skill. It is that ability to cover, mask, or shield ourselves from a virtually endless aray of social climates and environments that has permitted us to survive this long. It is deceit which allows us not only to survive those who might mean us harm but also to live peaceably with those we disagree with. In fact it is that ability to mask or lie depending upon our environment, which we often call manors or propriety. It is that lack of dishonesty and blatant authenticity which we call rude in the behavior of small children. The ability To be honest with those we trust, sarcastic with those we have comradery with but not intimacy with, and vulnerable with those we trust, is perhaps one of our greatest survival skills and tools. the ability to conceal or constrain certain truths and inner thoughts is not bad but instead vital for community.

People lie because they've been thought that their authentic selfs are un acceptable and that affection must be earned at the expense of our authenticity.

## AGENCY

Hidden In our innate proclivity to posit agency in the form of a god, lies both our desire to control that which we can't by the way of favorable divine intervention, and our desire to relinquish the overwhelming dread of our autonomous responsibility. There are things which we can not control which often cause us great suffering and thus we seek a reason for our pain in the hopes that we can mitigate it in future ventures. We try to gain the favor of some cosmic entity in the hopes of inadvertently controlling what we can't. Or at the very least relinquishing our responsibility for such consequences. It should come as no surprise then that Religion sought to deliver us from our submission to nature, in much the same way science seeks to liberate us from our captivity to doctrine. For religion despises that which comes naturally to man and beast alike, wile science detests religions social constructs of master and slave(suppression and blind faith). Religion seeks to vanquish, enslave, and indoctrinate,. science and philosophy on the other hand attempts to tame, study, and discipline, that which religion fears and hates.

Ultimately it comes down to what models we use to explain phenomenon in our environment.

## Evolution of philosophy and physiology

It is unfortunate that one still must argue for evolution as more then just "a theory" as if it is less substantiated than the spherical structure of the earth or its lack of centricity in the universe. But luckily for us the evidence for this origin of life as we know it is insurmountable.

For a start the fossil records provide a thorough progression of development or in other words there really is no "missing link" with very minuscule exceptions. Secondly the "infamous" dating of these fossils is in undeniable unanimity with one another. After all science does not seek validation for its claims but instead contently criticizes its theories in order to test their integrity in a earnest pursuit for truth. However thirdly and perhaps most prominently not only dose this "theory" continually prove itself thru evidence from the past, but it provides results in the here and now. Everything we know about genes and the human mind, all the psychology, physics, and biology, which has led to todays technological breakthroughs are due to/predicated on the data posited by evolution. It's the metaphorical wheel to the whole mechanism.

But what dose that say about god and these claims that god knit us together in our mothers womb?

Obviously this exert form the psalms is poetry, for as we can plainly observe thru the technology of X-rays, god dose not literally knit human babies together in their mothers wombs. But what about metaphorically? What are we to make of the intelligent designer who made air breathing mammals like dolphins and whales who live and make their living in the water, or who designed both the lion and the gazelle for equal competition with one another, when each's survival depends upon triumph over the other? Now in an ideal world Natural selection would choose traits which benefit the ecosystem as a whole as opposed to those which favor an individual at the expense of the ecosystem. But this is not the case. Instead organisms slowly evolve for survival in their corresponding environments and in relation to predation. And we observe this thru the allocation of recourses thru genetic modifications such as bright and bold colors/features directed at attracting potential mates in environments where predation is low or non existent. This is notable for two reasons. 1 genetic alterations are expensive. And 2 rather it is a virus or an dog, the primary prerogative of genes is to procreate. They don't really care about the organism as a whole. The dog, human, virus, etc.... Are just vessels for genes to propagate.

But wait a minute, if genes all originated from the same origin, why then are they incompatible and often in competition with one another? And what does that say about the hierarchical structure (if humans evolved from apes why are there still apes?)

The zoological "hierarchy" is not a progression of evolution but the conclusion of a derivation of species and particular traits. Apes are our cousins not our ancestors, we are both the epitome of our particular rung consecutively.

Fish and reptiles have more in common with other animals than they do with other fish/reptiles conservatively. This hierarchy as we commonly understand it today is problematic at best. The cat is not a higher life form than the rat, and nor is man unto apes. And this is true with the match up between the lion and gazelle as well. Both on the level of the gene and it's corresponding organism, this process of evolution is an literal arms race where the survival of the predator is just as much in competition with other predators as it is with the traits of its prey, and for this the rebel rules over the prudent predation. Even genes of the same species compete for survival and so tho some loins may be bigger and stronger then others, they forfeit speed and agility. For instance we as a species are not as fast as quadrupeds who deviated form a common ancestor, nor are we as physically strong as our biped cousins, because we reallocated the recourses for those traits into our big thinking brains.

Now I say "we" but truthfully it was our genes which did all the decision making (or perhaps more accurately our environment). In much the same way they did thru instincts before we became "Enlightened" or self aware. Much like The mammals who returned to the abyss/ waters such as hippos, wales, and seals, this derivation had much more to do with our progenitors environment then it did with a deliberate choice of the organism or gene. And in many cases "like that of the Galapagos Finches these changes are prompted by the formation and transplantation of organisms on islands (both of land and water such as lakes)

It is here where we see organisms adapt to their environment and subsequently why we find so many design flaws such as our backyards facing retinas, wings on flightless birds, and air breathing aquatic creatures with intact leg bones, because after all "GODs ways are not ours" (we were not designed) but instead retooled like a "bicycle which was converted into an boat" (as Dawkins would put it). We are the result of mutation not manifestation. This may be hard to believe considering the intricate detail in our anatomy because cells designing cells sounds just as preposterous as some invisible entity knitting us together in our mothers wombs and yet perhaps they are not so very different after all. Because In The words of Dawkins "that's exactly what you did in no more then 9 months time".

I think often science and religion end up at odds with one another because they both explain the same phenomenon in very different ways, but in much the same way, we as organisms share very distant ancestors with lions genetically, (and perhaps metaphorically in accordance with genesis 2). science and religion both derive from mechanisms of our genes. They are like two different programs for two different computers which are often incompatible with one another but are nun the less just circuits and wires, synapses and neurons, both attempting to answer the same questions and construct effective models for navigating our environment.

And I think it's worth noting that there was a time when this religious programming and faith based interpretation of life and the world was beneficial for the propagation of genes. Much like the seed of Abraham and the later Hebrews in Egypt, There does seam to be a correlation between faith and virility. So it should be asked Does nature/god know best?

## THE SPIRIT IN THE PSYCHE

In my first book (what it means to be human and what that has to do with the ineffable) I address the cosmological roots of this idea of a soul and how it relates to the union of our body and our breath (air/spirit) which was conceptualized as a form of ethereal essence of life.

This breath which was cosmologically equivalent with our spirit, intuitively expressed the correlation between the quality of the air we breath and our health (life).

This modern cosmological distinction between some ethereal essence and the unseen physiology of air or oxygen was not commonly acknowledged until the late 1800s.

And tho the distinction between the affects of air borne pathogens and the "spiritual" health of an individual is commonly accepted this day in age. The physiological ailments are still often attributed to the health of some kind of spiritual essence in many circles. Tho perhaps this is not far fetched, because everything we once attributed to this idea of a ethereal spirit is now better understood as a symptom of our neuro physiology. Or in other words in many cases the state of our psyche or mental health is directly dependent upon a physiological cause in our body.

Even things as philosophical as our mentality and empathy can be traced back to under developed neuro pathways and chemical Imbalances. Conditions which hinder or alter our behavior such as ADHD and autism or even psychopathy(which is very likely caused by a number of underdeveloped connections between the ventromedial prefrontal cortex (which is responsible for empathy and shame) and the DMPFC (which implements analytical process and rational).) This is important because it means that the traits often attributed to the morality of our immortal souls are actually caused and dependent upon very physical qualities in our anatomy.

Not only does this undermine our ethereal concept of a conscience, but Further more brings to question the validity of these eternal states and theoretical destinations for our "souls", as all our experiences of joy and sorrow, fear and peace, and especially pain and pleaser, are direct results of our physical nerves sending electrical signals to our bran, which then depending on the messages, produces and/or releases chemicals which make us feel pleasure or pain respectively. Or think of it this way, our body's are for all intensive purposes biological super computers which have become self aware. And yet without the necessary circuitry and physiological conduit, the electrical signals which comprise our programming, has no effect. And tho signals can be sent thru space like Wi-Fi, without a physical receiver the message cannot produce a reaction. Tho the signal may exist there can be nether pain or pleasure because those things are dependent upon a interface thru which it can materialize. Now Our body's function this way because our genes have constructed our bodies to reward behaviors which have proven effective in the past at replicating more (like kind) genes. So not only is the program dependent upon the circuitry(thought and experience is dependent upon our body's physical nerves and chemicals), but also this idea that our subscription to the illusion or belief in the sanctity of the feeling, is not a morally coherent standard by which to be judged.

Not even our sacred memories would be proven to be ethereal as they too are caused by sensations on our skin, thru scents in the air, or even sounds which trigger a reaction, mimicking and rooted/preprogrammed by the physiological reactions to previous experiences. Say for instance you smell a certain perfume which once was associated with extreme feelings of elation. This feeling was produced by the release of chemicals in your brain as a result of you'r genes primary prerogative being obtained. We call this feeling love and it is one of our genes strongest incentivizers as it is directly related to the replication and procreation of those particular genes.

This is also why rejection hurts so bad. It's literally our body/brain punishing us for failing at this task. The emotions we describe as negative are just as effective in incentivizing certain behaviors as those we associate as positive. Ultimately our genes don't control us but they do try very hard to. These memories and emotions are all directly connected to our physiology and without physical things like nerves, tissue, and chemistry, we would not feel pain or pleasure.

But what about dreams you may ask?

Are things as prolific and profound as dreams and ideas not proof of some form of ethereal soul?

Perhaps! However we see physical brain activity in correlation with our dreams and it should also be noted that Dreams don't actually predict the future but rather they simply amalgamate an virtual reality in our heads. This amalgamation then produces potential or possible outcomes for particular circumstances which often abstractly relate to our real world environment or experience. This is very likely some remnant of our instincts attempting to rectify the connection between our conscious and subconscious behaviors. This psychic amalgamation is ultimately attempting to calculate the best approach to a relevant obstacle by synthesizing circumstances and then testing certain behaviors. It's not so much predicting the future as it is predicting the probable outcome of a certain trajectory of behaviors. There is also ample evidence that our dreams are capable of drawing from experiences inherited from our ancestors thru our genes. Not so much complete memories but instead instinctual markers which have signaled things like Famines and droughts or even picking up on subtle changes in biometric pressure or the environment which foretell coming storms. These would not express themselves literally as an empirical image but rather as an instinctual feeling. This is because genes don't speak our language. They function more like the 1s and 0s in a computer program.

Ultimately that feeling in your chest when your intimidated or that tingle in your head when you are inspired by an epiphany. Is not the response of some ethereal soul but a biological response to some form of physical stimuli. That gut feeling is not god speaking to you but instead the voices of your genes trying to regain control of an autonomous creature. Ultimately the genetic algorithms we call our emotions, memories, gut feelings, and pleasure/pain which we attribute with some transcendent spirit is actually completely dependent upon our physical biology and anatomy in order to persist in existence.

But if things like psychopathy (a lack of empathy and moral dissuasion) are biological conditions then how do we address the problem of morality. Are those who exploit and abuse others for their own benefit to be permitted to continue their malevolence. Should complicity be accepted or should empathetic people be required to betray their biological wiring in their fight against tyranny?

Should evil be allowed to persist even at the expense of good itself? But perhaps more importantly if such psychological conditions are not a mater of one's soul or spirit but instead one's environmental development and biology then how can one rightfully judge and be judged? The question of morality is tricky because to some degree every living organism survives at the expense of another. From the parasitic wasp which impregnates zombified caterpillars with their larvae to then be consumed from the inside out. To the black widows that eat their male mates for no other reason then that they can. Nature has no morals. Humans on the other hand are not dependent upon nature to dictate our behavior. And thus we can push back against biological prejudices in the the name of virtue. This decision to develop some semblance of morality in relation to our development of complex emotions such as empathy is admirable. However it is also completely made up. And tho this code of conduct is progressive, the fact that it's basis for morality is dependent upon one's subscription to the cultural ideology rather then the actual virtuosity of one's behavior is proof that it is not rooted in some comic law but rather a man made model for congregation and cohabitation.

This is why it is not behavior but one's belief that validities one's existence culturally. It's predicated on their subscription to a form of orthodox indoctrination as a means of salvation. It's not based on some cosmic moral law but rather adherence to a man made set of rules or hierarchy.

All this phenomena which we once called god does not invalidate gods existence nor does it restrict the ineffable to the few and quickly fading gaps in our knowledge and discovery but instead gives god a face. God in a sense is consciousness, god is the meaning we assign above and beyond the rudimentary mechanisms behind love and awe, god is evolution and bicameral mentality, god is nature and the attempts of culture to nurture. We have seen that which we once called god and somehow decided that our new names for such phenomena invalidates it's past identity, god is ego but god is also self reflection. such things retain there meaning not because they are inherently meaningful but because we persist to attribute meaning and value to such interpretations. As we've already asserted it's all connected and thus the relationship between evolution and consciousness, meaning and mater, like the Hebrew nefesh (soul) is itself an entity (god).

We attributed cosmic value to our emotions and interpretations of them. And it was that attribution which then gave it value. "People stop at a red light, not because of a physical barer but because of an socially agreed upon value to the authority of a red flashing light".

We take the knowledge at our disposal and mistake it for the answers to the questions we face, often forgetting that the information is not some cosmic response but rather simply our attentiveness for answers to a particular question. And that's ok. In fact that is good. Because it got us this far. But when we find better solutions to the problems at hand we should not be reluctant to accept them because we have assigned some kind of cosmic value on the answers we made up first. And conversely just because we suffered things like poverty, debt, or physical/psychological abuse, doesn't mean that we should perpetuate those same things on the next generations.

Rather it's slavery or college debt, progress is unfair. Otherwise it wouldn't be progress.

Much of what we call "spirit" is simply an extension of belief. Rather someone is in the Christmas spirit or consumed by the "Holy Spirit" the deifying factor is the degree to which the individual believes in the sentiment. Contrary to popular belief it is not the "spirit" which determines belief, but rather belief which causes the "spirit" to manifest. Rather that be belief in Santa Claus, GOD, or even true love, it is the paradigm which then permits this "spirit" to manifest. However belief is not some ethereal power but instead an internal map and mechanism constructed by our VMPFC. Like empathy this mechanism generates an internal map or simulation of the world. However unlike empathy (which uses the external world to hypothesize probable outcomes in an Internal map) belief imposes an internal map on the external world requiring reality to conform to the hypothesis or internal model. We call this internal model a paradigm or world view. Now mind you, this phenomenon we call "spirit" is not strictly restricted to the VMPFC. As we shall soon discuss, this "spirit" has physiological correlates in everything from the activation of the sympathetic and parasympathetic nerves systems to the instinctual prerogatives of the amygdala. But for now the point I intend to be taken away is simply that belief causes spirit. Or more simply yellow is a happy color because the color yellow makes us happy!

# WHAT HAPPENS WHEN WE DIE

The after life serves the purpose of motivating individuals to sacrifice their lives for a cause or idea greater than themselves. A belief that our values supersede those of our enemies, and that our identities in an idea or belief, is more valid then our innate belonging and validity by simply being. the truth is we don't need A god to be good. I'll say that again "a god is not required to ensure morality" but instead more often then not has been employed to perpetrate far more evil by their authority then they have imparted moral judgment. Gods incite obedience out of fear of rejection form a tribe or community and not out of compassion or some semblance of morality!

And so what actually happens when we die? Well for starters Virtually All material eventually decomposes and takes alternative forms or states. Tho mater is not destroyed or created it does change in ways which are perceived as deconstructive (meaning it is no longer associated with those other particles which once defined their identity).

Our cells like all living tissue decompose or breaks down to more fundamental parts. These parts are then reallocated for other forms of life to form other atomic structures.

Much of what we call the soul or spirit(memories, emotions, and thoughts,) Are just electrical serges and chemical reactions in the right order to evoke a response.

Like data on a hard drive or a server, Once the physical computer (or device) is destroyed the program and data

164

stored there ceases to exist because like (consciousness) that program and information is a product of the physical circuitry. The electrical serges and frequencies which construct thought and data in ether the synapses of a brain or the hardware of a computer are a phenomenon caused by the physical components of that organism or device.

If you believe in a kind of love which fixes your brokenness, then you will be at peace only so long as you retain that belief, however if you believe in a kind of love which appreciates you as you are and not as a utility, then even in a loss of belief, you will retain peace even as you and your understanding of life grows and evolves.

How disappointing would it be to find out you spent your life validating the ego of a narcissist that doesn't exist (in an attempt to prove your utility) rather then enjoying life as it is thru acceptance with each new discovery and experience?

In the same way creation is done by dividing the whole into a hierarchy, chaos unites by destroying those illusions of division. What man calls corruption, nature calls adaptation and reallocation. What man calls death, nature calls creation.

Why do bad things happen?
Bad things happen as a result of neglect or as a result of natural processes. Rather that be the neglect to maintain the physical components of a car, house, natural environment, living tissue, etc..... or even the neglect of faulty or outdated programs (ideas and behaviors). Bad

things happen because things will inevitably happen and when we neglect to appreciate or respect those changes with due diligence their result is often detrimental to us and others.

With that said their are some creatures like the turritopsis dohrnii and the hydra viridissima which are (virtually) immortal. And in the case of the turritopois dohrnii, it is its ability to regress to an earlier developmental state when resources become scarce. This is key because for most creatures the defining factor of decay can be attributed to flawed or lost cellular data which occurs as we grow and evolve/adapt. What we call decay or aging is simply the repurposing of our genetic material in response to an ever changing environment.

## SUMMARY OF OUR BIOLOGY

On its deepest biological level the primary goal of life is not war, conquest, scholarship, or economic accumulation, but reproduction of genes in responses to their corresponding traits for survival such as physical strength and intellect observed in these activities. "Love" or the biological components of it are the primary driving force of life. This is not resigned strictly to the propagation of masculine traits but both male and female traits and their corresponding traits. Or in other words this does not validate the stereotypical male conquest and dominance but the reproduction of the species as a whole and this is controlled by both male and female compatibility with the environment.

In other words reproductivity is just as important

for women as it is men. The only difference is women expense recourses producing the organism and men expense resources obtaining a mate and so the female is typically more concerned with quality control and the male quantity on a biological level. However not all traits are strictly genetic. Many are cultural or learned and even some directly related to environmental factors. Both genders are concerned with attracting physically and intellectually competent mates for reproduction. (This includes philosophical and emotional capacity for communication)

However stereotypically the genetic male role is to provide ample recourses and the female is to transform those recourses into a healthy spawn and thus the male gaze is primarily physical attraction (signifying fertility) in their mate and power in themselves. Whereas the feminine gaze is typically more concerned with mental health in their case and prestige/compatibility in communication in their mate.

(Note: when predation or environmental threat is high, organisms favor evolutionary mechanisms for defense or camouflage. Whereas when environmental threats are mild or nonexistent, organisms typically develop cosmetics like bright colors or extravagant expenditures of resources oriented towards attracting potential mates attention.

## SOCIOLOGY IN THE CELLS

Healthy cells continue replicating until conditions are no longer favorable.

Cancer cells replicate regardless of desirable conditions.

The mark of a healthy cell is its boundary, it doesn't just produce frivolously, it is concerned with the quality of life and not just the quantity of its propagation. Healthy cells are wholistic and inclusive, cancer cells are dualistic, they make everything they touch like them, whereas healthy cells are all about diversity.

the mythic depiction of us taking that forbidden fruit and obtaining the ability to classify and specify division in life (man and animal). Articulates how the development of spoken and written word enabled us to congregate on a larger scale. It should then come as no surprise why In many cases we have a bias to dehumanize individuals who we lack the capacity to communicate with ether ethically or linguistically. this may very likely be why we have such a strong aversion to different cultures. However like the cancer cell which Is intolerant of diversity, isomorphism on a sociological scale is just as detrimental.

## ANATOMY OF THE BRAIN AND NEUROSCIENCE TERMINOLOGY

The most basic or fundamental part of the brain is the brain cell or neuron. We have approximately 100 billion of these brain cells. These neurons form complex circuits thru communicating with one another. In addition Gleo cells aid in this communication between neurons, as well as contribute to the infrastructure and insolation of our brains genetic make up. These gleo cells also store energy and aid in repairing neuronal damage. The proportion of gleo cell to neuron is approximately 10 to 1. Neurons in contrast to most cells are obtuse, barring a greater resemblance to trees then tiny uniform orbs (like blood cells). Also In contrast to the microscopic blood cell, Neurons or brain cells and their cables can extend many feet in length.

The anatomy of a neuron.

At one end of a neuron there are receptors (dendrites) which function as antenna and bare a striking resemblance to branches on a tree. On the other end are a different kinds of antenna (axons) AE white mater (in reference to the white myelin sheath which covers the axon). these axons transmit signals. The axons or transmitter of an neuron "connects" to the dendrite of another. Well not exactly but we'll get to that later. For now the axon functions like a mouth speaking, and the dendrite serves a similar utility to that of an ear.

How neurons do what they do.

On Both the Inside and the outside of the neuron membrane are ether negative or positively charged ions. When a neuron is excited (in an action potential), the ions inside the membrane are positive and those outside are negatively charged. However instead of holding a neutral charge when inactive (what we call a resting potential), the ion's charges swap, turning those inside the membrane negate and those outside positive. Now Half of the neurons energy is spent resisting or deliberately producing that negative charge. This neurological stop sign is not passivity but rather deliberate silence. This of corse is a simplified model. In reality it takes many excitations (approximately 30 mili volts worth) in order to reach the threshold necessary for transmitting that signal past the axon hillock. That is how neurons speak. But how do they listen? Well On the receiving end of the neuron (from the dendritic spines to the axon hillock) a neuron functions more like an analog signal which dissipates. But from the axon hillock down to the synaptic bulbs (tiny bulb like ends at the ends of the axons) it is an all or nothing signal (like a pressurized tube). It takes approximately half of an neurons positive ions to achieve an action potential (an ON or GO signal). However in most cases far less than half of one neurons axon terminals actually connect to half of another's dendrites. And so Instead of forming a chain, neurons form connections more a kindred to webs. This means that more then one neuron has to be excited in order to excite the next ones in line. This also means that any particular neuron can have a limited influence on any vast amount of other neurons. These divergent and convergent signal wirings allow for (and/also if) switches as opposed to the simple (on/off) signals of a binary circuit in the brain. To add to this complexity and intricacy the average neuron has approximately 10,000 dendritic spines, and just as many axon terminals. (It should also be noted that the axon hillock's threshold isn't fixed but can become more or less sensitive depending upon many (internal and external) environmental factors AE Hormones, access to recourses, past experiences, etc..) or in other words neurons that fire together wire together, and neurons out of sync fail to link. For example If you resonate with the lyrics of a song and it evokes a feeling of peace, then the sound of that song will stimulate the same neurons as it did the fist time and eventually will associate that stimuli (song/sound) with peace subsequently evoking the release of hormones that make you feel cozy or safe. The same happens with the alarm on your phone. Except where one (the song) is associated with tranquility, the sound of the alarm stimulates neurons which have been trained to evoke alertness.

Now One would probably intuit that the electrical signals would travel directly thru the axon and onto the dendrite of the next one in line. Forming something a kindred to wires. However this is not the case. The axon of one neuron does not actually touch the dendrite of another. Instead there are tiny gaps or spaces in between the neurons (axon and dendrite) called synapses.

Not only are the two (or 10,000) neurons not connected, the electrical signal (positively charged ions) don't even actually get transmuted between the two.

Instead we find Inside each of the 10,000 axon terminals are tiny sacks called vesicles. And Inside these vesicles are copies of a particular chemical messengers called neuro transmitters. The positively charged ions (electrical signal) triggers the release of these neuro transmitters which are then transmitted thru the synapses to the next neuron's dendrite. Subsequently exciting a completely different set of ions to become positively charged. These neuro transmitters are not all the same but instead are (molecularly) specifically shaped to fit the particular receptors of the dendritic spines. Much like a puzzle piece or a key in a lock. Moreover neuro transmitters are not univocal in their functions. Some excite positive charged ions to differing degrees, whereas others actually inhibit or cause ions to be more negatively charged.

(after transmission and reception of these neuro transmuters they are discarded, some being recycled and others being dissipated/consumed by enzymes). luckily neurons only produce one of these neuro transmitters, and so each neuron has a distinct neuro chemical profile. And because the neuro transmitters are specifically made/copied to fit particular receptors, the neurons (or cells) which that distinctive neuron talks to also has to be of the same (type).

(Note tho the chemical profile of any particular neuron is distinct, the same neuron can evoke more then one particular function/reaction in another AE both excitation and inhibitory.

Popular neuro transmitters.

Serotonin
Norepinephrine
Dopamine
Acetylcholine
Glutamate (+++)
GABA (-)

## HORMONES

Neurons can also release chemicals into the blood stream. These chemicals are called hormones.

Hormones, (like neuro transmitters) have specific target cells which they influence.

Hormones can change the function of proteins, alter genes (activated or inactivated), alter cell metabolism, and stimulate cell growth or atrophy,.....

Popular hormones include

Testosterone
Progesterone
Insulin
Estrogen
Cortisol (glucocorticoid)
Melatonin
Oxitocine

Inversely specific hormones often trigger the release of different hormones depending on the cells which have receptors for that particular hormone. And thus Blockages/alterations in those receptors can greatly effect the efficiency/effectiveness of those hormones.

## GENES AND PROTEINS

Neuro transmitters and hormones are made up of similar (puzzle pieces/key like) structures called proteins, which are then also made up of codes/links called ameno acids. The sequence or order in which the ameno acids form (in conjunction with environmental factors) decides the shape and function of the protein they make up. The possibilities for these sequences are virtually limitless. At least as far as this book is concerned. These sequences which make up a protein are (written or coded) by genes. (Like a program coding another program) Genes are made up of a sequence of 3 nucleotides. This small section/fragment (sequence) of deoxyribonucleic acid (DNA) functions as the code which is then copied and distributed for modeling specific proteins (sequence of ameno acids). This copy is then used as a model and is distributed throughout the body. Mistakes or alterations in the copping of a gene's sequence is called a mutation. We can think of this alteration as ether a misspelled word or a simple grammatical error in a sentence. If the first, then the message may still be received. However depending on the nature of the grammatical error, the meaning of the sentence may be changed to something entirely different. (In reality this is closer to the programming language JAVA then it is English grammar and spelling). Now Depending upon how advantageous or disadvantages these alterations/mutations prove to be, they may or may not be replicated (most prominently via sexual reproduction). This of corse is extremely simplified, but for our purposes it will suffice. Ultimately we find on the smallest level as well as the largest, it's about subtlety and nuance. Baby steps work best when it comes to evolution.

## WHAT A GENE IS, AND IS NOT

A gene is a sequence of DNA (Deoxyribonucleic acid). This sequence influences every defining quality in a living organism, from eye color (Determinative function) to height and metabolism (Influential function). Genes also influence behavioral traits of an organism. Genes do not however control an organism, they simply incentivize certain behaviors in response to certain stimuli. (They are likened to a line of code in a program). And thus rather it is the behavioral qualities or the influential functions of a gene, the gene is directly dependent upon its interaction with the environment. This means that the environment effects what a gene can and cannot do.

A common misnomer is this idea that genes primary goal is survival, this however is not completely true. A genes primary goal is reproduction and propagation.

And so a gene is more than happy to incentivize self destructive traits that promote propagation or preservation of offspring.

Another common misnomer is this idea that organisms are isomorphic. (humans have only distinctly human genes and dogs have exclusively dog genes.) This however is not the case. In fact humans have more bacterial genes in our body then we do distinctly human one's.

Organisms are not isomorphic but instead an large assortment of diverse genes.

Thus this idea of defining humans as a distinct species becomes increasingly difficult the further back we go.

Moreover a gene controls the shape and function of proteins. However only about 5% of our DNA is coded by our genes. Rather or not that gene is reproduced or transcribed (produces proteins via RNA) is dependent upon a portion of that other 95% of our DNA. Particularly a series of coding known as a promoter and its corresponding series called the TF. Genes are more a kindred to computer coding then sentient operators.

In addition to this code like nature of a gene, a specific gene does not necessarily create a particular protein but instead depending upon the number and nature of the exons and introns in the genes stretch of code and how particular enzymes splice or discard those introns and exons. For instance take a sheet of music with the same notes in each line but played in different keys depending upon where they fall out in the song. Now splice one line of music with another set of musical notes.

Moreover genes are not autonomous agents but instead lines of coding dictated by environmental factors both microscopic and cosmologically large.

Now my first intuition pertaining to the nature of genes was erroneous I'd admit. the phrase "we are vessels for our genes" may insinuate that genes are some kind of sentient homunculus controlling our bodies. This however is far from the actual nature of a gene. We are "vessels" for our genes in the same way a sentence is a vessel for a word or idea. Genes are lines of biological code which are then copied and applied to the construction of proteins. Proteins then make up everything from the tissue of organs to hormones and neuro transmitters. It is primarily in the form of neuro transmitters (in the synapses between neurons), and in the form of hormones (from neuron into the blood stream and to specific cells), that genes (code) via RNA and the proteins, then incentivize certain behaviors.

(The divinity of a monarchical blood lines are most probably a primitive idolization of these kinds of genetic traits.)

Diversity of life arose from mutations in genes

(alterations to the code thru errors in their transcription). The less complexity and redundancy in an organism (and its coding), the greater effect alterations have on an organism. And this Abomination (mutation) then, is how nature changes. These alterations or transcription errors are often effects of environmental factors. And so in short we change because our environments change.

## OVERVIEW

(Neurons/Brian cells are not strictly specific to the brain but rather can/and are found all throughout the body with dendrites found just beneath the skin tissue and axons reaching feet up and down the spinal cord and thru the brain stem. In addition, in conjunction with the limbic system the hypothalamus is responsible for regulating autonomic nerves system and hormones) the brain is not restricted to the brain but is made up of (in and throughout) the entire body.

Moreover as we've already seen neurons function lees like direct pathways (circuits) and more like interconnected webs with (and/if then) switches.

With that said neurons are organized and ordered by task and (role they play). This organization of neuronal type and function is how we divide the brain into regions. AE amygdala, PFC, hippocampus, brain stem, etc... each region specializes in particular tasks (the amygdala activates instinctive/emotional mechanisms, the PFC specializes in cognitive processes, and the cerebellum controls intentional motor skills whereas the motor cortex and brain stem are more specific to habitual/involuntary movements)

Despite this specialization, each region forms connections/circuits and networks with others. Moreover the brain on both the micro and macro level is not a binary but instead intricate and wholistic.

## THE BRAIN AND THE BODY ONE IN THE SAME

Much like the Hebrew word nefesh (soul) the mind and the mater which manifests it are indistinguishable. The philosophical phenomenon is an extension of the physiological. And (in our case) the program shapes the circuitry As much as the circuitry facilitates the software. The mind and the body are one. Not only is the mind connected and dependent upon the body as a whole, but the mind is not restricted to the brain, having neuronal circuits and centers all throughout the body. Many of these centers are often called chakras. Moreover everything we attribute to the spirit or the psyche, has a physiological epicenter or location in the flesh. With that said, the body, and more specifically the brain is separated into regions which specialize in specific functions. Or in other words Neurons that fire together (or facilitate a particular function), wire together (group themselves in sections or regions of the brain). And thus we notice specific neuronal functions in specific regions of the brain. For instance at the front of the brain is a relatively new region known as the pre frontal cortex or PFC. This PFC is then split Into more specializing functions such as empathy and belief which show greater activation in the ventromedial prefrontal cortex (VMPFC), and the dorsomedial prefrontal cortex (DMPFC) which has greater excitation during cognitive reasoning and executive functions. There are also regions which specialize in instinctual/intuitive behavior and emotion like the amygdala, and those like the brain stem which have pathways stretching all the way down our spinal cord and going to particular organs like the heart, lungs, bladder, reproductive system, and nerve endings. These regions in particular regulate involuntary responses and motor functions. We will go into greater depth into the functions of each and every one of these regions but first we will discuss the region known as the hypothalamus.

The hypothalamus (a part of the limbic system) functions as the interface between regulatory and emotional mechanisms of the brain (layer one brain stem, and layer two amygdala). In conjunction with the Brain stem, which specializes in regulating the autonomic nerves system or involuntary responses, the hypothalamus communicates and translates the instinctual or emotional signals of the amygdala into involuntary responses (like the secretion of tears or nerves causing us to tremble when we are anxious or frightened) and vice versa (physical stimuli effects our mood and emotions as much as our emotions effect our physiology) . Ultimately these mechanisms facilitate our bodies regulation via emotion as a representation of the subconscious in the conscious decision making (frontal lobe). Or more simply our emotions tell us when our environmental conditions, circumstance, or lack of recourses is unfavorable for homeostasis.

For more context here are the Metaphorical brain layers proposed by McLean...

Layer one autonomic and regulatory

Layer two emotion, instinctual.

Layer three neocortex, cognition, memory, sensory processing, abstraction, conscious volition.

Information is influenced and governed by the fist and second layers just as much as they are by the third. Or more simply we are influenced just as much by our emotions and involuntary responses as we are by our conscious thoughts.

Layer two (limbic system) obtains the majority of its information from olfaction (smell sensory organs) and visual sensory organs. And it includes regions such as the amygdala, hippocampus, sceptem, and habinula.

## AMYGDALA

The amygdala is the brains center for emotion and instinct and it's primary role is to translate and communicate the quality of internal and external stimuli to the rest of our body. However instead of waiting for the PFC to make a more accurate assessment of our circumstance. The amygdala often takes charge triggering activation of Autonomic systems (particularly the sympathetic nervous system). this (sympathetic) system specializes as a responds to stimuli evoking fear, fight/flight, and sexual erousal. The amygdala sends signals down the spine triggering the release of the chemical neuro transmitter norapenefren In the adrenal gland. the SNS or sympathetic nerves system then triggers the release of epamephrin/adrinalin. This illustrates the intimate relationship between our emotions/instincts, and our nerves. In contrast to the sympathetic nerves system, The para sympathetic system controls homeostatic functions such as relaxation and calming of nerves. SNS responses include faster heart rate, slowed digestion, and alert muscular responses, all of which are crucial for surviving threatening situations. Where as the PNS (parasympathetic nerves system) slows heart rate, increases digestive efficiency, and calms nerves, via the release of acedo coleene from corresponding axon terminals.(the SNS is triggered by inhalation wile the PNS responds to exhalation, faster breathing communicates panic, wile slower breathing triggers relaxation.) (note this is why exercises like yoga and dancing/jogging can exhaust (deactivate) our SNS buildup.

Because the amygdala receives signals faster then our frontal lobe, it often activates the SNS in responds to incomplete evaluations of our environment or stimuli.

When BLA (Basolateral amygdala)neurons correspond with amygdala responses to pain, the stimuli which triggers a response from the BLA will also trigger a response from the amygdala. This is learned fear or a trama response. And it is done by forcefully forming synapses (neuro pathways) between the BLA and the central nucleus (this increases the number of receptors for neuro transmitters in the dindridic spines. It is only when a stimuli ceases to correlate with pain, that our neo cortex can begin to negate those connections and the flow of amygdalae neurons. Note both environmental and social defense systems are governed by the amygdala.

Tho not exclusively, all Sensory information travels directly to the amygdala.

Notably the amygdala can respond to stimuli which we (via our corresponding cortical regions AE visual cortex, auditory cortex, tactile cortex, and neocortex) are unconscious of. This means that our instinct or intuition can pick up on stimuli too subtle for our sensory cortex's and cognitive mechanisms to register. And it can do so much faster then our neocortex can process the information. What this shortcut makes up for in speed and sensitivity, it lacks in accuracy. The amygdala also functions as the center for processing pain via an ancient core brain structure known as the pareaquaductal grey (PAG). In addition to external sources The amygdala also receives information from the insala cortex which is responsible for processing gustatory and social discust responses. Inversely the amygdala also communicates back to many of those same regions AE the pareaquaductal grey, insala cortex, sensory cortex's and the frontal lobe. The amygdala is triggered by stimuli, where as the hippocampus focuses on environmental context (or dispassionate rational). Now Typically the amygdala seeks the approval of the frontal cortex before mobilizing the body's motor cortex. however if frightening stimuli to the amygdala is sufficient it can trigger subcortical reflexive motor pathways directly.

An example of this procedure can be outlined below.

The amygdala via the bed nucleus of the stria terminalis (BNDT) (a region which functions as an intermediary between limbic and valence (visual stimuli) information) and hypothalamus, initiate hormonal release. it also activates the sympathetic nerves system inadvertently inhibiting the para sympathetic system. The amygdala then activates the locus cerilious which sends norapenephrin throughout the brain and neocortex.

The sympathetic nerves system is stimulated by excitement, rather that be eurotic or agritory (sex or aggression) which means as far as our minds and bodies are concerned love and hate are simply varying intensities of the same nerves responses and not opposites. The opposite of both being indifference (which is the goal of the homeostatic parasympathetic systems).

Or more simply put, the amygdala keeps us alive, however it is most fluent in raw primal languages and dialects like fear, anger, pain, and sexual errousal.

# PFC

A part from being slower and more accurate, the PFC is also more expensive consuming recourses at a much faster metabolic rate than our amygdala. And thus If the PFC is overloaded intellectually it will often respond by neglecting socialite, affecting morality judgments, and generosity. If a problem is overly taxing other processes of the particular mechanism will suffer. This also works inversely. If posed with a task which overextends the emotional regulation process (sympathy and empathy, typically functions of the VMPFC) then the analytical ones will suffer. Interestingly enough when a behavior is repeated thoroughly enough to transcribe to memory, the mental processes controlling that function are transferred to the cerebellum for autonomic responses or execution. This automaticity is also called muscle memory. (Note this may be one of the functions rem sleep accomplishes by reducing cognitive load (connectivity with the PFC) and processing simulations (dreams), and mental problems in the background with more ancient mechanisms like the (amygdala). This is key because roughly 20% of our caloric intake goes to fueling our big thinking brain.

Larger PFC's correlate with greater socialite of an organism. Subsequently Exposure to larger or more complex communities inadvertently increases the size of our frontal cortex. Inversely damage to the prefrontal cortex often results in anti social behavior, sociopathy, and a decrease inhibition, (note a similar disconnect between our VMPFC and DMPFC also occurs during rim sleep). Oddly enough damage to the DMPFC (a region of the frontal lobe responsible for analytics, planing and strategy formation) can become virtually incapable of regulating executive control of behavior and prolonging gratification, instead choosing the option offering the most immediate reward. (this particular DMPFC malfunction or diss function is often characterized by compulsive lying, and questionable moral and impulsive behavior). This is because The DMPFC controls impulse regulation, or more precisely it specialize in executive function/external application. Inversely Damage or lack of connectivity to the VMPFC often results in strictly utilitarian and rationally unemotional behaviors making a person unsympathetic or incapable of empathy. Those with damage or reduced connections to the VMPFC tend to care more about outcomes than intent. In the neocortex the DMPFC controls analytical rational, where as the VMPFC controls complex moral emotion or sympathetic behaviors. Oddly enough the two work very poorly (even in their corresponding specialty) without the other. The DMPFC or rational mind can articulate the meaning of something but without the VMPFC or emotionally mature mind, it can't understand the significance of that meaning. The DMPFC examines the world as a whole (zoomed out perspective) where as the VMPFC identifies and associates with the world on a macro scale (zoomed in and intimate).

The PFC (via a process known as antacedal reappraisal) can negate or inhibit the amygdala and sympathetic nerves system (and the pain response regions in the brain and receptors in the body). This is often referred to as the placebo effect. (Note tho we can sometimes reappraise circumstances "top down therapy". in many cases "bottom up" functions can be more effective.) this can be exemplified in the way somatic markers (internal simulations of feelings) in our limbic system are assessed by the VMPFC. This "bottom up" response can be utilized in therapeutic ways as well, We can calm our bodies via breathing exercises which engage the parasympathetic system, (slow mindful breaths which slow the heart rate and disengage the amygdala), or bottom up practices which include yoga, mediation, and even exercise which intuitively simulates the act of evading the threat by signaling to the sympathetic system thru an exhaustion of the chemicals produced for fight or flight responses.

Romans 7:15-17 (NIV)

> "For what I want to do I do not do, but
> what I hate I do. And if I do what I do not
> want to do, I agree that the law is good.
> As it is, it is no longer I myself who do it,
> but it is sin living in me."

(When our PFC assigns significance to a circumstance without also assigning value to it, it can result in us doing the thing we are deliberately trying not to do.) for instance the importance of touching the hot stove registerers. But the value (positive/do it or negative/do not do it) is not assigned. And so tho it is important that we (do or do not) touch the hot stove. Rather it is an affirmative or negative is unknown. This is why small children often do the very thing that are told not to do. If god really didn't want Adam to eat of the tree then god should have said "eat of all the other trees in the garden" .

## PLEASURE AND PAIN IN THE BRAIN

Physical pain can be characterized by When we stub our toe or bump our head on something hard, the neurons connected to our tactile receptors fire sending a "pain or pleasure" signal to the thalamus.

However much of Our conscious experience of both pleasure and pain are remnants of ancient biological equipment, who's responsibility it is to incentivize self preservation and propagation in an organism. The primary culprits for this are The mesolymbic mesocortical dopamine system and The ventral tegmentral area (located in the brain stem). We might call this social pain/pleasure.

In short These mechanisms secret and transmit the neuro transmitter "dopamine" to the nucleus acumbis as well as the amygdala and hippocampus, via the (mesolymbic dopamine pathway). The ventral tegmentral also send signals specifically to the PFC (mesocordical dopamine pathway). (Note in adolescence these releases of dopamine also aid in establishing novelty or culturally agreed upon traditions in and among piers AE the reason for nostalgia is directly correlated with dopamine enforced evaluation of preferences in environment and circumstance. Or in other words this accounts for the ariseal of tradition thru revolution in the adolescent brain. This is a deliberate testing of abstract or ambiguous boundaries).

The ventral tegmentral sends signals to other motor control regions as well. The ventral tegmentral also responds to positive stimuli triggering a release of dopamine. The stimuli which triggers this dopamine release can vary from visual signals associated with ascetic, erotic, or even intellectually provoking, to tactile sensations and auditory stimuli, (a cool breeze and warm fire or even a melodious song). This Subsequently evokes an biological response to everything from the smell of food to a beautiful scene. Dopamine release is also triggered in social circumstances where the greatest rewards come from corporation and the punishment of norm violations. Rather it is winning a competitive game, solving a complex problem, punishing wrong doers, or achieving sexual errousal, the dopamine release (or reward) is directly proportional to the scale of previous stimuli. AE if the reward for success is lower then the past experiences, then the release of dopamine will not be triggered or will be released at a much lower dose. "The first time is always the sweetest." We grow to fit our environment in relation to the viable recourses available. This scaling is mediated by the PFC. And thus This is where fasting and discipline comes in because it aids in equalizing that scale which when gone unchecked will become insatiable. "More will never be enough". this has positive connotations as well because it is what inspires us to constantly be learning and growing. It means we are constantly desiring to get, be, and do, better. This is made even more consequential when we consider that our Dopamine release is triggered more from anticipation for reward then it is from obtaining that reward. Which is why emotions like joy, hope, and happiness, are directions and not destinations. Epiphanies release dopamine for this very reason, because it is more about anticipatory (recognizing a pattern) then proportional to a direct reward (obtaining something).

Another important and related chemical produced in the brain stem (specifically the raphe nuclei) is a compound called serotonin. serotonin is distributed to the ventral tegmentral, nucleus acumbis, PFC, and the amygdala,. This chemical specializes in mediating and inhibiting impulsivity in conjecture with the dopamine systems by enhancing goal oriented behavior.

In short Our experiences of Pain and disgust (rather physical or emotional) are produced by our pareaquaductal grey, antigulal singulate, amygdala, and insala cortex,. The severity of these pain responses often depends on the activation and evaluation of activity in the VMPFC to decide the pain's validity in relation to homeostasis and external sources. Or in more simpler terms dopamine produces both pain and pleasure which are both just positive and negative reinforcements used to incentivize certain behaviors. The VMPFC is then employed to compare the sensory stimuli with our internal models to regulate the dosage for the release of these chemicals. (This is why small children often look to their parents reaction when they get hurt before assigning a value to their boo-boo themselves. This is also why a kiss is often capable of fixing many a bruised knee). Serotonin is utilized to modulate mechanisms like our mood, memory and learning, and reward and punishment, ect..

Ultimately this is why heartbreak and rejection hurt so bad. we are social creatures who require companionship and community to survive and thrive. The problem is that need has been weaponized as a tool to exploit and manipulate entire groups of people thru hyper individuality and divisive competition among piers. Competition and individuality are predatory tactics used to thin out the heard. Whereas the cure is cooperation. We need community not only to survive but in order to live. And regrettably a great deal of that need is due to the incentives our genes developed as a means of propagating. In part we need the romance and mythical fairy tales preserved in religion to make life bearable. In all honesty it is our irrationality that keeps us going, hoping, and enduring, when the rational option would be to cut our losses and give up. It is the way our bodies and brains deliberately exploit us, that then makes us so durable and miraculous. With that said the ability to recognize these manipulative mechanisms within our own neuro chemistry is pivotal to ensuring that other people are not able to (intentionally or unintentionally) exploit those same mechanisms to mistreat or abuse us. As often happens in toxic religious, familial, and corporate, setting. This self awareness also helps ensure that we are not (unintentionally) manipulative or abusive to others.

"Feelings (emotions) are one of the ways our bodies assess environmental conditions. And this if our proactive actions continually fail to produce productive outcomes, depression tells us to stop spelling effort and energy. In short pain tells us where the problem is."

## AND THE PURSUIT OF HAPPINESS

Dopamine (tho often associated with pleasure) is a hormone responsible for incentivizing the pursuit of pleasure and not necessarily pleasure itself. And thus it is secreted in higher/ more concentrated doses in circumstances where pleaser is possible but not necessary guaranteed, such as gambling or intermittent reinforcement AE hot and cold/on off relationships. This is why flirting works. Because it temps us with the possibility of pleasure but not necessarily the certainty of that said pleaser. And because this hormone's primary function is the pursuit of pleasure and not the obtainment of that said pleasure, the offs or absence of gratification, actually reenforces the repetition of that said behavior. For example if she kissed you every time you saw her each kiss would be less thrilling then it would if she only kissed you half or one third of the times you saw each other. This is also probably why prayer works so well. Because "if god" answers every prayer, then the turn of serendipitous situations would diminish in value. This incentivizing postponed gratification actually has many evolutionary advantages in the pursuit for food sources, but it is also why we have such a hard time getting over that summer fling and why we find nonchalant attitude so appealing. On a biological level what we really crave is the pursuit of happiness and not the actual obtainment of homeostatic states. And this is one of the biggest reasons we continue to grow, learn, and do better. But it can often be abused especially in our current economic and religious systems. Eternal life is in the persistent pursuit of knowledge and not in the obtainment of truth itself. The magic is in the mystery. It's the anticipation not the reward wince pleasure presides.

Desperation and depression are responses to futility.

Dopamine (the hormone we often attribute to feelings of pleasure) actually functions as an incentive for anticipatory stimuli, our bodies are designed to perpetuate. We are not designed for finality. Yes we need a destination but it is in our best interest that we never completely obtain homeostatic fruition. Depression and despair on the other hand are responses to futility. They are what happens when a voice is silenced and action is not reciprocated or stifled. And this is the crux of oppression.

Because people are designed to grow. And when that growth is inhibited, then we begin to atrophy and die.

As much as we need resistance or negative stimuli to incentivize our perpetual progression, if those attempts to alienate those same environmental stressors are not reciprocated then we atrophy, decay, or become emotionally unstable much like refined radium left unattended.

## SENSES

We can literally (tho not consciously) smell fear and confidence. The olfactory system (the sensory organs responsible for smell) tho greatly atrophied in humans, are none the less Finley tuned for picking up on the release of pheromones and delivering that information directly to the lambic system. But with a direct pathway to the amygdala smell and sight also evoke notable amygdala activation(particularly when smelling sweat secretion caused by fear) or when processing visual stimuli which corresponds with past experiences. We can also (primarily subconsciously) smell pheromones which signify markers such as hight, testosterone, age, sex, reproductive status, health, and genes,. These pheromones also alter our internal physiology and subconscious behavior. (Like the scent of cherry blossoms in the experiment accounted in chapter one).

As we can now see, Everything we once called spiritual, are direct responses to physiological conditions. Evil is not an ethereal problem but a physical one. This idea of the soul or nefesh discussed in the last book.

Not as a strictly ethereal or spiritual essence but instead pertaining to the relationship between the physical and the philosophical (the computer and the program, the circuitry and the software) is reinforced. Or as it was depicted biblically "we are breathed of dust".

This is also reinforced by the fact that altering our physiology AE standing in a proud superhero pose for 2 mins or forcing yourself to smile, can greatly effect your mood. Action alters emotion. Notably cultural beliefs and values also subsequently affect our internal neuro physiology (the structure and connectivity of our brain and synapses-axis circuitry).

Inversely exposure to lager or more complex/diverse social groups case increases in the mass and connectivity of/and between the PFC and the supieor temporal gyros (responsible for theory of mind).

Note (the medial orbital cortex is responsible for accessing the physical beauty of a person, and the moral virtue of their behavior or character, this often results in physically attractive people being judged as more morally good, intelligent, competent, etc… and vise versa.) this is the result of rejigging older mechanisms to accommodate newer tasks.

## SOCIAL RANK IN THE BRAIN

It was often believed that things like racism, bigotry, and sexism, were learned from our cultures. And to a degree the categories of race, sex, and sexual orientation, were applied to hierarchy, the behaviors themselves are innate. We know this from observing adolescent apes both captivity and in the wild. Those which were isolated from community exhibited the same behaviors as those which had been socialized their whole life, the difference between the two was the context in which those behaviors were exhibited. Those apes which grew up In isolation treated "inferiors" with respect and "superiors" with contempt or aggression. The behaviors themselves were the same, it was the context or categories by which the hierarchy was structured which changed or was learned from "culture".

behavioral motor functions are not taught but prewired. Inversely paternal and pier influence effect the context for behavior patterns (when, where, and to whom,) and not the innate behaviors themselves. Beliefs in hierarchy and status/prestige are learned so far as the application of those innate behaviors are concerned. This is most probably the primary role of social play AE games, jokes, dancing, corporative and competitive creativity, and physical communication.

This hierarchical training can have server physiological implications. It is now known that connectivity between and the integrity within particular regions of the brain are negatively effected by low socioeconomic status. This impairment to the copiscolosim (a bundle of azinhal fibers joining the left and right hemispheres) in conjuncture with impoverished frontal lobe components which specialize in working memory, emotion regulation, and impulse control/executive functions, has devastating effects. Low social and economic station also affects executive roles in the brain often subverting the authority of the PFC for that of the amygdala. AE creating more emotionally and instinctually centric individuals rather than cognitive ones. Or more simply put, if a person or animal is put in an environment where they are shamed, demoralized, or belittled, they will conform to fit that role. If a person or animal is beaten, violently chastised, and impoverished, they will develop mechanisms which favor emotion rather then logic.

On a positive note however research has found that a mothers paternal technique can alter (trigger or inhibit activation) of certain behavioral genes.

Interestingly enough our social identities depend a great deal upon our environments. Those raised in hostel impoverished environments like the dessert or barren wastelands are more prone to monotheistic hierarchies with strict authoritarian precepts, where as those who grew up in lush bountiful environments with a divers ecosystem are more prone to wholistic and inclusively corporative communities and social identities which value creativity and curiosity. Moreover in environments where the gaps in income and inequality are relatively small, people are more prone towards generosity and compassion.

Another component of socialite is found in the way our bodie's produce oxitocine and how it is then reinforces prejudice and allegiance consecutively. Or more simply put this chemical strengthens our divisions of us from them. This is made even more influential by the incala cortex responding to strange cultural practices with physical disgust. Ultimately prejudice is not a product of cognition (the PFC) but rather a instinctual mechanisms like the insal and amygdala.

We are prewired for conformity and instinctually inclined towards conservative mentalities. Moreover behaviors like compliance, impulsivity, conformity, prejudice, and adherence to hierarchy and tradition, are ultimately amygdala centric responses and come prewired in us.

## MORALITY, ALTRUISM, AND EMPATHY, IN THE MIND AND BRAIN

In collectivists cultures morality is based primarily on a social standard or function as a utility in obtaining a common goal. and to this accord collectivist cultures use shame to police the populous. As identity is found in correlation with the group and shame is a rejection form or by the group.

Whereas in individualist cultures morality is based on an inner law or code by which we hold ourselves to. In these cultures transgressions are punished via guilt or blame. Collectivist cultures care more about outcomes and are typically not as concerned with intent, whereas individualist cultures are more concerned with the motive or intention behind a transgression. And we can see this in their forms of punishment AE Shame is external, and guilt is internal.

Disrespect or an insult on one's honor deliberately injures one's identity within a group or community, this is closely tied to this social construct of hierarchy and prestige AE social rank or station.

The DMPFC, focus on more collectivist minds such as goal oriented, seeking the greater good or outcome for the community and Utilitarianism. Moral reasoning AE sacrificing the innocent for the betterment of the community.

The VMPFC, amygdala, and insula on the other hand tend towards a more intuitive morality, non utilitarian AE concerning intentionality rather than outcome. This form of morality is based on somatic experiences (or what we often call empathy). This intuitive form of morality oddly enough is not some form of primordial divinity instilled within us. nor is it simply reflexive, but rather the result of prolonged learned behavior. or in other words it is the end result of the firsts (DMPFC's) computations. On a neurological level morality is not innate but rather implicit. It's not intuition vs logic but much like the circuitry which necessitates the software, intuition is a result of logical reasoning. The goal however is not to subvert our learned intuitions back to cogitation. But also to continually examine those intuitions and their biases/prejudices. And renegotiate those models as the environment evolves.

Lying is a mechanism of the DMPFC and other logic based frontal regions of the brain. Compulsive liars have increased amounts of white mater (axinal cables connecting neurons) but less grey matter (neuronal cell bodies) in the frontal cortex. And thus Compulsive lying is characterized by both an atrophy of grey matter (membranes or cell bodies), and an greater development of white mater (connectivity, circuitry or appendages).

Lying also activates parts of the ventral lateral pre frontal cortex (VMPFC) and antereal cengulate cortex (ACC) a region associated with emotional and cognitively conflicting decision making.

The DMPFC has been shown to be causal in both resisting dishonesty and lying. Moreover honesty is characterized by a deactivation of the DMPFC, VMPFC, and ACC, AE instinctually learned and habitual like muscle memory or other autonomic functions and spinal reflex's.

This is why heroism is often not a conscious decision but habitual and instinctive. Past behavior programs future behavior.

Or in the words of the Greek philosopher Aristotl "We are what we repeatedly do. Excellence, then, is not an act, but a habit."

## EMPATHY

We are all empaths Becky!

In its simplest form empathy is one of two primary mechanisms of our VMPFC (a subregion of our PFC). The mechanism we call empathy is in internal simulation and somatic model constructed from our observation of the external world.

When we observe a painful circumstance AE a finger prick or emotional heart break, the parts of our sensory cortex (and possibly even motor cortex) which correlates (maps on to) with that physical (internal or external) body part, activates. This kind of resonance is characterized by the activation of minor neurons in the PMC (pre motor cortex).

The process of thinking, and doing, starts with Executive neurons in the PFC which makes an executive decision that is then communicated to the rest of the frontal cortex. From there this information is distributed to the PMC and then onto the motor cortex and motor functions AE nerves and muscle tissue where it is executed. Interestingly enough a significant number of the neurons in the PMC responsible for carrying out a particular activation or sensation also activates when observing an action or sensation. These neurons in the PMC which are directly responsible for converting "thinking/observing" functions, to "doing/feeling" functions, are called mirror neurons.

More importantly studies have shown that these mirror neurons respond both to a lager array of stimuli correlated with an action or sensation AE auditory, visual, olfactory, tactile... but also distinguish intentionality (context specific circumstances). In the case of empathy this process previously discussed is inverted using mirror neurons as an intercessor between the motor/sensory function and the internal model or simulation of it.

Or in other words We map ourselves into others subconsciously using them as a model or simulation for our own behavior. Empathy is an internal application of a social phenomenon. On a physical level this resonance activates our periaqueductal grey, PAG (a low level portion of the brains pain circuitry.) Whereas emotional resonance (sympathy) correlates with activation of the VMPFC and limbic structures. When "morality" is applied to the problem we see activation and coupling of the VMPFC, insula, and amygdala. Furthermore we see coupling between the VMPFC and the tempero periatal junction TPJ (a region responsible for theory of mind) as we develop and apply perspective to other peoples pain. More over we physically feel small degrees of other peoples pain and pleasure. Rather that is strictly philosophical (emotional), or a gustatory (physical) pain, it all happens in our head and is a very real physiological phenomenon.

(Note autism has been shown to correlate with impairments of mirror neurons) meaning their internal self model is incompatible with neuro typical models.

On a biological level this resonance (matching someones energy) is characterized by a production of oxitocine, and glucocorticoids in the antegla cingulate cortex (ACC) a region responsible for assigning meaning to pain, pleasure, etc... the ACC is also responsible for processing interoceptive information, turning physical or physiological responses into philosophical or emotional schemas/models. The role this frontal cortical region plays in "empathy" is similar to its role in conditioned fear and avoidance of circumstances, behaviors, and environments/ stimuli.

In conjunction with this mechanism of empathy in the ACC is the role of more cognitive mechanisms in the frontal cortex (particularly the DMPFC) which are primarily concerned with the causes and effects of pain and pleasure. These cognitive mechanisms function as gate keepers. Asking the question "does the responses match the cause".

Autonomic synchronicity with kindred AE synchronizing heart rates and Harmon production, activates the ACC. Whereas emphasizing with strangers activates more cognitive frontal cortical regions like the TPJ(which is central to theory of mind).

This need for commonality to initiate empathetic relation becomes evident when we observe the lack or significant decrease of empathy in those who are wealthy in contrast to those in poverty.

And thus Empathy or the lack there of, plays a pivotal role in socioeconomics. Or in other words if someone's socioeconomic station alleviates certain socioeconomic strains (particularly those prevalent in poverty) then their mechanisms for mapping with those models will inevitably atrophy and disassociate with the trials and tribulations of impoverished populations). Or more simply if a rich person doesn't have to worry where their next meal is coming from then they are less likely to empathize with those who do. Because the model they use for their own day to day life is virtually incompatible with that of destitute individuals.

Cognitive load AE the result of a finite amount of recourses fueling or feeding our very expensive frontal cortical regions and mechanisms, can easily explain why we are more prone to have compassion on an individual rather than on a large group. Particularly when we lack commonalities with that group. ("The death of one man is a tragedy, the death of millions is a statistic") (Joseph Stalin). The mechanisms engaged in first person sympathy (feeling someone else's pain) are distinctly different from those responsible for third person empathy (how they feel) this highlights an important role carried out by mirror neurons in the PMC. in humans mirror neurons are not as consequential in mimicking behaviors but instead specifying the appropriate context for instinctual behaviors and emotions. AE the application of hierarchy and social station (often instilled or imparted by paternal figures or piers). This Reasserts the evidence for learned altruism via a release of dopamine. Or moreover morality is conditioned.

Sympathy is feeling someone else's pain(emotional).

Empathy is understanding someone else's pain (cognitive).

Compassion is a proactive response to someone else's pain.

This illustrates the false dichotomy between logic based cognition and emotion based intuition, when it comes to empathy. More over we need both in varying degrees depending on circumstance. With all this in mind it's understandable how being empathetic (mirroring or echoing someone else's pain or pleasure can be just as counterproductive as a complete disassociation with them or it) AE the self gratifying quality of altruism.

## VOLITION AND FREE WILL IN THE BRAIN

EEG scans have shown a signal in the motor cortex signifying a decision (readiness potential) activating in the brain half a second before the consciouses decision to move was made. More simply put our brain may decide to act before we do (however the half second could simply be accounted for by the amount of time required to engage attention and interpretation of intent) but regardless of our brains decision to act we ultimately have veto power. To permit a behavior or to negate it. This of corse depends upon the quality of our regulatory mechanisms.

Some questions to ponder.

Does nature or nurture matter? We can't control our biology, but we can mitigate its agency.

How much does the biology or condition effect our autonomy or volition?

The incessant clambering of a scitzafranics intrusive thoughts vs the occasional intrusive thoughts of someone with the weak frontal cortical connectivity like a sociopath which then cause them to act upon it?

And how much does environment, access to recourses, habitual training, effect our agency?

The truth is we don't truly know.

To a great degree we are extensions of our environment and we are only capable of the the good we know.

# TRAUMA

When it comes to trauma we reenact it using the language we have until we understand and resolve the faults in our old internal models. When it comes to Trauma (especially religious trauma) it is often the result of a bad solution to unavoidable problems in our environment

Anger is not bad. In fact anger is very healthy. it's what tells us when our autonomy has been obstructed. We need healthy ways to feel and express our anger with hurting others, ourselves, or resisting/invalidating it's vital role in enforcing our well being.

What we call emotion is a biological systems check intent on achieving and retaining homeostatic states conducive for the propagation of the organisms genetic material. So tho it's goal is to motivate us to obtain healthier environments, all it has to go off of is the experience of past environments. We feel angry when we get hurt or when we see others injured because we are frightened of those same environmental conditions replicating. And thus unless god can be injured by us, then god has no reason for emotions of anger (which is a byproduct of fear). And if god hurts because they see us hurting, if the anger of god is kindled towards those who seek to injure us. Then why did god not care for our enemies pain, why did god not care to address the environmental strains which cultivated the need for violence in the first place?

More often then not the problem isn't the grief, sorrow, anger, fear, and anxiety, but rather the things which have suppressed and neglected those expressions to expend and be exhausted. Intuitively we blame the triggers without understanding why they trigger such responses. The why here is trauma.

What is trauma? Most simply it is an unresolved experience or a forced neuro pathway. However these definitions may be to simple to be accurate. As we shall soon discuss, trauma is probably best defined as a disconnect between our emotional, intuitive mind or components of our brain, And the rational, conscious parts of the same. More simply it is the result of our emotional, instinctual mechanisms mitigating our autobiographical one's. (Amygdala bypassing our PFC) as it is the corporation between the two which ultimately resolve (reappraise) that experience.

Trauma often results in us Dividing the world between those who can relate to an experience and those who can't. Us and them. And in much the same way trauma is like a finger drug thru the dirt creating a path for water to flow. The force which creates those pathways however, inhibits the healthy formation of alternative pathways until the forced paths which were formed out of necessity for survival are resolved.

Often the trauma of extreme experiences can numb any other (or even similar forms)of feelings. Until it is resolved.

Like muscle memory, the pathways forcefully formed in our minds (synapses) can case reactions to subconscious stimuli. This can result in subconscious behavior.

Trauma as a physiological response in the brain. It is not made up or illusory!

Trauma triggers/engages particular parts of the brain who's specific function is to account experiences as they are happening rather than to analyze or interpret those experiences into a story or autobiographical form. These particular regions of the brain include the amygdala (which controls the release of stress hormones cortisol and adrenaline, as well as nerve impulses, and autonomic nerves systems like the sympathetic system which specializes in fight or flight responses in our body's via the hypothalamus, as well as the entire right hemisphere of our brain (which is responsible for muscle memory and emotive impulses such as emotion and rhythm), and inhibits activation of more cognitive regions like broudmans atea 19 (a region in the visual cortex which specializes as a first respondent in regressing visual stimuli). this means trauma (or the triggers, visual, audible, or sensational,) activate automatic responses to immediate danger or threat. This is made more significant in the deliberate shutdown or negation of brain regions which control analysis (particularly the left hemisphere of the brain) and the bronchus area (responsible for the formation of language and words). This is why putting words to our feelings often helps resolve some of this disconnect and neurological diss function caused by trauma.

Some of these regions trauma disrupts or inhibits include key regions of the frontal lobe, the falamos, and the hippocampus.

The frontal lobe of our neo cortex controls inhibitions, temporal tracking and grounding, abstract and articulate communication, and assignment of meaning to abstract thought, reflection (what we call consciousness). The frontal lobe is also The seat of empathy as (mirror neurons) mimic and resonate with other persons behaviors or emotions. Without our frontal lobe we would be purely habitual or instinctual creatures.

The falamos integrates the sum total of our experiences (or the information transmitted from our sensory organs) with a story or narrative, converting the inputs into autobiographical data. From the falamos this interpretation is delivered to both our amygdala (emotional, habitual) and our frontal lobe (conscious, analytical).

The emotional (subconscious) amygdala receives information and responds much faster then the rational (conscious) frontal lobe.

Partially this is due to the fact that the amygdala mimics and reacts to the environment or stimuli, and the more cortical regions synthesize the environment and then run simulations thru a internal model. Via the VMPFC's belief and empathy mechanisms.

The Hipocampous and lambic systems work to integrate our amygdala (emotional and experiential) and our frontal lobe (consciousness, rational) parts of the brain.

Our body's defense systems (survival mode) is not compatible with connection and equilibrium in peaceful or safe/abundant environments. After all we are first and foremost communal creatures inside and out. From our mirror neurons to our autonomic systems, on an anatomic level we require communion with others. And so tho our amygdala developed mechanisms for surviving competitive circumstances and environments, our primary requirements are for connection and equilibrium (reciprocity) with communities. Isolation on an genetic level is death. And so when we have an inability to trust or more accurately a proclivity to fear others (rather that be in regards to familial bonds, cultural, religious, political, ethnic…) we become incapable of obtaining equilibrium within ourselves. Thru Our ability to relate to others experiences, we develop connection within ourselves for equilibriums states with ourselves and others.

"There is a part of me which loves deep intimate connection with individuals, there is also a part of me which gets anxious around large groups of people. There is a part of me which requires solitude, and another which craves connection with every person, critter, and tree. There are parts which love violence and passion and still too, those which desire peace and equilibrium. It is only when each and every part of our selves is heard and known, that we will then be whole." There is a part of us which hates those who have hurt us, and yet still those parts which will always love them. Those which fear our anger and those which are more inspired by the prospect of justice…. Our ability to honestly articulate and communicate with ourselves and others is the way we resolve the unspoken, emotional, expressions we calls trauma which quite simply describes (a deliberate disconnect between our emotional and our rational minds). it is only in reintegrating that connection, that making the parts whole, that we can then resolve trauma. Therapy is often simply learning to listen to our thoughts and emotions reestablishing equilibrium between the two.

Connecting thought and action, mind and emotion with body and agency.

We begin as synonymous with our experiences (hunger, temperature, tired, wet, lonely). It is only much later that we then become separate and autonomous from our experiences or emotions.

Dreams can function to reintegrate and connect these two mental systems (emotional and rational).

We have two forms of self awareness, the autobiographical self (which keeps track of ourselves in relation to the passage of time, linguistic), and moment to moment self awareness (which is sensory and experiential). We are not our pain, and yet our pain is a part of us.

"Our sense of agency is defined by our relationship with our bodies and its rhythms" (Bessel van der kolk, M.D) (the body keeps the score)

The autonomic nerves system can be broken down into two specialized sections, the sympathetic (emotive), and the para sympathetic (regulatory or inhibitory). Inhalation activates the SNS(sympathetic, emotive), exhalation engages the PNS(para sympathetic, relaxation) these systems synchronize with the heart. The activation of our SNS triggers a release of clucocoidacoids subsequently reallocating recourses used in homeostatic and reproductive purposes to those responsible for evading or obtaining/defending AE (flight or fight). Subsequently Slowing metabolic rate and increasing clotting in blood. (note the release of glucocorticoids can also effect personal altruistic behavior AE stress makes us more egoistic) moreover glucocorticoids inhibit or mediate empathy in the VMPFC. It is not that the release of glucocorticoids (like testosterone and estrogen/androgens) make us more aggressive or in women pro social within a tribe(us but not them) (via a greater sensitivity or concentration of oxitocine release),but rather that they amplify and reinforce the cultural or physiological behaviors or nature of individuals. AE if pro socialite is rewarded more then anti socialite then these chemicals will boost sympathy rather then aggression or defense. These hormones reenforce cultural conditioning.

Trauma can be resolved by ether consciously observing our own emotions and habits/subconscious behaviors (top down). Or by physically associating our emotions with healthier outcomes and deliberately creating healthier habits (bottom up). Essentially by telling our bodies different stories or physically adjusting our behavior for healthier environments.

## THE ABSTRACT IMAGE AND SYMBOL IN THE MIND

According to the neurons in our ACC (the part of the brain responsible for assigning meaning and evaluating expectation vs outcome) the emotional pain of rejection is indistinguishable from our physical pain.

The same thing occurs in the insula cortex which regulates and responds to disgust. This is activated by physically verseral as well as morally repugnant stimuli. This line between the physical and the philosophical is made even thinner by the effect of texture and force/weight on judgments of character. Such as a cold drinks signifying or inferring a cold personality, or a firm handshake or physically heavy book assigned a philosophical weight or integrity to the individuals character. The physical phenomena and environment influence emotional mentality AE harsh conditions, cultivate stronger or corser character. This is most likely a result of rejigging existing mechanisms which specialize in similar or corresponding tasks. The equipment we use to regulate physical disgust has been applied to moral disgust, the markers for physical pain and pleasure have been utilized in internal projections of that pain or pleasure in complex social settings.

Now none of this proves or disproves the existence of an intelligent/autonomous creator, nor is that a goal of this book. What this does prove is the character of such a designer (rather sentient or serendipitous). Which is the goal of this literation! Ultimately whatever wrote the code to life inscribed in our DNA nearly all of our human and cruel behaviors, and thus "they" programmed us to execute both our most virtuous and maniacal behaviors. The goal was propagation. We were the ones who imposed a sense of morality as a clause. Moreover we do exactly what we were programed to do.

## PRUNING PATTERNS IN THE PLASTIC BRAIN

Neurons that fire together, wire together. And thus it's understandable how we might lose certain autonomic functions as we begin to deconstruct harmful beliefs and behaviors that drove those beneficial skills. We can Often lose skills that once came naturally due to a natural process known as neuronal pruning. Neuronal pruning is the process of "atrophy" or intentionally deleting neuronal pathways that have become obsolete "like old beliefs and their corresponding behaviors".

The general consensus among neuroscientists is that certain skills are mapped onto corresponding neuronal networks in the brain (and this includes nerves and autonomic muscle memory). One of the main ways we learn is thru association. So it is not uncommon for a small child to call a cow a dog because of its quadrupedal nature and anatomic similarities. Value is based on Relation, a California 5 is a Arkansas 10 (excuse the crude objectification of a potential mate). And When it comes to learning we define in association with something els (similar or unseemlier). Swims like a fish but breaths air like a mammal, must be a whale (or dolphin). And so often these fast neuronal circuits are executed thru subcategories. AE animal, quadruped, not dog, lives on farm, .... Must be cow. But as we begin to challenge those categories and deconstruct many of these fundamental beliefs that once gave us a sense of identity, we can often lose many of the neurological shortcuts associated with those beliefs. Along with the deletion of those harmful beliefs also comes the pruning of corresponding drives that motivated or even facilitated those behaviors and skills. And so as we deconstruct ideas and resolve traumas that required coping mechanisms like a sacrificial work ethic (servants heart) or incessant need to control/know everything, we can often also lose qualities like our drive And curiosity which had been facilitated by those harmful beliefs and trauma responses. In some cases we can even lose access to emotions which had been facilitated thru the guise of spirituality. And tho the shortcuts or direct neuro hideaways are pruned or deleted (as they need to be) we can develop healthier networks which correlate with the same skills. Tho it may take more time and effect to solidify those new pathways.

I will warn my reader that tho knowing how and why won't necessarily make your life easier or happier and in some cases even make it harder and less "romantic", it will help you be and know better so that you can do better.

Tradition tells us who we are and where we belong but when we find better ways and challenge those traditions we were taught, When we abandon those stories, those parental figures and mentors who gave us those identities often feel like their assignments of meaning and value have been invalidated. Often these people were taught that their feelings and experiences weren't valid. And thus They feel cheated. they submitted to the traditions of their authority figures and yet those who are "their subordinates" are rejecting the traditional narrative. Because it's a bad story and we have found better ones.

These stories guide our choices which then becomes behaviors and eventually fate unless we choose to reevaluate and reappraise those stories.

Our genes developed instincts as a means of automating tasks. Our emotions are tools that our genes developed to monitor and then incentivize behaviors (we get angry "look what you made me do") consciousness on the other hand is a mechanism our genes developed to audit and optimize/reappraise those behaviors in real time. Beliefs are one of the tools our genes use to examine and update those behaviors. And so as we update faulty models and beliefs, we also often lose some of the automaticity associated with those behaviors. As we start to cultivate healthier habits our brains begin to prune or delete shortcuts to skills and impulses associated with the old harmful beliefs and behaviors. "When we don't use the fight or flight/freeze trauma response, we loose it". Emotions are a tool our bodies used to regulate our well-being in an environment. And so when our emotions are invalidated, our ability to regulate our well-being in our environment and then make the appropriate alterations to our behaviors and paradigms (world and self models) is inadvertently inhibited. Unfortunately many religions and social institutions have learned to exploit this inhibition of our autonomy/authority. This invalidation of the individuals experience and inhibition of their volition, then sets the authority figure in said religion or social institution up to dictate the identities of an entire populous. Far to often this new assigned identity designates the individual to no more then property who's value is reduced to their utility.

That's why the role of god in the creation of ancient cosmology as we see in the genesis poems is not a claim of manifestation but of assigning meaning and value. Gods role is to literally tell people who they are and are not, to define their identity. And tell them where they belong and what their utility/purpose in life is. This is the narrative underlying this rhetoric reducing our identity to no more then "sons and daughters of god".

## THE HOPELESS ROMANTIC

I wanted her to fix me, because that's what I thought love was. I thought we could only be made whole if we fit the mold.

I believed them when they said our value is dependent on an identity. But That was before I realized the only thing separating us from them, and me from you, was those beliefs in hierarchy. Identity gave us a place to belong by separating the whole into parts. Thus Chaos only threatens to destroy those divisions which separate the whole. This claim that Unless we validate that ego we to will be lost to the flames. Like a father taking credit for their children's success and yet blaming their child for their resemblance to them. Such a creator has no moral footing. Love is not concerned with receiving glory and praise, all wile manipulating thru shame. Love admires diverse expressions and validates experiences of joy and pain.

Love is not a price paid to counter its own intolerance. Love does not see you as broken, love recognizes the mold as flawed.

I wanted her to fix me because I had believed that i was broken. But it is only thru appreciating all life's parts that love makes us whole.

# THE NARCISSIST

Because we will refer to both god and Jesus as narcissists in this book. it is important that we outline the specific characteristics which denote that labeling or diagnosis in accordance with the personal information at our disposal.

The text book narcissist is a person who has an excessive interest in or <u>admiration</u> of themselves. However more profoundly it is a byproduct of someone who was raised in a world of survival rather then love. Narcissism like any diagnosis is simply the name we give to a particular set of mechanisms (or programs) which were developed to understand and function in the world. And so more often then not the narcissistic tendencies derive from those who had to prove their worth to exist. These are people who see themselves as a utility. And thus these people learn how to obtain love from others thru manipulation. because they are incapable of truly loving and appreciating themselves. these people who our gods are modeled after were most certainly the ones who paved the path to our present world of plenty by surviving the persistent perils of the past. They are not bad or wicked they simply developed mechanisms for survival in a world of scarcity and are thus simply Ill equip for appreciating qualities that developed in a world of abundance. There are of corse exceptions to this rule, those who truly believe they are inherently superior to others. However for the most part this constant self focus is rooted in an inability to truly appreciate one's self as they are. And thus it is not enough that good be done. But that like the ancient Israelites and even the modern evangelicals "that good must flow thru and from them" otherwise what makes them sacred or necessary? I know this because I myself still am in part the narcissist who is never satisfied with themselves and thus requires the approval of others. And so my dear reader the goal which I profess are those not even I myself have obtained. (For they are not obtainable but must simply be accepted). For it is only because we continue to resign ourselves to these titles that then makes us inadequate.

Sadly the closer we observe the love of theses archetypal identities of god and Christ, the closer they resemble that of a narcissist. Tho god is more transparent in their requirement for glory and praise, Christ's claim that people are unlovable and they only have value because they (Christ) loves them, is a blatant representation of a person who users their "love" to then necessitate others reciprocation. however the worst part is not that the narcissist never truly loved someone els, but rather that they are in fact incapable of loving themselves, that's why they require the love and adoration of others. until we learn to love ourselves we too will require the love of others to fill that void. Like the egotistical god or the narcissistic Christ who assigned value to the lives of people only in relation to those peoples praise and worship of them (god and Christ). Until we are capable of appreciating our selves, we too will sell our souls to be loved by others. And sacrifice our self worth as part of the whole to belong with "the chosen or holy ones" in isolation from the whole. A Narcissists definition of love is rooted in utility and survival/scarcity, whereas the alternative is rooted in appreciation and acceptance/abundance.

"Satan doesn't whisper, "believe in me"
(seeking his own glory like god does) he
whispers,"believe in yourself"

(Matt smethirst)

## CONCLUSION

The individuals contribution to society should not be the primary purpose for their development. A Society's purpose is to serve the individuals who comprise it and not the other way around. This is where the communist party went wrong. They crucified the people who did not serve the system which was originally instituted to liberate those oppressed by feudalism.

Feelings are instinctual reactions to stimuli and data points which resemble past experiences like the smell of a certain perfume, a particular sound, or even a familiar behavior or environment. And thus in many cases our failures or success in relation to a feeling has more to do with the confidence we have approaching said experience then it does some mystic form of superstition.

If we go far enough back we find all living organisms on this planet originated from a singular common ancestor, and tho some species are closer then others, and others have been here longer then others. We all evolved alongside one another, which means we (humans) are not more evolved then our piers but simply differently evolved. Every tree, virus, mammal, and insect, is our distant cousin and not some lesser evolved organism. They are our equals, they simply specialized differently. Another interesting point is that no organism is isomorphic but instead consisting of a large assortment of different cells. In fact humans are made up of more bacterial cells then distinctly human cells. And yet every living organism on this planet survives at the expense of another organism. Rather that is a virus attacking a body, a carnivore hunting it's prey, or a plant developing toxins to combat its consumption by herbivores and insects. All living creatures depend upon the death of another in order to survive. And so the question arises "where do you draw the line" this brings us to creation. Because creation as it is described In genesis is a cosmological claim that "this is not like that" it's the deliberate act of separating man form beast, and "us from them". And it was this distinction which has condoned the genocide, rape, and slavery, of those people who didn't belong to a certain tribe or set of traditions. As on a cosmological level faith in said traditions and assignments of meaning and value is what distinguished "us from them". It is no wonder then that the very act of transcending those traditions or claims of creation, is described as apocalypse "the end of a world" in its revelation that all those distinctions which separate us form them are entirely made up. Chaos is a world undivided by order, and it only threatens that which divides the whole into parts. This is why our faith in a set of traditions gives us identity. It promises a sense of belonging at the expense of expulsion form the whole.

The truly interesting thing about this revelation is it's need to mimic nature. You see because we as humans are capable of negating our instincts and thus we must establish a basis for morality which is not something other organisms have to worry about. We have the capacity to question and even mitigate our behavior and subsequently empathize with our "prey".

And as horrendous as we've been. We have slowly but surely argued on behalf of more nurturing representations of nature and more compassionate conceptions of creation (division of species). We still survive at the expense of other organisms but we are conscious of their suffering and studious in establishing more humane ways of farming and killing those creatures who permit our survival. The point is these lines we've drawn distinguishing us from them are just as arbitrary as our claims that one land is holier the another. All life on this planet is more then connected but integral to one another.

Rather it's the fire in our chest when we see an injustice done, the satisfaction when social norms are punished, or the fires of love when we fall hard for someone, it's all our experience of adrenaline being dumped into our system. This is also why dominance and fear play into eurotocism. And why jealousy rather of a god or of a man is a result of insecurity and a fear of abandonment. In the same way passion is an incentivizer to mate, jealousy and heart break are just our bodies way of incentivizing/punishing our failure in obtaining the prerogatives of our genes (AE a mate and progeny). Ultimately rather it is the voices of "satan" in our ear, or the voice of "god", they are nether abominable nor the result of divine intervention but rather completely natural.

The evidence in this book which has displaced the

ethereal essence of some supernatural spirit does not disprove the existence of god. Nor is that its intent. What it does do is fill in the gaps of what was once unexplained and thus attributed to some supernatural divinity. When the fact of the matter is (that which we once called evidence for a supernatural god is in fact a very natural and empirical phenomenon which is ordained no omnipotent authority.

Rather it is gods love or gods anger, it resides within us and not some external entity!

## POEM

Are not even the trees our distant relations.

Do they not also have bodies and minds?

Do they not commune and converse with one another?

With grace and solidarity trees grow tall, blocking out the light in their might. And yet distributing their wealth and knowledge thru their roots.

For even malevolence and compassion are not foreign to the trees. Even trees believe in some form of morality.

Are trees not alive, do they not breath? do they not know wisdom, do they not grieve? What more is required for one to bare a soul?

For are not even the trees our kindred.

# CHAPTER 3

————— ◉ —————

# THE IDOL AND THE IDEAL

It is only belief in hierarchy which makes malevolence moral, makes one life more valuable than another, and separates us from them. hierarchy is the only thing separating gods, form men, women, and beast, .... Chaos is only violence because both sides have been convinced that one is good and the other evil, rather then equally valid and vital parts of the same whole. Chaos only threatens to destroy that belief in hierarchy which separates the whole into dichotomous parts. It is love which recognizes they were one from the start, love is not a savior who saves us from ourselves and fixes the parts which do not fit the mold. Love appreciates the diversity. For love sees what hierarchy obscured and repressed. And yet every living organism survives at the expense of another. we needed a belief in hierarchy to make our malevolence, moral. It was that malevolence which found sanctity in scarcity and sought war on the abundance of chaos. The divine right to rule of conquering kings pathed the way for saviors who sacrificed the self for an idea and inadvertently made such ideas idols.

With the development of the frontal lobe came the ability to write our own stories and create our own unifying (and identifying) narratives. However it also paved the way for consciousness. It meant that we were no longer shielded from our behavior. And thus we devised the concept of morality. We were capable of empathizing with our prey. And thus we needed a hierarchical identity to separate us from them. The beast lived in darkness and was driven by instinct, but man had seen the light and needed the idea to guide them. This subversion of agency to an idea is what we shall here call the ego. It is the part of our body which gets defensive or feels threatened by different perspectives, contradictory interpretations from our own, and dichotomous beliefs or ideas. It is the result of our identity being defined by an idea, and the validity of our existence being decided by that identity. It is a result of our existentialism.

In the ancient cosmology ideas, revelations,and epiphanies, were things of celestial principalities. It was the zeitgeist or muse which granted man sight and by their authority that Conquest and indoctrination was condoned and even commanded. "Our enemy is not of flesh and blood"(Paul).

What we now attribute to genes, ancient man called genies, or angels. God in this context is not one who made the world but rather the one who has been shaping it. Or perhaps more accurately the one who recognizes a pattern and then uses that divine knowledge to then manipulate its functions for their benefit. God is a an assignment of meaning and values on reality. And thus Righteousness is about being in right relationship with the world and it's societal utility.

This may explain why "the fear of the lord or god" is so important. because it bypasses our frontal lobe (conscious autonomy) the mythical "knowledge of good and evil", and reverts to the instinctual amygdala. We look for god when we are frightened because our genes are more evolutionarily adapt at surviving threat and danger. This state of fear or awe would understandably be idealized for its proclivity for innovation. It's also worth noting that the two attributes most commonly associated with the "Old Testament god" and ancient gods in general (fear and anger) are primarily consecutive functions of the amygdala.

This idea that the spirit will guide us, the muse will entice us, or the genius will inspire us.. is perhaps the root of this "indoctrinated belief in hierarchy"

Note (our beliefs and cultural habits influence our biology via (testosterone, cortisol, and Glucocorticoid, often in the form of placebos.) as we saw in the last chapter these hormones reenforce cultural conditioning.

## THE EGO AND THE INSTINCT

The ego here as I am referring to it is our innate proclivity for existential identification with the idea which then subverts our autonomous agency in much the same way our subconscious instincts dictate our behavior and habits. They both rely and function thru traditions and habits.

The defining quality of Humanity is our ability to negate our instincts and decide for ourselves our own nature. This trait bestowed upon us the image of god.

But If your identity is more dependent upon an ideology than it is your own autonomy then you have relinquished that authority as an "image barer". for your religion has become your god. Ultimately The utilitarian idea is just as unfit as a god, as our instincts and nature was. Our mutability is not an iniquity, it is rather our saving grace. our ability to adapt and change, to empathize and emulate potentially positive traits. But Any representation of god which is more concerned with a patriarchal kings claim of divine right to rule and adherence to traditional standards of hierarchy, then it is with the autonomous authority and honor of the least of these, is not the image of an compassionate creator but a malevolent master seeking to justify their conquests.

There is this illusion propagated by our instincts that we are some how permanent or binary. when in all actuality we are made up of a cacophony of heterogeneous traits facilitated by their corresponding genes and whatever we were the day before has no baring on what we shall become in those to come, the parts that defined us in the past will inevitably decompose and be reallocated for new forms of creation. But not even these genes are homogeneous but alterations and rejigged variants of older simpler genes. Or in other words our erect back is a modified variant of our ancestors quadrupedal spine. We have both predatory traits and scavenger traits in us and so why would we assume that we could effectively identify with one set of traits or idealized habits. The point is you were never meant to be reduced to the confines of an binary ideology. You are not ether theist or atheist, conservative or progressive. The epitome of humanity is not to revert to a set of traditions which themselves are mutations of our primal instinctual behaviors. The whole point is that we decide our own nature. Our Instincts do have one major benefit in this regard however and that is their temporal permanence. that is they are only capable of focusing on the here and now and so they lack the wholistic potential for variation. Our conscious mind on the other hand is subject to the existential reality of our impermanence. And thus we crave The illusion of our association with a false sense of permanence and perfection as if we are the ego or façade. But we are not our world view or corresponding ideology we identify with. Originally this trait of relating, empathizing and even emulating with qualities was intended to aid in the propagation and sustainability of genes. So of corse it's understandable why in the light of an existential awakening we would revert back to the prerogatives of our instincts to govern our behavior in the form of identifying with an empirical set of values and narratives.

But the whole point is that we get to grow beyond natures plan for life and it's meaning.

## THE IDOL OR DEIFICATION OF THE OBJECT

Early in our development we developed a social mechanism for arranging hierarchical stations for individuals. This social construct or mechanism is often termed prestige and it can also be observed in the social circles of primates and wolves. More commonly this Prestige is obtained thru the accumulation of wealth and objects which signal to the group that this individual not only has ample skills in obtaining the necessary recourses for survival but also an surplus allowing them to spend energy on trivial and even futile endeavors. This is often how we signal success in a dominance hierarchy, however this is not a good indication of our value as individuals. This inflation of our identity with objects is where we get the term idolatry. Assigning the value of a dynamic creature to that of a stationary object. Tho This may explain why in ancient societies artisans would perfect their craft of immortalizing figures and the corresponding ideas in correlation with prominent people and animals thru the meticulous molding and carving of precious recourses into likenesses. Because These objects functioned not only as signals of one's devotion to an idea thru the futile assignment of vital resources but also functioned as an incentive for one's adherence to an law or social code thru the exemplification of an idealized individual and traits.

I would note that Tho one may be immortalized thru identification with a stationary idea, it is ultimately that stationary quality which strips them of any semblance of life which is anything but static. this immortality comes at the expense of death to the dynamic of an eternal or ever changing life. To identify with the idea is to be resigned to but part and subsequently surrender one's belonging in the whole.

## THE IDEA

The personification of an idea.

Resigning our identity to an idea and our autonomy to the authority of an ideology is by no means a new concept but rather a rejigging of those same mechanisms which we inherited from nature for identifying with our environments (like a toddler identifying with their pain or pleasure or an infant identifying with their mother). Tho the idolization of the idea makes sense especially when those ideas have given us a sense of stability, security, and even purpose. they often come at the expense of innate liberties. The idea functions as a means of obtaining vital resources and serving as a defense against both the physical and the psychological threats that may pose us harm. Tho this static stencil supplies us with protection and direction it also often severely hinders our autonomous motion and growth. Ultimately this Existential crises we fear is often the revelation and epiphany we need as our salvation. To die to our identification with an ideal so that we can truly live, grow, and change.

Which is the false teacher, the prophet or the prophecy that revealed itself to man. Obviously it is the one who reads the runes and not the runes that lay as they always were meant to! And yet why must god speak so cryptically when "she" wishes to say a thing of importance?

Should you not trust god? This is why the profits asked for signs when god told them to contradict their sacred writings! So again I ask, which is the false teacher, the preacher who condemns a man or woman's sexual orientation, or the person who loves who they were made to love?

According to Christian scripture we are not supposed to test god? (Deuteronomy 6:16)

And yet every person god called, tested god in order to insure that they had heard correctly and that the voice they heard was really of god. It's more then simply obeying but having faith that the supposed voice is one in good faith "god's".

Now in the secular world this phenomenon can be expressed as societal reform for as any good teacher will tell you " half the journey is unlearning what you thought you knew" The idea gave us an image to model ourselves after which oddly enough was what we had just been liberated from thru our obtaining autonomy over our instinctual nature. Never the less the idea gave us a direction in the endless possibilities we had just found ourselves in. And as counterproductive as that may sound it was actually incredibly useful in helping us conserve vital resources and energy, obtain shelter and perimeters for security, and give us a goal to aim for. Because as restrictive as these ideals which have shaped our societies and individual identities are. prior to the achievement of that goal they where progress. They gave us a place to belong and a purpose in life. They functioned as an answer to the cosmic questions in life. But like a rest stop on a trail or a check point in a video game these ideals are not final destinations but simply directions and waypoint markers. And that is where reform comes in. because often we arrive and find the thing which had once inspired life has begun to die. Because the idea is not the full picture. Imagine if you had told the ancient Greeks that the world was round and the stars weren't gods but flaming balls of fire. Imagine trying to explain evolution to the cainanite tribes of the ancient levant.

The partial scope of the idea is crucial for its application. Because just like the processes for eating food, we have to break down information into smaller parts that are easier to digest and then allow those baby steps to build on each other.

> "The sign of an educated man is the ability
> to hold an idea without accepting it"
>
> (Socrities)

Whom chooses who, the human or the idea? Is it the muse who inspires man, or woman who entices the genius? To whom does the other belong?

It is my solum belief that Inquisitive humility is the only proper way to aproceh the unknown. An open mind has served us much better in our development and pursuit for truth then our defensive aversion to alien or contradictory evidence.

The religious warn that Man shall always be susceptible to temptation, however I would like to point out The "sin" there is the lie which tells us that temptation is a wicked thing and not the echoing voice of our inquisitive muse beaconing us to grow and change.

Now there are some temptations which are not friendly but rather malice, but in many cases the unknown wants to be discovered and seen for its beauty. It is a sad fact that the malevolent master has taught us that this enticing genius is an evil creature weather then the way to life. They made us hate the serpent in the garden in the hopes that we might see our autonomy as a curse, our curiosity as a vice, and believe the miracle of life is incurred with some insurmountable debt. The ideal took its solution and made it a problem. Rather than surrendering its authority to our autonomy and dyeing so that we could live. The ideal declared itself god in the form of religion. "I am a Christian or I am muslim" Are all infringements of our autonomy and authority as humans to submission of an identity in the idea. We were made to be partners not slaves. And true life is found in honoring diverse depictions of the divine image, and any voice which tells us to vilify those who look and see differently then we do, is most certainly the voice of death.

## THE STORY

The story rather that is a tv show, movie, comic, or novel, gives us an series of scenarios and variables that provide a framework for testing moral and causal outcomes in a given circumstance. It does this by giving us archetypal characters (one or more of which we can identify with thru their development in progression of the story). Where as the idea gave us a direction, the story gives us much more.

It gives us an amalgamation much like our unconscious dreams or the somations of our VMPFC. They offer us a venue to test probable outcomes in potential or even ridiculous situations. The story also gives us important information about the pros and cons of certain character traits in any given scenario.

The story is an indispensable recourse but it too can become corrupted. We can be given stories which present us with an abusive authority as the hero and savior, and convince us that the outcast skeptic and heretic is evil because they don't look like us. We see this all through history as the tyrannical victor rewrites history in their favor describing their enslavement of indigenous peoples as salvation thru indoctrination which completely justifies their horrendous crimes of rape, murder, torture, and theft, in the name of a honorable cause. And as horrendous as these stories are, the truly horrific ones are those still playing out today in the lives of those who have been taught that they are problems that need to be solved. Because we all have stories we tell ourselves. We all have stories we were told growing up. And we all have stories we have come to believe about ourselves and the nature of the world we live in. But here's the thing. These stories and ideals serve a vital role, yes. But that role is not to define us but rather simply to help guide us in defining ourselves. And so for this reason, the unifying narrative too makes for a bad god.

Among the epitomes of our humanity is not only the ability to negate our instincts and decide our own nature, but more distinctly to entertain an idea without accepting it. Or more precisely to relate and empathize with a perspective and yet not identify with its ideology. To be human is to reject the static binary and to appreciate life's expressions holistically. To be nether theist or atheist, conservative or progressive, happy nor sad in mutual exclusivity, but to be all and at the same time none. To no longer be subject to the whims of man or beast, muse or instinct, but to bare the image of the divine in our autonomous authority as human! To recognize the prolific and profound wisdom in the intuitive heart and the rational mind, in emotion and sacred text/tradition. But regardless our obtainment of autonomy and proclivity for mutability which permits us to empathize and emulate potential traits, is not a problem to be solved but our saving grace. Ultimately understanding the impermanence of any given state aides us in appreciating our fleeting feelings of joy and sorrow consecutively but not dualistically as if one invalidates the other.

## QUICK NOTE

Any story or narrative which presents its audience with an absolute good and evil or a purely dualistic conception of the world (us vs them) was most probably constructed by the conqueror who painted themselves as the hero or savior as a means of retaining power and control over a divided and ignorant people. These stories are not meant to promote growth or wellbeing much less truth but rather to instill fear and complacency or even hate for anything that might inspire growth.

## SOCIETY

Humans are social animals. But then again most animals depend on a kind of community to some degree.

But often I think we forget just how much this quality (need for community) influences us. In a recent study done by grad students attempting to incentivize conservation of electricity thru fliers outlining the individual economic, environmental, and moral benefits of not running the ac units during the summer. In virtually every case this form of incentive had no effect. However when told that their neighbors where behaving in an more environmentally conscious manner, there was a noticeable decrease in power consumption.

Another example of this can be observed in the behavior of a small child showing their parents or guardian a bruise, cut, or scrape after encountering a mishap. Now of corse this is done as an attempt to be comforted but especially in younger children (prior to the development of complex linguistic skills) small children will show their guardian the injury to see how they respond before they themselves (the child) evaluates the extent of their own injuries. As if they (the child) needs the discerning judgment of the guardian to distinguish the extent of the injury (to validate their pain). Tho the injury may case physical pain, in many cases this is not as important as the parents reaction to the injury. If the parent brushes it off and kisses the wound, in many cases the child will run off and play as if nothing had happened. However if the parent reacts critically (as is sometimes necessary) the child will mirror their parents reaction.

On a psychological level if someone's trauma or emotional pain is validated, the individual is more likely to address it themselves as an actual problem, where as if someone's unresolved emotional experiences are discarded as inconsequential or invalid (regardless of the symptoms and behaviors that stem from that unresolved emotional and psychological injury) it will go for the most part untreated.

Not only is our behavior (generally) regulated by our sociological environment and community, but in many cases our perception of ourselves is dependent upon social pressures and signaling.

From the gods we found in the stern reproof of a father to the nurturing embrace of a mother, and the immaculate beauty of a lover or fearful abrehinsion towards a stranger, and even the regal reserve resigned to those of sage governors. We have been given a foundation for such things as gods. To be un moved by beauty, prestige, or fear, is to be free of the malevolent will of gods. It is that same desire to assert one's authority and to be seen, to receive attention, to be admired and to ourselves admire, which then Permits the phenomenon known more commonly as god to arise. And thus to be unmoved by emotion is to bare such images of god for oneself. for only one thing is as vexing as to be ignored and that is to be adored.

What if god is just a projection of our ego in response to cultural paradigms. more concerned with honor then morality?

Both hitler and Joshua perpetrated horrendous genocides in accordance with their beliefs. For hitler like Joshua, Moses and Abraham before him acted out of obedience for an nationalistic god? Is morality and racism any less universal when done on behalf of a "creator" and is the unchanging god of old not subject to the moral standards of today? For not even Hitler would have stood out in Joshua's time.

In virtually every culture we observe our perception of god is a direct projection of our own identity thru a world view or paradigm or an anthropomorphism of our instinctual impulses (transcendent feelings and emotions). And thus In many cases baring a loving image of god require us not to be instigated by our instincts and our cultures.

The meaning making mechanism and the prediction programing in our genes are interdependent. And it is these traits which recognizes patterns and then interpret them in preparation for causal phenomena in our environment that has permitted us to survive and even thrive this long.

Our society then is shaped by our models of god. tho Christianity has most definitely shaped our society and culture in the west, our society doesn't look anything like the god, Christ and even Moses before him sought. Our capitalistic Christian society is built on values like nationalism and economic wealth at the expense of the working class. And this is because unlike the movement practically spearheaded by Paul who used the apocalypsest Jesus as the model figure for. (This movement more commonly known as the way.) Christianity is the result of Constantine's nationalistic version of this movement. Interestingly enough this is also where we get our economic system which depends on a slave or working class to serve the gods(bourgeoisie). We inherited this economic system from England who inherited it from the remnants of rome.

An example of this may be observed often in the quality most valued by companies, When a boss says I'd rather have someone doing their best and get less done then have someone do the bare minimum and get more done, it's not about profit but instead about asserting power over someone.

In many cases making someone behave is not about acting morally but instead exercising one's authority over those deemed lesser.

The big businesses buying up property in order to monopolize the market is more a kindred to that aristocracy of the lords virtually enslaving peasants (feudalism), then the ideas of our founding fathers who sought the ability of anyone to obtain wealth in an open market (this of course was still depended upon a slave class). but the slave class "was not their own". Obviously we see just how selfish and morally ignorant our founding fathers where by today's standards. And yet capitalism (like the communism of Soviet Russia) requires a class of slaves for the elite to profit from. Because the system is designed for the "gods" who regard people as resources to profit from. This is why things like higher education (which benefits the society as a whole), healthcare (which is a fundamental human right), housing and food(which are basic human rights and completely feasible this day in age) are not given freely but instead used to exploit and profit off of people.

The evil we see in society today is not the result of some cosmic devil but instead the result of the miss figured Christian god who looks nothing like the holy hobo Jesus or the social activists Moses, but instead the nationalistic Roman Caesar and Egyptian pharaohs.

As of 2023 the top 1% own roughly 43% of the wealth in America wile the bottom 80% of the nation's population poses only 7% collectively. And contraries like Denmark have better working conditions and fairer market then the us.

Example

The price of a burger at a prominent fast food chain in America cost $5.81 wile the employees only make $9.00 an hourly. And get no benefits Compared to Denmark where the same burger form the same fast food franchise costs only $4.82 but the employees are paid $22.00 an hour with full benefits like insurance, vacation, and even a pension.

In America many companies buy up their own stock and then burning that stock as a means of increasing the over all value of remaining stock. Instead of paying their workers a livable wage at no loss to their bottom line . Some major companies could award their employees (as much as a $5 increase) in hourly wages without effecting their profits, but instead choose to spend a whopping

$7.6 billion yearly buying up their own stock so that they can then burn it. This practice is not legal in other countries because it gives shareholders unlimited potential to manipulate stock prices and inflation.

As grim as this may appear, I assure you there is hope. we (gen z, millennials, and gen x) know better and thus we can do better. We already are pushing back against the divisive tactics used by past generations to exploit and abuse marginalized communities. And so as bleak as this capitalistic hells scape looks. There is hope. And all we have to do is keep learning, growing, and challenging the status quo. Rather it is single celled organisms or a workforce, the greatest tool we have for combating predation, is cooperation and collaboration.

## BIBLICAL ECONOMICS

Religious institutions like the church have often aligned themselves closely to those in power like the czar and the queen of England, not as a means of speaking for the people but rather so as to profit themselves form the economic systems which exploited so many.

However Jesus and the movement that he started were members of an economic rebellion and would be more likened to the hippies and hipsters who are selling their possessions and moving into vans as an active boycott of the monopoly big companies have accumulated thru buying up property and controlling the housing market then those conservative Christians profiting off of the miss fortunes of others.

Here are some statistics for those who like that sort of thing.

As of 2022, Christians accounted for 55% of the worlds wealth that is more then Muslims and Hindus combined (5.8 & 3.3 conservatively) and the elite Jews who were vilified by the nazi party for their excess of wealth in the 1930's make up a staggering 1.1%. In a country where an estimated 580,0000+ people are homeless, 38,800 acres of valuable property in California (the state with one of the highest rates of homelessness) is owned by religious institutions and establishments. The Catholic Church owns approximately 177 million acres world wide. And the Trinity church in Manhattan owns approximately $6 billion worth of stock and assets. Which is pocket change compared to Mormon church with over $100 Billion dollars in one of its (nonprofit reserve fund accounts).

And yet even the malevolent god of the Old Testament despised those who profited off of other peoples debt by collecting interest.

Exodus 22:24 (JPS 1917)

> "If thou lend money to any of My people,
> even to the poor with thee, thou shalt not
> be to him as a creditor; neither shall ye lay
> upon him interest"

Deuteronomy 23:19-20 (NIV)

> "19 Do not charge a fellow Israelite interest, whether on money or food or anything else that may earn interest. 20 You may charge a foreigner interest, but not a fellow Israelite, so that the LoRd your God may bless you in everything you put your hand to in the land you are entering to possess."

Leviticus 25:36-37 (KJV)

> "Take thou no interest of him or increase; but fear thy God; that thy brother may live with thee. Thou shalt not give him thy money upon interest, nor give him thy victuals for increase."

Not to mention the Hebrews god instructed a complete resolve of debt to occur every seven years in the year of jubilee. Subsequently the freeing of any Hebrew slave in bondage, returning of all land back to its tribal heritage, and even giving the land a sabbath or rest form it's utility in the economic system.

Even Jesus commands his followers to sell their possession and live in community (where there are no slaves or masters). To find life outside of the empirical hierarchy of the Roman economic system and in reaction of its tyranny. Jesus's or perhaps more acutely Paul's movement (of the way) was more a kindred to socialism then the Christian capitalism of today. (Tho very much resembling a cult, this movement still sought to exploit its members).

We see verses like Leviticus 23:22 and Deuteronomy 24:19-22 instructing landowners to leave edges of their fields for the poor. And even passages like Deuteronomy 23:24-25 which permit the sojourner or those with homes to eat their fill of a crop (this term sojourner may even extend as far as to those who simply do not own property but instead rent or borrow shelter, as we see this practice observed by Jesus and his disciples in Matthew 12:1-8. And We know that Jesus and his followers would often stay in the spare rooms of his wealthier benefactors.)

The social economic system idealized in scripture was not the one constructed by capitalist Christians but instead one where things like housing, healthy food, healthcare, and education are free and un encomberd by the need to validate a hierarchical claim and authority. Because even in the archaic times of exodus they realized Peoples health and well being should not be profited off of. and those who do monopolies off of the misfortune of others are despicable. The truth is things like Learning and work are fun! when not restrained by a systems need to assert authority and superiority over the working class. Education of the entire populace befits society as a whole and it should be distributed freely. Knowledge should not be the sovereign property of the gods.

For work can be done more effectively and efficiently or at less toil and torment at the hand of the labors if not for capitalisms need for a hierarchy and dehumanization of "the slave class" by which it depends. For after all all work and pursuit for knowledge can be fun and pleasurable if not for the interjection of omnipotent authority. work and learning is the very object of child's play.

But without slaves, there are no gods.

And "knowledge makes one unfit to be a slave " Frederick Douglass

yet we would rather follow a path made by gods rather then tread our own. We would rather be slaves in a system for gods than blaze the unknown on our own. Like birds in a cage or chickens in a hut safe from the snares of cats but willingly submissive to those who intend to devour us all the same.

We would rather erect our idols behind our white picket fences then tread up the mountain and join nature in partnership as creators.

however for far too many these white picket fences or even a roof over one's head at all are entirely out of reach.

With that said it should be noted that according to the parable in luke 19:23 god has no problem with charging interest so long as it is "HIM self" who is profiting from it.

## THE ECONOMY OF GOD, MONOTHEISM VS PANTHEISM/ NATIONALISM VS GLOBALISM

In my last book I discussed monotheism from a philosophical perspective as a cosmological model describing the movement away from dualistic world views which saw nature as divided against itself. similar to the way the cheetah competes with the gazelle for survival at the other's expense. But on a larger scale both are part of a more wholistic cycle. However in its original context monotheism was more a kindred to nationalism and racial prejudice/supremacy then it was a more wholistic and inclusive world view. On a biological level this monotheism is an economic and political imposition of values and racial traits in favor and even opposed of/to those of other cultures. Thus it is virtually no different then an organism favoring the traits of its corresponding genes and subsequently its own familial sect/species. And if god is simply the way ancient peoples understood their genes then this would make perfect sense. However this ideology far exceeds the simple parameters of biology. because this sanctification of a nation or people group deliberately exiled itself from trade and corporation with neighboring tribes on the bases that wealth and blessing would not exclusively flow from and to members of its own population. Meaning if they broadened their horizons to tolerate and even appreciate diversity they might lose business or glory in and among their own people. For them diversity and adaptability/mutability of different cultures and even their own younger more tolerant generations was seen as abhorrent and abominable.

Interestingly enough The power of a god was directly connected to the breadth and depth of the peoples faith and value of that god and it's corresponding traits (similar to a nationalistic attribution of values to genetic and philosophical traits). This too affirms the stance of god as an anthropomorphism of our genes and their corresponding traits/incentivizes. In many ways The monotheists of Israel are virtually no different then the conservatives today who fear foreigners will steal their jobs. They viewed tolerant kings like Solomon and Ahab as wicked for the same open mindedness which is vilified in progressive politicians today.

## THE POWER OF GOD

Ultimately the concept of god on a social/national level is primarily about controlling people. In this way Religion Is not concerned with doing what's right, but instead being right, obedience not morality, adherence not truth. It's about controlling people and thus

Religion is how societies like Rome, England, and now America, control people. But as we've seen people want to be controlled or governed. Rather it is subordinace to instinct/intoxication, or compliance to a religious ideology. We choose the idol and master over partnership and responsibility every time. We chose the security of acceptance in a community or social system rather then liberty and justice for all.

The same conquests we observe in the "supposedly waring nation of Israel" can be seen in the holy war and conquest of Europe by England during the crusades. And tho Rome most certainly did not begin as an Christian nation. It did hinge heavily on its captured peoples submissiance to its Roman gods. In fact this was the complaint Rome had with the early movement of the way (which would later become Christianity) it was that this movement could not be controlled economically as it refrained from sacrificial practices all together. And thus it was not until 312 ce when Constantine realized Rome could monopolize off of this movement (in much the same way paul had many years before him) if he'd simply convince people that Rome was a Christian nation and inadvertently infuse Christianity with a form of Roman nationalism which would be inherited by its later English progeny in the form of the Catholic Church of England and much later by its American counterpart. All of this infusion of nationalism and capitalism into what is now known as Christianity, laid the foundation for the The godship of corporate ceos and career politicians similarly to those of Roman governors and isrilight merchants of old.

## SIDE NOTE

(Contrary to popular belief The historical Jesus would not appear to be cross culturally/ethnicity tolerant. the acceptance of non Jewish peoples was one most likely constructed by Paul. Jesus the king of the Jews most probably considered those of other cultures like the cainanite women to be no more then animals and not even people in the eyes of god!)

See Deuteronomy 23:1-6

## POEM

I think we feel broken because we were taught to fear the unknown. And thus we fear ourselves when we cease to resemble "our own"! But our hearts crave the inquisitive sentiment of life!

Our curiosity is not a problem to be solved, we are not broken things in a box. We are beauty itself. The box is the thing that is broken because it separates us from everything within that doesn't fit its mold. Sorrow and grief are not independent of joy. You belong to so much more than one idea, emotion, or expression of life.

## SOCIAL CONSTRUCTS OF GENDER AND SEXUALITY

If a man finds that his value is restringente on his production and ability to provide then he will seek those who can appreciate him for who he is. and the same is true with women, if they find that their value is solely dependent upon their aesthetic and utility as a womb they too shall seek out those who appreciate them for their minds and emotions. This is not to denote the genuine physical attraction to those members of the same sex as prompted by these individuals genetic makeup. but rather to acknowledge that life finds a way. Especially when these gender roles in society have proven to be severely lacking in their ability to appreciate those of the opposite sex apart from their utilitarian role in the relationship. Which is not healthy. Rather it is men or women, we crave attention and connection. and those things should not be restringant on one's adherence to a given role.

Until women can appreciate a man's vulnerability and men can appreciate women for their minds, the prominent example of masculinity and femininity will be primarily toxic and perverse because they lack the capacity required in the other to appreciate their whole expressions. That which was meant to obtain connection has instead produced a detachment from members of the opposite sex. This is not to say that men are completely masculine and women primarily feminine but weather to bring light to the fact that each have both masculine and feminine qualities in them to varying degrees.

Jung spoke of the feminine soul of the man and the masculine spirit in the woman that seeks to express their selves as if we are not one or the other but both.

> "What about masculinity? Do you know how much femininity man lacks for completeness? Do you know how much masculinity woman lacks for completeness? You seek the feminine in women and the masculine in men. And thus there are always only men and women. But where are people? You, man, should not seek the feminine in women, but seek and recognize it in yourself, as you possess it from the beginning. It pleases you to play at manliness, however, because it travels on a well-worn track. You, woman, should not seek the masculine in men but assume the masculine in yourself, since you possess it from the beginning."

> (C.G.Jung the red book)

A notable points the writer Alan watts makes is how unfit the institution of family is for the industrial culture. And I think this is becoming more and more apparent as more an more women enter the workforce. In the 1990's singles accounted for roughly 23 million people in America, by 2022 that number has exceeded 37 million and rising. In a culture where people spend the majority of their waken hours working to making stock holders richer, having a family isn't feasible. Families make work well for agrarian lifestyles, but not for capitalistic and industrial ones. This is particularly devastating as the greatest defense against predation is community, which is why hyper individuality is such a useful tool for those seeking to exploit a populace. Now don't get me wrong, there is nothing wrong with singleness, and in face it will be the death of capitalism in the way a lack of families leads to a decrease in manpower and an increase in the electoral power of individuals as we saw in the 17th century following the plague and famines.

As gender roles become less and less rigged, the narratives used to exploit and abuse people will also lose their power.

The thing about humans is it doesn't matter if they are gay, straight, bi, man, woman, trans, non-binary, black, white, theist, atheist,… they are still human and their experience, identity, and worth, are just as valid as anyone else's. It's only when one requires the "eradication" (genocide) of all other populous, that that validity comes into question. It is only when one seeks to deliberately disenfranchise, exploit, and abuse any other people group, that that "identity" loses its validity. It is those who are inhumane towards others, that are then themselves less then human!

# MODERN SYSTEMIC INJUSTICE

There are many modern forms of systemic injustice which clearly resemble those of the dark ages and Greco Roman eras from big corporations buying up all the property in order to monopolize the market like lords hoarding their power over the serfs. To the rampant discrimination of people of color, gender, and differing culture in the workplace. However the goal of this particular discussion will be to get to the root of these injustices and how they relate to our concepts of gods.

In 2020 the government substadised business for the sole purpose of mitigating unemployment caused by the virus. to cover the payroll of working class employees. these PPP loans (public private partnership/ paycheck protection program) were gifted to a large number of big businesses. However of the 100 billion dollars intended for working class employees (88% of which was forgiven by the government) 75% of these loans were pocketed by the employers instead of using these substadies to pay employees during quarantine.

In 1989 the ratio between the average workers income and that of the CEO was 59(CEO) to 1 (average employe). By 2021 that gap has just increased exponentially to a ratio of 399(CEO) to 1(average employee). According to inflation and productivity minimum wage should equal $27/hr (a livable income) as opposed to its current $7.25-$15/hr(bellow poverty yearly income). All this to illustrate that the working class is not a leach but instead deliberately and unapologetically exploited by these economic systems.

It is no secret that those who profit the most from an corporation often do the least. And those who actually produce the bulk of the service or product is shortchanged for the benefit of the managerial class of a corporation. This is primarily because the quality of the product or service is dependent upon the employed whereas the employer's role is public relations. Like a god who is judged by the loyalty and diligence of their followers, the CEO of a company is only a face (or idea) of the company and not its actual substance. Now their may very well be exceptions to this rule as there are those who actually care about the service or products they advertise. however for the most part the image is the primary concern of the head or higher ups of a company. This need for a slave/exploited working class is fundamental to capitalism and even one of the biggest political components behind the American civil war as well as the American Revolutionary War before that. Each had parties who sought the exploitation of a class/colony deemed lesser. (it should be noted that in both cases a level of exploitation was retained and required for ether party to succeed) for instance tho the American colonies wanted civil liberation from England they still relied upon the exploitation of slave laborers (indigenous people and Africans). And Tho the northern states fought for individual rights it was at the expense of states subservience to the government as a whole. Where the south who sought to exploit the individual did so in favor of the autonomy of the state independent of the governance of the nation as a whole. In both case one sought to exploit the other. I will clarify, I am by no means condoning the atrocities of slavery and degradation of humanity. What I mean to point out is that in both cases the compelling argument at the time was not moral but economic. As far back as ancient Egypt the hierarchical structure of society was constructed by (gods) at the expense and exploitation of the working class. This is the heart behind the profundity of the ancient Hebrews concept of sabbath which essentially functioned the same way our modern day labor unions and organized strikes do today. (Tho this peaceful alternative is being threatened by legislation passed by Congress banning strikes nov 2022). despite what the gods, kings, pharaohs, government, CEOs etc... claim, authority is not given by god or nature but by the allegiance and complicity of the people. This is why the egotistical gods are so concerned with their public image because tho they may rule to a degree, their power and authority is dependent upon those they rule. The problem with Communism is that it forgot that The sabbath was made for man not man for the sabbath.

It goes without saying that a populous which is taught to treat every situation as a test, will inevitably be much less likely to fail said test. However what kind of life is dependent upon nothing more then the utilitarian values imposed by a judgmental god who is constantly watching and waiting to see if you are worthy of their very conditional and abusive kind of love. My advice is not then to simply do whatever you want to but nether to spend your life being tested. Your value is not dependent upon your productivity/ the extent of your exploitation by the gods(and their systems).

God has been used to immobilize our critical thinking skills and propagate ignorance so as to establish an omnipotent hierarchy for utilitarian reasons. Or to vet divine right to rule of the ruling class.

I am personally appalled at this representation of god. A god that is more concerned with retaining patriarchal traditions and conquering kings claims of divine right to rule then they are with each individuals autonomous authority as image bareres. A religion more concerned with controlling people rather then appreciating each individuals diverse representation of a world and/or a god that made them that way. To think after all this time those same traditions "intent on liberating people from slavery, looks more like the oppressive abuse of innate human rights then they do a good god. This imposition of a law which does not seek liberty and justice but instead imposes an insurmountable cosmic debt which enslaves. All so that the god who made that law can deliver us by grace.

Regardless of the unbearable incumbrance of the law which not even Jesus could keep, Jesus goes a step farther and makes one liable for their thoughts and not just their actions. This attempt to illustrate the inherency of "sin" doesn't free mankind from the "gods judgment" but instead solidifies a insurmountable debt dependent upon one's submission to a god that made us inherently evil in order to necessitate their intercession. Ultimately this so called "grace" functions as a cage, or a form of "spiritual Stockholm syndrome". And so it is no wonder that Christianity and capitalism (tho original completely

253

counter intuitive) have become so intertwined in there ability to create a perpetually indebted slave class to exploit and profit off of.

Rather it's capitalism or Christianity, deliverance from a problem they caused is extortion not salvation.

The rhetoric of the Bible has often been employed to reinforce this subscription to those earthly gods particularly in the form of quotations such as Ephesians 6:5 and Colossians 3:22 which was used to ordain the submissiance of slaves to their "masters" during the civil war. And yet in Deuteronomy 23:15 people are instructed to protect runaway slaves from their masters.

There is a big difference between laws that protect peoples liberties as a whole and those constructed to control people, those that inspire, and those that manipulate people. The first is out of respect but the second is out of a desire to fabricate and obtain power. The bad guy is afraid of losing something and holds on too tightly, the villain is someone who has already lost everything and has nothing left to hold on to. The villain here was exploited and then discarded. And in many cases the villain ends up being the hero the oppressed and marginalized need but also the one who deliberately opposes the authority and exploitation of those in positions of power. This is important because as we shall soon see those who seek to liberate and empower the oppressed are often vilified and demoralized by the political and religious powers who profit off of the exploitation of the poor and ignorant.

God can't tolerate sin especially when it concerns people profiting financially from their toils. Take for instance the New Testament account of Ananias and Sapphira who where struck down for lying about their donations to the church. The question tho is why is god so concerned with our monetary wealth? This is an important point the biblical Jesus made in his command to relinquish our monetary wealth. Presumably he recognized that Rome and the systems which would follow would be too powerful to revolt thru physical violence for their injustice and exploitation of the people. but it was susceptible to the peoples mass financial boycott of their economic system.

Interestingly this form of economic warfare has been adopted by countries in modern times. (See The destabilization of Russias economy hindering their ability to fuel their invasion of Ukraine in 2022.)

However it should also be noted that tho the apocalypsest Jesus supposedly required his followers to sell all their possessions and give the money to the poor. Tho He often stayed in the guest rooms of his very wealthy and influential benefactors houses. And suddenly this movement looks more like that of a modern cult then an community of social activists. Tho it should be noted that unlike today where the execution of unarmed protesters is frowned upon, Rome had no problem with nailing a rebel rouser to a tree for good measure.

The truth is there were many such apocalypsists who were martyred for their beliefs. And more then a couple who's names where Jesus as it was a very popular name in its day and place.

## ECONOMY

The only way the existence of multi billionaires can be considered moral is on the presumption that capital is not a finite resource. And yet if capital is not a finite resources then why is there such a scarcity that billions of people live paycheck to paycheck?

Ever since 1971 the US dollar has not been grounded in any physical standard. And this is the way our brand of capitalism works by creating money which the government then loans out to banks to invest. The government charges interest on those loans, much like banks do on their loans to the common people. That Interest is entirely comprised of imaginary money, because the only real money is printed by the government who then loans that money out for an interest fee. The "debt" (much like its spiritual contemporary) is entirely made up. To the point where if every debt was paid in full, theirs would be no money. In fact because the debt is incurred thru interest on the loan of real money, the debt is greater then the physical money to pay it. It is the very reason that people live in such disparity that then makes capital a nonfinite resource.

The money changers in the temple have made the earth a hell.

We have created a generation who could argue ethics with god themself, but we still worship the gods who continue to build their empires on the bones of their humble servants.

Now The concept of life after death was one which gained popularity only after the Christ didn't return in the New Testament authors foreseeable lifetime. However it was with The promise of eternal life that then justifies the egregious taking of life in the form of genocide. And the reason they need genocide is to snuff out all the (trans, unconventional, gay, and minorities,) who are disproving the narratives written by the oppressors to keep their slaves submissive and complicit. They have profited so much form the lies that they cannot allow the truth to see the light of day. That's why they seed doubt in science and vilify people groups who disprove their narrative. Because they know when everyone is equal, they will cease to be gods.

Similarly capitalism justifies the exploitation the working class by claiming that if you work hard enough, you too can reap the rewards of other peoples toils. But that's not the case is it? In fact our pedigree of capitalism may be more classist then even feudalism. And what is so abominable about the claims of socialism is the breakdown of that hierarchy. Socialism claims that everyone is endowed fundamental rights like access to clean water, healthy food, affordable housing and adequate education and healthcare. However what is most outrageous about socialism is its claim that the worker should reap the full value of their own labor. Whereas capitalism claims that the egregious majority of the profit of one's labor goes to the supervising/managerial class. Much like grading in schools, the capitalistic system cultivates medocraty not innovation. You really want to incentivize people to contribute to society, you don't cultivate a false sense of scarcity, you give them access to the professions which inspire them. As of 2016 Americans under capitalism are working more hours across all fields, getting less of the over all value of their labor, and are paying more just to get by then any previous generation. Unpopular opinion: we don't need more investment bankers and businessmen to instigate how much they themselves profit off of other peoples work. What we need are people who actually contribute to society. Those are the people who should be revered. People like nurses, teachers, factory workers, cashiers, cooks, construction workers who actually built our cities, and all those city workers who keep everything working smoothly like clerks, trash men and women, portapot cleaners and sanitation workers…. We don't need more cops who use their position of authority to act out their aggression and insecurities. We need EMTs and firefighters, and we need good cops who actually ensure the safety of the whole community and not just the privileged elites. We need mechanics and architects, we need engineers and computer scientists, we need biologist and physicist who are not constrained by profit driven fields of science and grants. We don't need higher management or wall street brokers. At least not as many as we have. But our system deliberately profits those who contribute the least (if anything at all) to society, and it penalizes people in fields who are most essential for the well-being of our communities at large. The people building the houses should get paid more then those selling them. We don't need more idol gods to worship, we need to start appreciating the lowly peasants who actually contribute to the growth and wellbeing of society at large. We don't need more politicians who infringe on our rights nor those who use their defense of our innate rights to lord over us. Because here's the kicker, the government doesn't give people rights! They take them away. People have rights.

It is worth noting that everything we were taught to fear or despise about socialism is being exemplified 10 fold curtesy of capitalism. From the increased cost of goods and services, resources shortages, and lack of access to quality and innovation in medicine and technology, to the utilitarian brutality of police on civilians. Capitalism is making the very hellscape it prophesied to protect us from a reality. This is where trusting in gods plan got us.

Innovation is not exclusive to capitalism. Greed on the other hand is a prerequisite for survival in a capitalistic system. Innovation is a feature of evolution and by extension life as a whole.

Ideally the majority of the wealth should go to those who contribute to the infrastructure of society like the plumbers, pipefiters, electricians, teachers, and sanitation workers, not only are these fields the last to be allocated to robotic kiosks and AI. They are also those which are most indispensable in maintaining a society. However what we see under capitalism is those jobs which harm society at large like landlords, politicians, CEOs, and bankers, which consolidate the majority of the wealth. Land lords contribute nothing to society but instead commoditize basic needs. Politicians explicitly serve the interests of the corporations which funded their election campaign rather then serving the people who elected them. And banks cater to major corporations and not individuals and the general populous at large.

I'm not really concerned with the states rights, I'm concerned with the individuals rights to exercise their autonomy in a community regardless of wether or not they identify with an orthodox ideology. Man, woman, trans, non-binary, gay, straight, black, white, rich, poor, young, or old, they are people who are endowed fundamental rights simply for being. And it is not the state or the nations right to infringe on the authority of the people they are supposed to be serving and not the other way around. I believe that all peoples are owed the truth, innate liberties and access to justice. This book advocates for those willing to traverse the mountain of god and exercise their executive function in partnership with the ineffability of life rather then reman at the foot of the mountain erecting their gold statues and ideological idols. Ultimately we make our gods, or at least our conceptions of god, in our own image. For the Europeans Christ was white even tho the historical Jesus was Palestinian. And there's merit in that! but there in is also the root of the problem, because The biblical god is modeld off of primitive peoples and parents who exemplify particularly poor character by comparison to present day patrons. And so there is no real reason to be concerned with the biblical gods who are of no real consequence, for such a small egotistical narcissist who is so insecure and fragile that they require the validation of their slaves. Is a poor example for the modern day.

## THE CORPORATION NATION THE ECONOMY
## AND BUSINESS OF COUNTRIES

The very fact that nations identity by economic system is a sure sight that countries are in fact gigantic corporations with their own capitalists or socialistic business models. The former resembling an co-op where distribution of the profits are governed by those actually producing the goods and services and not by capitalists who function as a form of upper management. To put this difference in perspective let us look at the $50 trillion stolen from the working class, that $50 trillion of corse is justified by labeling the egregious percentage redirected away from those who actually produce and into the form of corporate profits passed directly to those owners and CEOs. The point is the directionality of resources and respect. In capitalism (much like Christianity) it is the idol god and those in positions of authority and power who are owed the glory for the work done by their subordinates. Whereas Socialism makes the outrageous claim that the ruling (civil servants) should serve the masses and not profit from their egregious exploitation of the working class. And yet much like a dog barks at other dogs in defense of their human owners, far too many subordinates have been taught to fight tooth and nail to defend the very systems that abuse them all in the hopes that they might be thrown a bone. But the rich are just getting fatter and the working class are being starved. The directionality of this economic system strongly resembles that of the Christian gods claim that those with nothing owe those with everything.

Hyper individualization is a byproduct of the need to exalt the idol gods. But the thing about idols is we learn to emulate them in hopes of obtaining adoration (or at the very least acknowledgement and recognition).

when cows are being herded they inter into a form of trans state or zombie mode where they move as they are led to. However they aren't moving towards something but rather away from certain pressures from behind. This state is synonymous with that of a trauma response in which people disassociate or zone out. In humans this particular state is induced by sensory overload and thus is often treated by "scrolling on social media" playing video games, watching porn, or even running/exercising as a form of regulating our emotions/nerves/instincts… but what's interesting about this is we see similar behaviors in cows grazing. Intuitively this passivity is key in processing vast amounts of data all wile conserving energy by inhibiting more cognitive mechanisms. However in the case of the cow, it is used to cultivate compliance (in the same way punishment or "the rod" is used to train reactionary behaviors) this is why so many of us shut down when we encounter confrontation. Traveling in herds of 10 to 100s and weighing up to 50lbs, it is not unreasonable to deduce that a heard of gazelles could effectively deflect any attack from a single lion so long as the gazelles corporate. In actuality however the lion triumphs over it's prey because the lion is only competing with the weakest or most feeble in the herd, whereas the gazelles are not only in competition with the lion, but also with themselves. And so instead of standing in solidarity they sacrifice the weak. This spirit of competition is key because it cultivates a survival of the fittest mentality by strengthening the individual at the expense of the community. And that is why the decisive two party system of America resembles such a stark contrast to the unified protests of France and Israel. We are in competition with our own rather that be republicans vs democrats, or leftists vs liberals. Now of corse the tactics used by the French won't work for our American citizens because we live in a police state. But we don't need violence. Those in power are not afraid of us owning guns. They are afraid of access to knowledge, and us uniting with one voice for all. that's why they pass legislation regulating the books available to children in schools that teach critical race theory and gender equality/fluidity. They are afraid of alternative narratives which contradict the decisive capitalistic story they sold us, that's why they push bills like the restrict act of 2023 or the 450 anti trans/gay bills. We win by learning to corporate not out compete. We fight back by boycotting their industries and narratives used to divide us. We win by striking. Because we produce all of the goods and services. And we are also the consumers.

Interestingly enough Anarchism is disproportionately common among anthropologists (the people who study human civilizations across cultures). And perhaps this is because The rigidity of hierarchy prescribed by the authoritarian gods and the societies they inspired contrast any real ecosystem in nature. In fact the earliest forms of theism are animism, henotheism, and polytheism. And tho we intuitively posit meaning to circumstances in order to extrapolate the cause and effect, forms like monotheism didn't begin to surface until much much later. This natural hierarchy is seasonal and not authoritarian or supreme.

Anarchy here is not the absence of authority but rather the equal distribution of authority among autonomous agents, it's not diss order but rather a minimal imposition of hierarchy. It's power distributed more equally and diversely.

Now truthfully Hierarchies do arise naturally but these natural hierarchies are not static authorities but dynamic and seasonal, the teacher transferring their wisdom to their student is a good example of this dynamic.

Power as described by those in positions of power is the authority over individual's autonomy and agency. Power is those who seek to be gods among men to control or force certain behaviors. Gods do not seek the power to control themselves or their genes but rather to control other people and things. And thus We live in an age where men seek not the power to govern their own agency but rather to rule over others. Thus we live in a world where laws dictate how certain people act, dress, and love, but that condone the egregious exploitation, abuse, and dehumanization of the same.

> "I am not afraid of an army of lions led by sheep; I am afraid of an army of sheep led by lions."
>
> -Alexander the great-

Lions that lead sheep, lead them to slaughter, because in the end lions eat sheep. And as long as they can convince the rest of the sheep that the black sheep, or the scrawny sheep, or the sheen sheep, deserved to be eaten, the lions win. There are just as many Caucasian men who are ignored as there are white women and black men. Now women of color and trans people are among the worst treated people in America right now and that can't be overlooked. But the point is as long as it is us vs them then the lions win and continue to abuse the majority of the people. Black, white, male, female, trans, non-binary, straight, gay, bi,…. We are all on the same side. And the sooner we realize that, the sooner we can start to combat our own predation and genocide.

It's not about anti semitism, it's about anti humanism (anti people group)..

Hatred is simply ignorance when people truly understand and know better they do better. Ignorance and wickedness are often just faulty VMPFC models. In order to heal trauma we must first get to a place where we no longer need that trauma to survive. Rather that be access to resources, a healthier environment, or inspiration, knowing better and doing better often depend of external factors. But Everyone can be saved. not form themselves but from the behaviors that their socioeconomic station, from the biases and prejudices of religion, and from the environments and climate which cultivated scarcity.

## THE CAPITALISTIC CANCER

The closer you are to the actual production of a good, the less you profit from its distribution.

The government doesn't belong to the people, the government (by way of its politicians and lawmakers) is owed by billionaires and their corporations.

Now admittedly the government often gets blamed

when tragedy befalls people. But in all honesty it was not the government which caused the problem, but rather the corporations which caused the problem and then employed the government in enforcing peoples compliance with the corporations exploitation of people.

In the 1970s scientists employed by prominent oil companies like EXXON MOBIL accurately forecasted the effects burning fossil fuels would have on the environment and global warming and yet the company continued to deny the validity of such claims, instead spending an excess of $30 million dollars to "debunk" the ecological effects of their product.

In the year 1890 and lasting until roughly 1921 a series of conflicts and alterations between disenfranchised workers and prominent members of the coal industry took place in Colorado. These events later got the label "the coal wars". By the 1890s coal companies housed their employees in company owned houses and towns where all animates were furnished by the companies. In turn employees were paid not in US currency but in script which could only be used in that particular companies town. This meant that the companies had a captive audience and thus they started price gouging diving the workers into debt essentially making them slaves to the company. In retaliation the workers attempted to unionize, to which the companies responded by bribing law enforcement agencies to arrest, falsely charge, and even kill, workers suspected of unionizing. This eventually lead to all out war between the workers and the union busters which climaxed in September 1921 at the battle of Blair mountain where 10,000 miners fought off 3,000

Union busters. At which point the national guard was called in to enforce compliance to the companies terms. Tho the workers put down their arms out of respect for the uniform, the US government at large sided with the companies resulting in a strong diss trust in the government ever since.

In the 1920s and 1930s women workers who had suffered adverse effects from prolonged exposure to hazardous materials in their work environments (more prominently radium poisoning from the paint used to paint watches). Brought this issue to their employers attention. However even After women developed server health conditions from this prolonged exposure to radium (like radium jaw or necroses) the denied the adverse effects of their working conditions and environments. These companies even bribed physicians to deny the cause of these ailments. These companies launched smear campaigns to invalidate these women's claims. These girls (called at the time ghost girls) later become known as the radium girls and played a big role in the enforcement of quality control in workplace conditions in the present.

As a result of (*alleged*) neglect and greed In 2023 a Norfolk southern freight train derailed in the town of Palestine Ohio. This derailment exposed millions of people to a hazardous chemical payload of hydrogen chloride which was promptly burned and sent into the atmosphere resulting in the deaths of many fish and wildlife as well as the unknown health defects which will ultimately befall the communities effected. As devastating as this tragedy is, the most nefarious part is the way it escaped public broadcast because the company who neglected to preform upkeep on the tracks, and who owned Norfolk also controls the majority of the stocks of virtually all major networks. Black rock is the majority stock holder in virtually all the major competing companies across all the major industries. From android to Apple, form Pepsi to coke, form fossil fuel to solar, and everywhere in between.

However the fluctuation of stock prices in these companies do not make or break blackrock's bottom line. Because At its roots the 30 year old company (blackrock) is an administrative investment advisory firm which receives the vast majority of its revenue from advising on profitable investments. It more prominently deals in inviting ETF's (exchange trade funds), and index funds. These funds are funded by the majority of the money we pay into retirement ($10 to $7 trillion dollars worth). And so the majority of the money blackrock is investing (using to buy stock) is our money. Or More precisely the money we pay into our retirement directly funds blackrock's consolidation of stocks. As of 2023 US retirement assets totaled approximately $33.6 trillion dollars. That's the money we are collectively paying into retirement. And that is the money being used by companies like blackrock to invest in stock and become majority shareholders across virtually all companies and industries. To put this in perspective. The biggest commercial bank (JP Morgan) controls only $3.2 trillion dollars in assets. And the S&P 500 (the sum total of the 500 largest companies) controls only $36.5 trillion dollars. Whereas as of 2022 individual investors made up 22% of the market accounting for over $7.2 trillion dollars worth of stock. But because companies like blackrock control roughly 40% of the assets across the market, they can vote for lower wages for workers, all wile increasing wages for CEO's (General Mills CEO income exceeds $15 million dollars a year wile the average employee will be paid only 52k -120k a year). Now the majority shareholder of Black rock Is vanguard, and ultimately all the decisions made in this "free capitalists market" are instigated by the members of these two companies who have a literal monopoly on the world economy. Tho their names are hidden the owners of vanguard are no different then the Caesars and gods of old who omnipotently preside over the populous at large.

when the 99 exists to serve the 1, their lives are forfeit. But If god is willing to exploit and abuse their subordinates then they never really loved them at all.

The truth is You will work your life away and still be poor, but far too many of us bought the lie that we could be gods too. And thus we become slaves who live to work all so that the real "gods" can live in leisure.

If they were truly god then they would not require worship for love, even dogs can love unrequited. if they were truly rich they would not require sacrifice for blessing, not even Samaritans demand beggars to surrender more. If they were truly good then they would not require obedience and loyalty to teach and lead, their merit would prove fruitful. if they truly offered liberation then they wouldn't need the wages of sin and an unobtainable law to enforce peoples compliance. if they were truly a savior then you wouldn't still need saving. If they were truly powerful they wouldn't need others to fight their battles for them. If they were truly present then they wouldn't have to remind everyone every week to have faith. If they were truly god, they wouldn't need us to worship them!

People a-range themselves much like brain cells according to roles. and tho PFC mechanisms are more expensive then the lesser brain regions, the more auxiliary functions often actually perform the majority of the tasks more efficiently.

The cool thing about capitalism or any faulty economic system for that mater is we the people don't have to do anything. because they will burn their own house down and all we have to do is not put it out. Boycotts and strikes are how we win and how their corrupt economic system and its egregious exploitation meets its demise.

At the peak of the Great Depression in the year 1930 the average individuals yearly income was $4,881.00. (Which when we account for inflation would equates to the equivalent of $88,888.86 yearly income in 2023).

The average yearly income for an individual working in 2023 is $32,000.00 (which is less then half of that brought home by a single individual at the peak of the Great Depression 1930.)

But here's the thing, this egregious disparity is actually detrimental to the very economy that the same billionaire class that caused it, profited form so egregiously in the first place. Because if people spend all their time working just to put someone else's roof over their head. Then they aren't going to waste their money on products. And without consumers, our economic system crashes and burns.

## THE SOLUTION?

I will preface this by saying I am not an economist. And when it comes to viable economic solutions which are equitable (in pursuit of equality and over all profit) I am certain there are more competent sources then myself.

We don't have the luxury of protesting like the French in 2023. The top 1% own the police. The top 1% have privet armies and technologically advanced fortresses. The billionaire gods have the lawmakers in their back pockets. And tho the top 20% are the land owners, the bottom 70% produce all the goods. We are the farmers, the electricians, the programmers, the nurses…. The top 1% contribute nil to society. And the only thing that gives them any power is the imaginary money which has no real value apart from that which we attribute to it. The US dollar isn't even grounded in gold anymore. We (the bottom 70%) produce all the bread, produce and distribute all the electricity and water, actually BUILD all the buildings and infrastructure, comprise the supply chain, drill for oil, plant the gardens, invent the cures and administer the aid, the bottom 70% provide all the services and goods. ALL THE GOODS AND SERVICES are provided by those who live under the poverty line. So what if we just stoped taking their money? We produce the bread, we purify the water and design the infrastructure for the distribution of electricity and goods. We can't fight them like the French. But what if we didn't have to eat the rich! What if we could just starve them, the way they've starved us. Now I know that would destabilize the capitalistic system (or more accurately the neo feudalistic system). But the CIA has been destabilizing the socialist systems for decades. What if the socialists destabilize the capitalists who have exploited us for far too long. What if (the people, the workers,) actually profit from their own labor. What if we owned the goods that we produced. What if we stoped paying homage the the idol gods and their imaginary money?

We can't beat them with might. We can't win with a show of force. We win by boycotting their exploitative system. And we do so not by eating our own or our enemies. We do so by simply reaping the benefits of our own labor. By rejecting their capital. By quenching the flames of hell which heats their heaven. This is how we revolt, we stop worshiping the gods who have exploited and abused us. We win by being our brothers, sisters, non-binary's, …. Keeper. We win thru solidarity, by forming more compassionate, comprehensive, conducive, and compatible communities. We stop emulating the idol gods and narcissistic saviors. We plant the female trees to absorb the excess pollen and produce fruit for the poor. We honor our neighbors authority as autonomous agents with their own valid perspective and ways of expressing their needs. We construct and continue to critique more comprehensively compatible models which are more conducive to values like objective truth, justice, and healthy growth (prosperity for all). We win by rejecting their dualistic systems. We win by joining the heavens and the earth and recognizing the taboo and common as holy and a vital part of the whole. We win by healing the hurt and addressing the underlying need,

If America was still a democracy we could solve the issue democratically. But members of congress on both sides of the aisle proved that they do not represent the people who elected them, nor are they working for the best interests of the people. But serve only the interests of the corporations which bought them like slaves to be their political puppets. The same corporations which have exploited and abused the people of both this nation and many others.

When American congressmen pressure (communist) china to censer the information American citizens can see in regards to national policies and world affairs without the filter of American propaganda. Then it may be too late to save any remnants of our corrupt economic and political systems built and vetted by the Christian god. This is particularly frightening because from the laws being passed to police the teaching of racial injustice in schools, to the anti trans/gay bills which outlaw a whole group of peoples experiences, and finally with the fervent veracity by which elected officials have sought to squelch the freedom of thought and speech (S.686 RESTRICT ACT) (an attempt to control the dissemination of information) America is well on its way to becoming the next fascists nazi regime. The up side is that these radical bills are the final attempts of capitalism and Christian nationalists to regain control because they know that theses authoritarian systems days are numbered. They are trying to force the broken pieces back together and resuscitate the decaying corps of their god. If they cared about the people they would be addressing the corporations which have stolen $50 trillion from wage workers to pad their own pockets. They would address the school shootings and the increase in suicide as a result of these corporations egregious exploitation of the populous. If they actually represented the people we would have affordable housing and healthcare. If they cared about child safety they would be banning churches who endorse the millions of sexual assaults perpetrated by their leaders and ministers every year.

The goal should never be the destruction of the old world. But we also can't let the old inhibit the new.

There's a good reason why revelation (the kingdom of god, the world made new) is synonymous with apocalypse (the end of the world as we know it, the day of the lord/ judgment day). With a change of perspective also often comes a challenge of hierarchy and a redistribution/ negotiation of power and authority. Revolution doesn't have to mean revolt and rebellion. But it does require change (growth/progress). And that is often met with contempt and a show of force.

(Note: the apocalypse/revelation is not a singe linear event but rather an inevitable reoccurring process realized in all times and among all peoples. It is a perpetual death and resurrection and is as much a law of life/nature as gravity it a law of physicists. It is the process by which life pursuits and the inevitable way life finds. And thus what tradition calls corruption and perversion, life calls adaptation/evolution and mutation. Some of which are understandably corrosive and faulty. Most ideas are bad, but some are vital.)

If we want to live in a world where the debt is forgiven, and children are not held liable for the sins of their fathers. We must sacrifice our gods. For as long as there are gods There shall be slaves. For there can be no gods without slaves to serve them. But any heaven built on someone else's hell, is oppression not salvation.

Ultimately rather it is the gods of old or modern industry, their power dies when the people cease to believe their lies.

## THE CRIME OF POVERTY

My dear reader it should be apparent by now that poverty much like murder, is very much a heinous crime. Not in the since that to be poor is a crime (tho in America by law in many ways it very much is) but rather that like the one who is murdered is the victim of a crime and not the perpetrator of it, so too are those forced into poverty (which is far too often the case.) for poverty is the plight not of the idol aristocrat but of the working class who produces the wealth for the bourgeoisie. In this age of society every basic need is more then obtainable if not for the insatiable greed of those who profit from the labor of the impoverished. The fact that education, and housing (tho imperative for survival in a society) are then held far out of reach by so high a rate of interest that only those who already poses them are then capable of obtaining more. Then denotes the crime of such a fundamental level in the restructuring of our modern society at large. If it were not for those who craved exaltation over man that then cultivates such depravity and scarcity for the other 99% of the nation and even the world. Such poverty would not even be possible. And thus The criminal here is the lobbyists and the corporate stockholders who exponentially profit (over $50 trillion dollars worth in 2020) off of the work done by those who are left without. (This is theft) Here I speak for the factory worker and blue collar laborers working 80-100 hours a week, and for those teachers who come out of their already starved pockets to ensure that those who come after are equip with the skills to bring about a healthier and more inclusive future for all. For these are the victims of those idol gods who regard those very souls as slaves and not even human. Like finely dressed men and women who leave those who fashioned their luxurious garbs naked or in tattered rags. Poverty is a crime which our whole society is guilty of. For we permitted these gods to rule with such malevolence.

Now my dear reader the point here is not to demand that the world cater to your every need like those aristocratic gods but nether to devote yourself to their service but instead to find what they cannot, which quite simply is the capacity to love yourself fully with no need of the praise of gods to validate your purpose. For it is only when the multitude are no longer fooled into assigning value to the extent of exploitation they have permitted, that the idol gods are then no longer granted honor for their exploits. Of corse people have to work in order to survive, however living to work is completely counter intuitive. And for some devoting themselves to their job is a worthy cause but it should not be a requirement. The truly sickening part is that practically every company expects people to devote their lives to their job but refuse to pass on the excess in profit to those devoted employees. Leaving those who actually produce the goods to remain in poverty. The average hunter gatherers would have worked only about 8-12 hours a week in order to live relatively comfortably. However with the development of agriculture and later the Industrial Revolution that average more then quadrupled to 70-80 hour work weeks. And tho for a time in the late 80s these kind of hours actually profited people. Since that time the cost for higher education has increased by 169% and housing more then doubled wile wages stagnated. Poverty isn't simply a tragedy but rather a deliberate systemic crime which has only increased wile the more violent crimes have decreased exponentially. The benefits of agriculture and industry paved the way for curing world hunger and disease (which are very feasible now, tho not accessible to far too many). We are certainly within a dozen years of a vast majority of manual and intellectual labor (roughly 80% of all jobs) being more efficiently performed by AI. But unless we find a solution to the economic system at present, those who are exploited will be replaced and discarded (with no place, property, or potential for obtaining resources by which to survive by) by those with power. This is a reality which far exceeds those Babylon myths of the igigi creating man to serve them. It is important to remember however that These aristocratic god are not evil but also merely products of a system. As we saw in chapter two.

"One should not be held guilty for the evil done in good faith. For nether the unconscious beast or the indoctrinated man is liable for what they know not better.

To incur a cosmic debt on one's indoctrinated ignorance is no more just then to incur the same on one's nature."

## RELIGION

"What is god and even Christ, if in the extent of one's whole life in the here and now (even after pursuing fervently) the saviors in question are not once encountered, then of what consequence must they or their existence truly be. If not for the burning of witches, hell hath no fire. What good is heaven if after 2,000 years it has still yet to be realized."

Where the tribes of Israel and Canaan sought to slay and even vanquish the beast in man, the indoctrination of Rome and England sought to enslave the beast in man. And yet is perpetual torment any better then a quick death?

For the most part religion claims to be concerned with morality but instead cares far more about authority than it does morality. It seeks power not justice and mercy. And that is why it is a sad revelation when we see far Too many people learned to love from their religion, because religious love is not love. The religious "love them as they are, but don't allow them to stay that way" never really loved the individual as they were. Religion is not tolerant of other cultures or our autonomous expressions of personhood, and that may be why it is so concerned with sexuality. because religious love isn't about appreciating someone as they are but rather about controlling people, it's about social power. And that's why it is so concerned with converting people and also why it has such an animosity for things like science and nonbelievers (heretics). ultimately religion is about validating an ego (I am right regardless of the insurmountable evidence against my beliefs and all the physical and psychological harm these beliefs have caused so many people) it's about being right and not doing what's right, that's why it's so concerned with faith and belief as opposed to truth and honesty. Tho this most definitely has its drawbacks, religion in its modern context is not concerned with understanding ourselves or the world but instead about providing a sociological groundwork for community. And thus perhaps it's greatest tools are Communal pier pressure and enforcement of an individual's potential to mate and congregate. (in fact this very point is made by the animosity many missions leaders harbor towards societies such as Swedes who take care of their peoples needs, subsequently inhibiting the religious mission team to exploit the poverty of the Swedish people in order to necessitate conversion to their religious beliefs.) "Self-sufficiency is the enemy of salvation. If you are self-sufficient, you have no need of God. If you have no need of God, you do not seek Him. If you do not seek Him, you will not find Him." (William Nicholson, shadowlands) or more simply it is not enough that peoples needs be provided for, if that provision does not then permit the exploitation of people for the glory of god. A big part of this communal quality in religion is its approach to governing our inherently instinctual behaviors which are still very much submissive to our genes primary prerogatives of propagating and preserving life. This is not a cop out to our autonomous authority as humans but instead an acknowledgment of an fundamental disposition in all living organisms to both seek out and procreate with potential mates and to then protect and defend that progeny from danger. And thus we are wired to react to people both erotically and defensively based on certain qualities and circumstances. In the mythic depiction of genesis this is illustrated in Addams blaming Eve for his reaction to her (pier pressure). This is addressed later by the rabbi Jesus who claimed we are responsible for our own emotions and reactions to other people (Matthew 5:27). the point is yes people will evoke reactions within us which to a degree we will never be able to control. But ultimately we can't control other peoples actions (as religion tries to). But we can control our external reaction to other's actions. You are not responsible for the way you "make" other people feel, any more then they are responsible for the way their actions "make" you feel! You can't blame Eve, or addam, for your actions or reactions. And tho yes ultimately it is directly related to our genetic wiring, we can't blame our instincts, animalistic impulses (or mythological serpent) ether. In many ways these evoked responses in ourselves make us aware and conscious of our own responsibility for the reactions of others. Which is important because after all we are communal creatures.

By now it should come as no surprise that this social influence is so compelling because on the most fundamental level the primary prerogative of all living organisms is to duplicate or procreate. And thus community is integral to our purpose for existing (on a biological level anyways).

We experience this rejection from community and society as a whole, as a form of hell which is a punishment for opposing or aggressing the social norms. Hell is not the result of our exile from some invisible cosmic entity but rather our divorce from communion with all other living organisms or at the very least those of like kind. This daunting experience of the lack of certainty and mythological (egoic) meaning is just a side effect of our disconnect from the security of a community. And perhaps that is why so many people seek solitude and isolation this day in age. Because with the intermixing of diverse cultural norms and the predisposed enforcement of social norms, we feel safer alone than we do around such a broad array of potential behavioral transgressions and their subsequent social rejections. we fear genuinely expressing ourselves might offend that ambiguously broad social construct. In many cases Our need for solitude and exhaustion from social interaction is a response of our inauthenticity and constant masking in social circumstances. The ego often mistakes the reflection in the mirror for the individual themselves, or in other words we often mistake the story we've told ourselves(or been told about ourselves) for our own reality and identity as humans. The shadow cast by an object can be helpful but ultimately it is inaccurate and only a representation of the subject and not the object itself. Here that which religion exemplifies is not the object but instead the shadow and dim reflection.

The ego doesn't like taking responsibility or exerting autonomy but instead prefers to place blame and acting habitually. Ego follows orders from ether an ideology, culture, or our own instincts(emotions).

"Perhaps the greatest evil to be perpetrated is to act both as an autonomous agent and yet at the same time in subservience to our instinctual prerogatives. Tradition then is the subconscious progeny of the unconscious habit and instinct."

The modern incarnation of this sanctified entity driven by the divine voice of nature thru intuitive instinct may very well be depicted as the modern day artist!

Note: Rather you are in defense of a god or in defense against a god, I have probably stirred up some anger in you. But that anger is important because not only does it help us better understand the context of these narratives but it shows us how our agency has been usurped. Ether surrendered to an idea of god, or obstructed by the institutions employed to enforce that idea of godship. after all (a god) is just an entity which subverts our authority for their own. In ether the form of a king/queen, or as an idea which rules over some else. It must be pointed out however that The righteous anger is that which defends those who's authority has been subverted, not that in defense of an authority which usurped that agency in the first place. For the Israelites who were subjugated to more mighty empires, the wrathful justice of their god (the day of the lord/judgment day) was seen as mercy and love, it was that of a savior liberating them from oppression and tyranny. The rhetoric we see in the Old Testament is that constructed by marginalized communities to critique the biases and prejudices of their ruling nations myths. And Tho Christians still often see themselves as the victims and minority in their narratives rather then the oppressors. This is far from the reality. 50 years ago roughly 90% of Americans identified as Christian. As of 2021 that number dropped to 63% (still a majority). Only 44% of people abandon the religion they were born into.

Matthew 7:13-14 (NIV)

> "Enter through the narrow gate. For wide is the gate and broad is the road that leads to destruction, and many enter through it. **14** But small is the gate and narrow the road that leads to life, and only a few find it."

in part this victim blaming may have to do with the way Christian's are oppressed by their own abusive god. Or idea of god. like a sociological form of Stockholm syndrome. The problem is that we have been convinced that it is those without, who owe those with excess. This is not the case however. In actuality It is those who profit off of other peoples work, that are the freeloaders. And when that authority and power is redistributed among those who were displaced and disenfranchised by those omnipotent authorities, it is not then those authorities who's surplus of authority and power is reallocated to those deprived, that is then a victim.

A truly virtuous god is not one who fights for his own people, but rather one who stands up for the rights of marginalized communities which will never represent him. Such virtue is exemplified when cisgender

Christians speak out in favor of equal rights for gay and trans communities.

We should seek to gain power over ourselves, not over others.

Too regain our own authority, not subvert someone else's.

## THE EVOLUTION OF SACRED TEXTS

In many cases the scribes who copied biblical texts and scriptures by hand would alter or change the text in favor or even defense of/ against an particular theological interpretation, or even in an attempt to obtain a sense of unanimity and harmony between conflicting accounts and claims. We know this because of the various assortment of these scriptures which do take certain liberties in their transmission of the message, we see differences from copy to copy. And thus contrary to popular belief even the distribution of these texts are not in fact inerrant as many would like to claim. For this reason the common rules of thumb among scholars is to prioritize the texts which lack congruity as the original or at least closer to the original.

Critical analysis of the ancient texts construction, formation, and transmission, verifies the errancy of biblical scripture.

However Contradictions aside, these ancient New Testament text account events 10 to even 100s of years after their occurrence. And so not only are These accounts not eye witness accounts but not even originals and instead copies of copies. These accounts come to us $3^{rd}$ and $4^{th}$ hand with numerous accidental and even intentionally theological alterations.

This most certainly produces stumbling blocks in our attempts at retaining antiquity.

However not only do we not have eye witness accounts or original copy's but instead we find conflicting copies of copies. Contrary to popular believe this supposed inerrant word of god didn't manifest immaculately but rather evolved slowly and went thru many alterations. Just like us. "Perhaps evolution is just a prerequisite of being god breathed".

Not only does god and Jesus not live up to their own standards of perfection and virtuosity. nether does scripture it's infallible claims within its self. But even those standards have become antiquated and malevolent by today's standards. As we've already seen morality on a physiological level is not a stationary standard but rather an ineffable quality we must continually wrestle and contend with (update and reevaluate). Morality is dynamic and ever evolving.

More over This claim that all scripture is theo pneustos or "breathed of god/spirit" in continuity with its usage in other late $1^{st}$-early $2^{nd}$ century writings such as sibylline oracles and porphyry ... most probably referred to "life giving" or " to bring life" rather then "inspired by god". this idea is articulated in more detail in John Poirier's (the invention of the inspired text) but it is worth noting that even in their time these "omnipotent standards" for our behavior were not as restrictive as they have become this day in age.

When we approach the Bible and scripture not as a doctrine or creed we must subscribe to, or believe in, but instead as a historical account of ancient civilizations cosmological interpretation and evaluation of their experiences accounting relatable phenomena. we are given a deeper understanding of this phenomena as it pertains to its fundamental motives. Now we know why we feel such strong impulses, whereas in ancient times we simply made a note of how they made us feel. Moses was not there in the garden and Adam did not see god speak light into creation. These are all made up stories describing our conscious thought emerging from primal instinct and distinguishing between the heavens above and the earth below. These stories and poems describe this extremely relatable phenomena pertaining to our human experience and then attempts to ascribe cosmological meaning to those feelings and outcomes. These ancient traditions recognized the correlations between the way we breath and the activation of our parasympathetic nerves system (a phenomenon still accredited today as evidence of gods presence and attestation of the ground's holiness).

The truth is In their time, the writings of Paul and even the later gospel accounts were like those of the modern writings of C.S.Lewis, rabbi Abraham Heschel, and J. Richard Middleton. In their original context these letters of instruction and parables for introspection were not divinely authoritative (tho some claim to be). This means that there is no more authoritative weight in the words of Paul or the "supposed apostles" then that which is in the words of a modern writers who offer introspection and rhetorical commentary on the theological claims of "what is now considered canonical scripture". In the traditional respect of both the old and New Testament authors (which both critique the ideas and claims of those who came before) the educated writings of Rob Bell or John Milton, hold just as much (divinely authoritative) weight in their criticisms and renegotiation of the terms of the texts meaning (and or application) as the former.

## GOD IS WHO "HE" SAYS HE IS, BIBLICAL CONTRADICTIONS AND FALLIBILITY

The sacred text is not the word of god. The word of god very well may have come to the profits and sages, but the words in documentation we have deemed scripture are those of man and not god. If God is static or stationary like words on a page that must then be interpreted, then god is quite small indeed. but if god is instead the interpretation itself then that cannot be infallible but most certainly ineffable.

By now we know that a lot of the phenomena attributed to god in the Bible or in religious circles is explained by science as very natural and not supernatural. However If god in fact devised the laws of nature and physics, why then would god require derivation from his own precepts? If we can't find god in nature (the artist in their masterpiece) then we have a big problem.

However in the same breath we are faced with the fact that if god (the creator) is restricted to the laws of creation. Then god is not greater than, but part of creation.

Is God simply what we call our instincts in light of our consciousness?

we now know that It is our genes that replicate thru us, and produce motivational impulses compelling us. We are their vessels for replication. And so is god a gene which we have rebelled against? To some degree I believe there is some truth to this conclusion. so far as that which is often described as god intervening in religion and scripture is more accurately explained thru science as a phenomenon associated to our biology and genes. However the idea that there is a god gene is absolutely absurd. That's not how genes or "gods" work. However it can't be overlooked that the majority of what we describe as a "spiritual" experience is inevitably a phenomenon we encounter or feel in our physical body. and thus most certainly a result of our biology and not some supernatural entity. This doesn't discredit or invalidate the existence of god but simply questions the definition of spirituality. Because after all.

> "If our beliefs tell us one thing, and the needs of real people tell us another, can there be any question of which we should listen to?"
>
> (Claudia gray)

If the "ethereal" does not relate or pertain to the ephemeral, then what purpose does it serve?

Unfortunately many of our theological claims pertaining to the identity of god is not based on an honest observation of nature but instead The unnatural exploitation of Conquerors who convinced us they were saviors. However these supposed saviors are in fact so distantly removed from us that we can hardly call them saviors or Christ anymore. Them and their ideals function only as outdated models for our behavior. And thus this is the reason that faith in the priest's prestige is so important. Because these gods incarnate have become no more then internal social projections.

How do you love someone you've never met. Because everything we know about Jesus and or god comes third and fourth hand. How do you have a relationship without real communication?

# CHAPTER 4

## THE INTUITIVE SPIRIT, THE NATURE OF GOD OR THE GOD OF NATURE?

(prior to the a Babylonian exile in 360 BCE) The original usage of the Hebrew word (nefesh) for soul, was not a strictly ethereal essence. But rather a way of describing the integration of spirit and body.

Deuteronomy 11:13-15 "listen to god and god will listen to you, to listen is to love" (authors amalgamation)

Society is our attempt at mimicking the interconnectivity of nature in an ecosystem. and so it makes sense that we would shape our societies in a uniform manor. As we recognize that nature is wholistic and not dualistic. However what we often fail to realize is that wholistic does not then mean the same nor does it negate the necessity for our diversity. In fact that dichotomy Is the whole point of an integral network. We thrive because we have both holistic and autistic brains, men and women, and progressives and conservatives.

(Albert enstine)

> "A human being is a part of the whole (universe). He experiences himself (his thoughts and feelings) as something separate from reality or nature. This division however, is an delusion."

The goal is not to nurture this delusion of duality but to dissolve that dichotomy until the diversity of the whole is realized and appreciated.

In any given ecosystem you will find a food chain which, is often mistaken for a form of hierarchy. This idea that a lion is more important then a fly in regards to the ecosystem as a whole is a folly however. in much the same way the jungle does not exist in solidarity of the desert or grassy plains, so too does the lion and the fly make up two very small parts of the same whole (ecosystem). As we have just discussed in the previous chapter there are many pitfalls to a stationary picture of a dynamic terrain. Because as you are probably well aware. Environments change, life functions as cycles and seasons, and the smallest insect is just as consequential to any environment as any apex predator (if not exponentially more so). And so a good rule of thumb is If the map doesn't match the terrain you change the map not the terrain. It is not sciences job to conform to religion any more then it is the earths responsibility to accommodate the cartographers mistakes. You don't shame "what is" in favor of what "should be".

In this chapter we shall examine a number of the phenomena we once (and even still) attribute to god as evidence for their existence. And tho the obvious conclusion is not the result of some omnipotent deity, I myself am still inclined to presume the existence of some cosmic form of mental agency posited in the universe and nature. We are not however inclined to reduce that agent to an egotistical personality. Especially one which is prejudice against cultural, ethnic, neuro orientation, (that is left brained or right brained) or gender/sexual specific temperaments. any divine creator which harbors a spiteful distain for traits which are posited by our genes such as physical attraction, sexual orientation, and gender not to mention those of mental disposition and ethnic demographic, is most certainly not an example of a wholistic progenitor but instead a dualistic perversion of that source. Further more Any god who is racist, homophobic, or misogynistic is itself a perversion.

And so without further to do let us begin.

Believe it or not Our ancestors still speak to us thru the genes we inherited from them (as we saw in chapter one "epigenetic trauma in the mammilan brain"). And tho the memories we inherit are not like those we recount in daydreams or flashbacks form our past, certain behaviors and subconscious memories still present themselves many generations later. It is true we don't literally hear the voices of our ancestors or see what they saw. For these (ancestral messengers) don't often exhibit themselves in our conscious cognition but instead in subconscious gut feelings and prolific intuitions. Tho some may diss credit the validity of such ancient mechanisms. They none the less played a vital role in our survival up to this point. And it would be foolish to discard their prolific utility even now. despite our advancements since obtaining autonomous cognition, our genes and their instincts are well equipped for survival and even "thrival". And so if this subconscious programming (instinct) as we understand it today, is so commonly referred to as "the will of god" or "the spirit" in ancient texts. Which they most probably are! Then these ancient practices of prayer and meditation with its prolific messages should not be discarded flippantly. As they often facilitate these intricately woven neuro networks.

Now the first thing you may think of when you hear "the will of god" is fate and predestination. Or as I always use to put it. "It will all workout the way it's supposed to."

But what does that really mean especially as far as free will is concerned?

Many scholars will conjecture forever in regards to the topic of free will and fate. Making arguments like " you're genes or some cosmic deity decided your outcome long before you unwittingly decide to agree with them" .

However for the sake of sanity let us simply put it this way.

Your instincts/intuition have been programmed with directives such as, danger, shelter, procreation, preservation, protection, and comfort/ adequate recourses.

This is your nature in accordance to biology. You inherited certain traits from your ancestors and many habits as well. This is ultimately no different than any other animal on this planet. However much like birds that fly south for the winter or apes in captivity having inherited a cautionary response to snakes, there is some very compelling evidence to show for the validity of those instinctual inclinations. This however does not mean they are always correct or without error. And this is where free will comes in. because instinct aside, you get to chose how to react to any given situation. Tho on occasion your subconscious fight or flight response will kick in. Even that can be rewired and reprogrammed thru updating models in our PFC. Ultimately you decide your own nature and no religion or science textbook can take that away from you. Fate then quite simply is the result of continually making the same choices over and over again. And so of corse it makes sense that As long as we remain submissive to an idea, ideology, or unconscious instinct, we are doomed to their evil or benevolent designs.

But is it really this simple? and if it is. would we really ever know it? The answer is unequivocally yes! You are not evil and nether are your instincts.

"Between the intuition of the amygdala and the nucleus accumbens, our psyche most likely does hold profoundly prolific information pertaining to signs and cycles in our environment. And this information was inherited from as far back as our most primal single celled ancestors. What our subconscious experiences (often expressed as emotion and abstract imagery) are an amalgamation of this prolific ancestral memory."

## THE INTUITIVE SPIRIT

The famed psychologists Carl Jung spent a good portion of his career studying the conscious and subconscious states of our psyche. And he noted how our dreams most probable function as an intercessor between these two states as an attempt to process and quantify the vast wealth of information our subconscious had acquired and accumulated in the background. He also points out that it is for this reason that the abstract expression is often far closer to reality or at the very least has a greater degree of accurate data then our empirical attempts to quantify or describe those phenomena and environments. (the compass is a better judge of the terrain than the map because the terrain is much vaster than the small map). The compass is also prevy to cosmological forces the map could never even imagine. This is why it is important to listen to what our emotions are trying to tell us as opposed to attempting to suppress or repress them. Ultimately for Jung the goal was to obtain equilibrium between the two states where we were not unconsciously controlled by our instincts nor complete ignorant of their knowledge but let them function in tandem (Rational mind and intuitive heart). In chapter two we saw how these cognitive appraisals become autonomic behaviors that too must be reappraised.

Mythically this is depicted as the garden where culture and nature co exist In harmony as opposed to the mythic depiction of nite and day, which expresses a duality between the two. (day is man's domain and the nite is where the wild beasts roam). This actually makes a good deal of since because in light of what we now know from our study of evolution. There is a vast wealth of information posited in our cells dating back as far as the dawn of creation. From us crawling out of the primordial waters (depicted in virtually all the ancient myths) to the emergence of consciousness (the light filling the darkness) that trove of information is coded in our subconscious.

This is why we feel art. rather it is music or a painting there is profound wisdom instilled in those expressions.

So the question arises then. How often is gods voice simply our subconscious eluding to what our rational mind is unaware of? And does that then undermine the authority and even existence of god? But perhaps a better question may be why should that change anything?

Why should we not seek those prolific messages hidden in our instincts? should we not pray/speak candidly with our subconscious and meditate on that wisdom posited in nature? does this revelation detract from the profundity of its wisdom or the virtuosity of its claims, as it is in accordance with nature and our ever changing environment.

But is that it? is god just our instincts manifesting themselves? And if not, how do we distinguish between our own anthropomorphism of our ego or doctrine and those prolific messages posited in our intuitive cognition? Or in other words how do we distinguish between the man made god of doctrine and ideology and that of the muse and stroke of genius which shaped man? A great deal of the phenomena we attribute to transcendent encounters and experiences are actually expressions of our intuitive nature or inherited instincts.

Now I will grant that this idea of god being inflated with nature or the universe, can be disconcerting. but at the same time Why should we be averse to a god posited in nature? One which is unhindered by ego. A god which formed all mankind in its image and is not restricted by any one culture, that is a god not made in man's image. But a god sown into our very genes and present in all living things. So why is this conception of god so uncanny. This is a god at work in the slow and gradual process of evolution and thus one devoid of any real prejudice against its creation. It should also be noted That In virtually all the observed universe, it is mater which manifests a mind, and not mind unto matter. And Tho the mind does shape matter, it is matter which posits the potential material for mind to manifest. the idea that some form of formative consciousness was the result of mater and not the other way around should not come as a surprise to us.

Perhaps the most compelling reservation to this prognosis is in regards to the virtuosity of nature as a good god. As if the narcissistic gods of religion are any better then that of a nature which programs lions to hunt and kill, and posits snakes with venomous defense mechanisms. Perhaps the true reservation against nature being god, is that it doesn't discriminate against those who look and think differently than us. Nature, unlike religion, and culture, is not dualistic but wholistic and inclusive. (Note the serpent is only evil because it is not defenseless).

Is nature not nurturing, is the cosmos not compassionate, is culture not more cruel than the wild creatures? Perhaps not! The story evolution gives us is of mindless micro organisms slowly evolving more efficient bodies and more comprehensive minds which then allow us consciousness and the capacity for compassion for other life forms. We don't need an intelligent designer because we have the process by which intelligence itself was designed. This however does not by any means demean those prolific messengers which guided life's progression. That is we don't need an entity to manifest the material, but simply one to mold the mind and mater into more precise mechanisms and materials. And so Who then is the muse, which after all is among the oldest conceptions of god in antiquity. What divine agent decided compassion was favorable to conquest?

If we are honest most probably natural design (which was not simply a decision made by the ruling class but rather by the masses) is that muse. It is instincts interpretation of the environment doing the very same thing our conscious PFC does when it reappraises habitual behaviors and models to fit that very same ever changing environment.

Regardless of what you call this miraculous muse, our instincts main goal remains, to be ensuring that we survive and even thrive in our terrain. and so It would make perfect sense that we would then look for signs in our environment to guide us. because after all that's exactly how that mechanism works. It subconsciously calculates signals and cycles in our environment as an attempt to predict events (subsequently positing agency in the process).

Like those in touch with nature, with dirt on their feet and rain dripping from their hair, the sage profit seeped themselves in the stories of those long past. they listen to the past to find the rhythm for the future. to be in tune with creation and nature. Not that it was created but instead contentiously being created thru the process. The artist studies the strokes of the paint and led caping their fingertips or smeared on their faces. Similarly Those who live their messy chaotic lives are the ones who know the steps to life's dance. We learn to be emersed in loud noise so that we may feel the silence of the song. The point then is not to disregard your instincts and emotions but rather to become more aware and conscious of their subtleties.

Let us revisit the study done by Brian Dias and Dr. Kerry Ressler (of Emory university school of medicine) which shows how fear to certain stimuli can be inherited by offspring. The researchers repeatedly exposed male mice to the scent of Cherry blossoms and electroshock stimuli in concordance. Understandably the male mice developed a aversion (trauma response) to the scent, however what was interesting is how that fear was then inherited by offspring which was not exposed to the negative electroshock stimuli and never even knew their fathers. This generational trauma is transmitted by genetic changes in the DNA prior to the conception. And can be passed down up to three generations. Tho these inherited memories are expressed as behaviors and not as abstract images in mice. it is possible that the same process could be communicated thru subconscious expressions and imagery acted out thru our dreams and meditations.

Is god "god" just because that's the name we have given that experience. Did we anthropomorphize those agents in nature simply because we intuitively deified our ancestors and progenitors for their discoveries and assignments of meaning and value? Would a god by any other name still be a god? Do we worship god simply because we ourselves want to be seen? In Christianity today We worship the celebrity of Christ, the idea, not the human. For that's all we know of Christ. We only know what we've been told third and fourth hand. None of us have had a personal relationship with a figure who has been dead for 2,000+ years. And yet the Christian hates the Buda who looked very much like the Christ. Why is this? Truthfully Any image that is alien to our own culture will appear demonic. This is because we've idolized the tree and forgotten the soil it grows from. we worship the mountain and have ignored the rivers and oceans which shaped them. We are unmindful That the wind that shakes the branches and the ocean waves alike is in fact the same wind. And yet not even the wind is ineffable for even behind the wether is a unseen change in biometric pressure in our atmosphere. And so as we saw in chapter two, we have biological biases and prejudices towards qualities which contrast or even resemble but are distinguishable from those we naturally formed during adolescence thru a process often referred to as (nostalgia)(see pleasure and pain in the brain chapter two).

## SPIRIT OF GOOD OR EVIL

Much like those martyrs of the tyrannical Roman empire, today we are flooded with People who would rather die then live in the evil world the Christian capitalist story created. And understandably so! Oh How often this need for an compassionate consciousness at the heart of creation result in us submitting to the malevolent master of religious doctrine and dogma. For Such a god requires us to be nothing in order for it to be anything at all. but we are not nothing. We are part of it. Not all of it or the penical of it all. But a vital part of the whole none the less. The point is not looking up at the stars to remember we are insignificant but miraculous, not a drift in chaos but belonging in the abyss.

For both Bible and science agree that all life crawled out of the primordial chaos waters.

The only question that remains is what is that spirit hovering over the abyss?

Man has been searching for god since the dawn of time and yet god has yet to come down and show themselves to humanity but instead has left it up to us to decide their nature thru the way we bare gods image. For some, god is a tyrannical conqueror, for others, a malevolent master, and even some, a compassionate intercessor. but in all cases it has been man who gives god identity and never god who shows themselves to humanity.

Among the oldest quandaries in prosdarity is that pertaining to the nature of certain cosmic agents and how that nature has influenced the development of life as we know it today.

There does appear to be some force propelling life towards more comprehensive and cognitively aware states. as well as a muse inspiring more compassionate forms of consciousness in the process. For the more we understand ourselves and the world at large, the more compassionately we are capable of appreciating and entertaining both ourselves and nature. The problem is we often have a proclivity for anthropomorphizing that cosmic agent thru an egoic persona often modeled by our parents and cultures. But this dualistic conception of the cosmos restricts this god to a corresponding culture and religious tradition. That is why This restrictive attempt at quantifying the ineffable often undermines the relational connection as a source for all of creation which by expression would resemble that source.

Perhaps the question should be asked " why are we as a species so inclined to Endow these agents with dualistic and egotistical traits"? I think perhaps it has to do with our insatiable craving for meaning and significance in life. Somewhere along the way we got a very scued view of hierarchy and decided that the lion was the king because it could devour all other beasts. And so to attribute credence to a lesser beast seamed egregious. We want so badly to look like "god" that we never stoped to ask what does that really mean? Is god an ego? Is god petty and jealous? And frankly how big can a god who takes credit for everything their children accomplish but none of the blame for those traits that their children inherited from them really be?

Does god really look like a king taking all the glory for himself and leaving his "children" to squabble in the dirt like peasants?

The Deification of ancestors, parental figures, and those in authority is a intuitive mechanism used to retain the values, ideas, and discoveries posited by those individuals wile also functioning as a coping mechanism for the loss of those loved ones.

This idea then of a god that gets jealous and picks favorites, is a god that hates humanity because we look like them like a parent with a temper, who then resents that trait in their child. and thus these depictions of god are most probably just our inflation of god with bad parental figures. And It makes logical sense that the cosmic progenitor would resemble our own caporal ones by which we retained all our wisdom and knowledge of our environment form. of corse we explained the unknown with an example we do know. It is no secret that both man and women are often left to grapple with the sins of their parent's parents. And thus the best thing any man or woman can do is resolve such scars in themselves, so as not to pass them on to those who come after. It then is of paramount importance that we as a people wrestle with all that which we once called god.

God is our answer for why things happen. and that makes sense. because it is hypothesized that this bias of mental agency is posited by our psyche as a means for predicting the causal relationship between phenomena in our environment and the probable outcome of those events. Or somatic simulations in our PFC amalgamating probable reactions to our behavior in complex social circumstances). However This does not then in any way demean the autonomy of environmental agents as they do function as a cause to a equal or greater reaction. Ultimately on the deepest level there are still phenomena who's causes still elude us today. And so it is completely reasonable to deduce that things happen for a reason unbeknownst to us. And further more to posit a form of mysterious mental agent to account for such phenomena especially when those seemingly unfavorable circumstances result in serendipitous outcomes. However it's worth pointing out that a great deal of the mystery has to do with the fact that my model doesn't necessarily match your's exponentially outward. for this reason The question is not " is there a god" but rather "what is that pivotal agent underneath it all" and most importantly is it moral?

Carl Jung recognized that archetypal images weren't universal but instead could mean very different things to different people depending on the relationship the individual had with the image. Similarly many people have claimed to have heard gods voice telling them to act on faith. However the claim "god told me to" has also been the answer given for some of the greatest atrocities perpetrated by humanity. And so the question arises. Did god tell Abraham to kill his son, did god demand the genocide in Joshua 6 and did god condone the crusades? and if god did command such things why then does god speak only to a small select few and hate such a vast majority of "his" creation?

My reader may interject here that Nature is no more moral then religion! the lion has no compassion for its prey by which it is sustained. nor is the elk merciful unto the grass, or the virus unto its inhabitant, and yet we see things like adoption in the wild such as a lioness adopting a baby oryx or a pod of whales adopting a dolphin which demonstrates that Nature in just the same way does not favor any one culture over the other (as religion does). the same lion which devours the elk shall return to the earth and cultivate the soil and grass of which the elk will graze.

Well okay then nature may have some semblance of morality but god is miraculous and supernatural!

Nature is supernatural! from the catapulted becoming a butterfly to the clownfish changing its gender. Every day we discover the nature of life is not what it appeared the day before. from the expanse of the new heavens to the four dimensional reality of the once flat earth. Miracles are innate to nature.

Is it really that crazy to deduce that our subconscious psyche, instinct and the priory programing of nature which after all did design us gradually and meticulously, is what we call god. That instinctual curiosity and muse which picks up on the prolific messages in phenomena in our environment is that mental agent often depicted as god! God was associated with our ego and quantified thru our ideologies but perhaps was actually just our subconscious psyche communing with our rational mind. Ego after all is a facet of belief (one of our VMPFC's primary mechanisms)!

The ability to distinguish between yourself and others and the environment is a mechanism not even fully developed in adult humans.

Empathy is not a human trait but observed in other apes who grieve/morn their dead, care for those in their tribe who have lost a familial member like a sibling, parent and even on some occasions a child, they also tend to living companions who have been wounded or hurt.

If we are driven by our desires and our desires are incentivized by our genes then the fault lies in our designer.

If the artist plays the wrong note then the fault is on the the artist and not the instrument or song!

How can god blame man for being inherently evil when that evil is not only inherent but also command by god.

The biblical answer for this of corse is we are to blame because we chose to discern between good and evil (took of the forbidden fruit of knowledge of good and evil) and yet our ability to reappraise our habitual behavior is the only reason we are even capable of morality at all.

## FREE WILL OR SLAVES TO OUR GENES?

From the ideological predestination of Calvinism to the unconscious instinctual prerogatives of our genes. the question of free will is a tenuous one. However I think it should be noted that The more conscious and mindful we are of our instincts and subconscious programming, as well as those subliminal messages posited in the habitual subjugation of doctrine and tradition, the more authoritatively autonomous we become. Or more simply our ability to question our habits and traditions rather than mindlessly submitting to the whims of ether ideology or instinct is what then permits any form of autonomous agency. But the argument can still be made "does our ability to observe our own thoughts and actions, then habilitate our autonomy?" Are we not still subject to the outcomes of our environment and the limitations of information at our disposal? I mean are we not all doing the very best we can with what we have? Is it really fair for us to judge the biblical hero Joshua by the same standard we do the notorious hitler for their racist form of nationalism and genocidal extremism? And should the fact that both acted on behalf of an "higher authority" be completely dismissed? Obviously our ability to judge the actions and motivations of past figure heads in relation to our modern standards for morality, is what gives us some magnitude for autonomy. And it is clear then that Our autonomy can only be obtained by rejecting our proclivity to identify with our desires or beliefs. But is that even possible? How do we act in contrast to belief and desire. Is not our choice to skip a meal or choose a less flavorful but healthier alternative not the result of some belief which simply seeks a different object of our desire? Can we really exercise "free will"? I think we can. For I would argue that a person who had been taught to believe in hell, can deliberately act out of compassion for their fellow man/woman, subject themselves to the immanent consequences of their rejection of doctrine on behalf of another's wellbeing. For this act is perpetrated deliberately despite their beliefs and in pursuit of someone else's well being even at the forfeit of their own desires. Genes seek to procreate like genes. And so when an individual sacrifices themselves for a stranger (one who does not look or think as they do) they exercise free will because they do not act out of some need to be extolled or remembered or out of any programing to preserve genetic traits. However I would argue that one does not even have to go that far. because any attempt to derivate from the status quo and question our beliefs and instincts is a form of self control and exercise of one's agency.

Now part of this questioning is asking

"Who is the muse?" Who's voice is speaking? And where did it come from? Surprisingly enough we find many answers to this question in accordance with science.

One of the most notable is that of the bicameral mind proposed by Julian Jaynes which explains the transition form animal to man.

In his 1976 book "the origin of consciousness in the breakdown of the bicameral mind" Julian jaynes make a number of compelling arguments for an intermediary state of human cognition adjoining the once purely instinctual creature to the rationally self conscious man. Jaynes does this by not only observing the functions and limitations of the brains hemispheres (left and right hemispheres) but also and perhaps most notably thru scrupulous examination of anthropological development observed in ancient literature like the epic of Gilgamesh and the Ocyrus myth. However I believe some of the most compelling evidence in favor of this theory has to be the entomology of ancient Hebrew words such as (chemah) which often means anger or hot, illustrating this anthropomorphized projection of early humans own anatomic reaction (getting hot) onto the environmental phenomena which evoked that sensation. (Anger as we've seen activates the sympathetic nerves system which in turn prepares the body for a fight. This flooding of adrenaline and cortisol causes our body to heat up). Another example can be found in the Hebrew word (shema) to listen, hear, or obey, which in the ancient man were indistinguishable from one another. To hear the voice of the gods, was quite simply to obey. It is quite apparent that this voice or dialogue we have in our heads which we use for analyzing situations was understood by ancient man as separate and alien. And yet at the same time it was him/her. In much the same way that Jung concluded that man first acted habitually, and only later questioned why "we do what we do",jaynes illustrates that even as ancient man observed their actions, they did not resonate or associate their own autonomous agency over their behavior. In the same way the Greeks believed that the "muse or genius" was separate from them, ancient men felt anger and jealousy within themselves and attributed it to the external stimuli which evolved it. "The anger of god burned hot" in the hearts of men who could not distinguish between themselves, their emotions, and the external stemuli that evoked that anger. In many cases this is man before we learned how to reappraise our autonomic habits.

This was man walking with god in the garden or as depicted In the ilid the gods voice was that which modern man replaced with conscious volition. "It was thru word that the darkness gave way to light"

*M.R.Holt*

Or in other words Consciousness (light) is restringente on the construction of language. And so God in a bicameral sense is the lack of consciousness, or perhaps even the disassociation with consciousness.

Jaynes argues that this mechanism originally formed thru the individuals awareness of others presence in their absence. Like a illusive version of a loved one in our head, or the grieving memories of the deceased, the god kings spoke to primitive man. And this is still observed in humanity today, thru the idealized versions of potential mates amalgamated in our dreams, to the religious "relationships" with figures who died thousands of years ago and who we have never met yet claim to know. They are projections simulated by our VMPFC to amalgamate potential behaviors of those we know (or know of) played out thru somatic scenarios which then help construct and adjust our internal models). or more simply put, as we interact with people our minds construct internal models of their behaviors. We then use those models to simulate hypothetical situations with them as a means of "testing out" our own behaviors around them. If the version of these people In our heads are excepting and descriptive then we "know" we can be open and trusting with the real life version of these people.

The ancient man was not an autonomous individual but subservient to the will of the gods in much the same way animals are to instinct. Their actions were not of their own volition and thus they no more bore the blame any more then they did the glory for the fruits of their labors. And thus this idea that god/kings are glorified for the actions and sacrifices of their servants and slaves finally makes sense! Like a artisan who's muscle memory no longer requires the artist to be mentally present. The sage who had meditated on the traditions of old long enough, no longer acts on her own volition but in accordance with the will of god. But the heretic who questions such habits and behaviors is the cause for folly. This is very likely what we observe all through the Old Testament, especially In the events of the 1st-2nd kings and the profits. We are almost certainly watching the breakdown of the bicameral mind and ancient societies attempt to regain control thru fear and shame as a means of reasserting the divine right to rule of patriarchal authorities. for most of humanity was still illiterate until quite recently In our history. However it's hardly worth calling these ancient texts literature, at least in our modern sense. Because both in tradition and linguistic formation, these ancient texts where rhythmic poetry or songs. The words were abstract and ambiguous, more like the muses of a song bird or the hums and squeals of apes. For Poets were profits and songs were the proclamations of gods. Now the lack of empirical rational does not denounce the prolific importance of these ancient texts or traditions. For after all, like the primordial chaos waters we emerged from, We too are made of ancient relics of primordial times. And we owe those remnants their reverence thru prayer and meditation.

We are communal creatures and these ancient traditions recognize that.

Or As Dawkins points out, religion is in many ways

just a meme or means of social signaling so as to attribute a form of prestige or establish a biases for communication.

And consciousness is no exception to this, because after all (as we've already stated) Before man was self aware he was aware of others, before she had an internal self, she imposes an selfhood on others as she observes their behaviors. (These traits are now probably best described as personalities.) and more familiar in the lyrics of songs describing mental amalgamations of past loves in our heads or observed in hypothetical arguments we have in the shower or laying in bed at 3am. And so it comes as no surprise that The effects of religion often mimic those emotions. Many times even transposing the passion and vigor of sexual romantic love upon an idea or deity to the extent that we "turn to see god" and "swoon during worship songs" these traditions deliberately manipulate our nervous systems to evoke these reactions. I mean after all What child doesn't look on their parents as gods, and rebel as they become more and more like them. "How vile must god be to despise even the beautiful thangs that resemble them." And in this way not only does religion idealize ignorance but vilifies transparency and anything which threatens the ego. However this is exactly why obedience does not equate to morality. Retaining one's religiosity on no more then the bases of obtaining meaning for life, is not evidence based, but synonymous with saying I am only a Christian because I want to be. This kind of faith is more dependent upon fear then it is morality and in many cases it is even opposed to morality. Perhaps Morality may in fact be consciousness to the intricacy of an wholistic world view. And an appreciation of its diversity.

Our extent of Consciousness is directly proportional to the complexity and precision of our vernacular. and this may very well be why our greatest tool for healing trauma (here referring quite simply to any experience which has not been permitted to express itself, be realized, or energetically exhausted,). And thus the remedy is often self reflection. Therapy after all is just the process of learning to listen to ourselves and developing the musculature/ capacity to tolerate and even appreciate all of ourselves (especially those things which society shames or neglects).

This brings us to the subject of shame, or as it is referred to in religion (sin). Contrary to popular belief, sin is a social construct devised as a form of signaling taboos in communities. And it is directly related to our awareness of others which then evolves our awareness of ourselves in relation. Because after all Before the knowledge of good and evil, there was no wrong, because there was no right to judge it by. No set of social standards to acknowledge our vulnerability and produce a feeling of nakedness or exposure. But this is not the only revelation modern psychology shows us in ancient texts. Another profound point is languages ability to perceive and keep track of time.

We observe time in a dialogue and so it would track that ancient man with their ambiguous linguistics would account millennia in the span of a week and lifetimes as hundreds of years with no definitively clear vernacular by which they could keep time with. (They also may have inflated themselves with their parents for many generations AE Tom son of Tom was an extension of Tom his father, Tom bar Tom).

This brings us to the catastrophic chaos flood depicted in virtually every ancient cultures mythos. This flood functions as a cosmological return to a primordial state and a breakdown of society. it can be equated to the modern day apocalypse. However in contrast to the epiphanic revelation of apocalypse, the flood illustrates a regression to more abstract understandings of the world as the distinctions and divisions of creation crumble. Rather then a reintegration of the now expounded part back into the whole, as revelation and revolutions attempts to do thru reappraisals. (For as I stated in my last book "creation quite simply refers to the construction of boundaries which distinguish the heavens from the earth and define the separation of land and water") the flood is the result of an absence of law and or order. However this breakdown of social constructs plays a pivotal role in the development of human cognition, as we see immediately following this cataclysm and its subsequent erection of paradigm (cosmological ark/vessel specifying creatures and their kinds, mammal, reptile, and marsupial,) we see a rainbow fill the sky. A symbol which today still intuitively represents unity in diversity, or religiously as a promise that in all it's chaos, life still has meaning (even when our definitions for that meaning appear to dissolve. I find it then not surprising that the next mythic event we see is the erection of a hierarchical tower and the subsequent incompatibility of differing world views (the Tower of Babel) Because "to the poet the prolific protestations of the mathematician sounds like babble" In a system or hierarchical social structure that omnipotently elevates one over the other, there will inevitably be confusion.

314

It should be noted that tho animals were sacrificed to god before the flood, it was only after the flood that man was permitted to eat (cooked) meat. Which then allowed the body to allocate more resources to our big thinking minds. Another crucial point is the significance of animal sacrifice which functionally provided both sustenance for the group and signaled allegiance to the community, also symbolically illustrated the death of man's animalistic nature in favor of the deified idea/god. The bicameral mind describes this (deified idea) as an internalized voice of some god/king figure who should then govern the will of man.

But like jacob wrestling with the ineffable, or Abraham and Moses arguing with god on humanities behalf (Abraham on behalf of sodom and Moses on behalf of Israel) we too are invited onto the holy mountain to join in partnership with god. rather than remain subservient to our instincts and ideological egos as masters and lords(bal). And this is the role of humanity and the goal of the messiah.

> "Whoever, at any time, has undertaken to
> build a new heaven has found the strength
> for it in his own hell."
>
> ~ Friedrich Nietzsche

Before the unifying narrative of the sacred text we were ruled by instincts. For Both are subject to interpretation, adaptation, and adjustments. In ether case mind did not manifest but instead evolved slowly, and as physics and psychology have shown us the mind did not metastasis mater but it was mater which provided the material for mind, there is no software or program without circuitry and hardware. Regardless rather nature or the muse/idea one would be foolish to extol the creator wile in the same breath berating the innate design of creation. To critique the masterpiece is to question the artists skill. And yet there is no ideology or instinct which has developed without the aid of mutation. Mutability is the way of life! this inerrant version of perfection, is the path to atrophy and death. Relinquishing our autonomy to ideology is just as counterintuitive as submitting completely to our instincts to dictate our behavior.

## Consciousness

To say something as ambiguous as consciousness is god, is a cop out unless we are willing to define consciousness itself.

Now The common consensus on the emergence of consciousness (or where it came from) is as an evolved mechanism of our primal instincts in response to or alongside the development of our more complex linguistic musculature. Ultimately the primary function of this particular mechanism we call consciousness is to amalgamate a fluent model for the complex and rapidly evolving social structures of culture. This model allowed us to construct comprehensive versions of both ourselves and others so that we could then test hypotheses and internally synthesize situations and their corresponding causes and affects. This also served as an aid in reading social cues, as this mechanism gave us the ability to impose our self model onto others in order to understand why they act or react certain ways. We often call this trait empathy. This is important because tho We don't have a concrete definition for life, we do have a rather solid definition for death, which is quite simply the loss of mind or consciousness (brain dead).

Physiologically all the matter is still there. and similarly on an anatomic level all life on earth is made up of the same stuff just in different configurations. The thing which denotes death then is not a physical change per say but rather that of spirit or consciousness. Tho we will go into much greater depth on this topic of spirit later on, for now we shall simply define Spirit as that which describes how something feels. And so tho science tells us what something is and why it does what it does. For things like spirit, we are dependent upon more abstract venues such as art and religion.

Basically Consciousness Is evolutions answer to virtual reality, it allows the individual agent to observe their own behaviors both in relation and in solidarity of others and their environment. This mechanism then produces a working model for predicting potential outcomes and reactions to certain behaviors. (Thus like the prolific quality of dreams, meditations, and prayers,) Consciousness functions as an internal amalgamation of the exterior world.

But like any muscle or mechanism, consciousness is not some static object but instead a spectrum with no foreseeable extreme. And so as Jung noted "we often act first, and only much later ask why" consciousness and instinct/intuition are two sides of the same coin. consciousness only grows by becoming aware of what instinct already knows (unconsciously). This is where it gets interesting, because not only is obedience the primary process of instinct, it's also its only approach. Instinct obeys and behaves accordingly. For instinct there is no difference between hearing and doing. Consciousness on the other hand is attentively disobedience. Consciousness asks why, and how will this behavior effect ourselves and others. However consciousness also requires discipline because it is not inherent. Tho we are conscious of pain and injustice to a degree. We still naturally behave instinctually and not rationally.

It is no surprise then that Obedience is not a sign of knowledge, but disobedience is. Or in the words of genesis 3:5 consciousness (the knowledge of good and evil) makes man like a god.


At its roots Consciousness is a social mechanism. And tho there are many facets to this mechanism. It is more commonly proposed that our awareness of others is the primary source of our consciousness of self or "self consciousness" if you will. It is a tool which developed to aid in our social interactions. as chiefly social creatures, our ability to hypothesis how other "people" and creatures in general, will interact with, or react to, us and our actions, is Indispensable . And thus we developed a mechanism with the potential to produce a variety of intricate amalgamations of individuals and circumstances in order to synthesize our social groups reactions to our behavior in certain circumstances. Tho this day age these scenarios are often criticized for causing unnecessary stress and anxiety. They still serve a vital role in acclimating us to a community. And Further more (tho it is often a mocked trope for insanity) the versions of people we have constructed in our heads, and then proceed to converse with from time to time, is in fact exactly how consciousness is supposed to work. Those "imaginary" interpretations and conversations, actually function quite well at helping us prepare for reality. And tho yes, the versions of people and "potential events, we have constructed in our heads are often flawed and incomplete, they are still extremely helpful and necessary for our survival in society. moreover what I am telling you is that talking to ourselves is actually healthy and even exactly how therapy works. Because after all, more often then not, therapy is any practice which teaches us to listen to ourselves and our bodies. Consciousness is how we process information. and so the ability to process that information in its entirety and without the addition of biases, (or listen to ourselves) is understandably helpful. it comes of no surprise that our first depictions of consciousness are brought on by social pressures from members of the opposite sex (the mythical Eve/Adam) which then results in us becoming self consciousness of our nakedness or vulnerability. As well as the sudden separation of us from the omnipotent authority of god/ prerogatives of nature by way of instinct. We also become aware of this disconnect in the way that masculine models obviously don't match the feminine mental models of ourselves in relation to our environments. The psychologist Julian Jaynes articulated this evolutionary development of instinct into deity and then into autonomy in his theory of bicameral mentality. His theory illustrates the phenomenon commonly observed in small children (particularly infants) where they become self aware or conscious that they are not part of their environment. On an anthropological level we see this development take place gradually thru the poems and narratives of ancient writers. Once again take the Hebrew word "shema" which means both to listen and to obey. In part this one word articulates Jayne's theory. For In much the same way a dog or cat is compelled to chase any moving object without the slightest inclination to ask itself why? we too heard the voice of nature and obeyed its commands. for to hear the voice is to obey. Jaynes outlines this development in great detail thru the carful examination of ancient texts and their representation of the evolution of personhood (the species itself discerning between our own autonomy and those involuntary responses to external environmental agents). This explains what is being described in the "famous fall of man" in the garden depicted in genesis. as man questions and then defies the will of nature or god.

The instinctual animal is not responsible for its actions for it nether asked why it behaves as it does nor is it cognitive of the correlation between its actions and its present outcome. For the instinctual animal simply submits to the prerogatives of its genes. We on the other hand need an space where we can hypothesize circumstances and amalgamate potential outcomes for our actions and reactions in a community and in the world. we who are capable of rewriting our own nature are thus responsible for our sumbisiance or amendments to that nature. And thus we need a space where we can test our theories. For we who took of the forbidden fruit can no longer walk the garden unawares of our vulnerabilities. nor can we blame the animalistic impulse (serpent) or fellow man (Eve) for our behavior but must instead sacrifice the animal in us, in favor of the divine idea, and shame the human for the traits she inherited from her heavenly mother nature. We stole fire from the gods in order to light up the nite. For Our consciousness required answers to our instincts blithe. And we could no longer linger with the wild beasts in the darkness of nite.

This hi-lights another pivotal component to consciousness which is our ability to observe and critique our own behavior. Or in other words to deliberately question and negate our instinctive prerogative.

It is no secret that We would often desire to be spared of this autonomy ether by hallucinogenic substances or religious dogma. Because consciousness (tho an profound evolutionary development) is not comfortable or in many cases enjoyable. In fact it is only our consciousness which subjects us to suffering. And that's why the mythical depiction of this phenomena is described as a curse or punishment. That "edge" or "anxious nerves" we try to numb or silence thru alcohol and drugs, is actually a symptom of our consciousness to our instincts. And so Like a dog eagerly pondering an inanimate stick, we too often do things that we know in themselves are not real, or constructive. In the hopes of inducing a chemical reaction (release of serotonin). Because after all The phenomenon we call Pleasure or happiness is just a mechanism our genes use to motivate us. In the same way the feeling we call love on a biological level is just our genes primary prerogatives for procreating similar genes/ (duplicating) themselves. Thus its When we do what we don't want to, when we exercise our inhibitions, when we don't act out of desire or necessity, that we then exercise a greater degree of our autonomy then when we are subject to the whims of our biological processes and habitual instincts.

Perhaps this is one of the most profound distinctions between us and other animals. for we are capable of illuminating the illusion, un like our animalistic counterparts and modern cousins we are not condemned to be fools to our imaginations but instead aware that our play is in fact that, a play or action by the prerogative that our genes have orchestrated in order to incentivize their propagation. With This said, the reality of such a consciousness is often particularly unpleasant. We need the illusion to add luster to life, the mythical to retain some semblance of magic in the mountainy of the every day. We crave that unconscious state of play to perform the tedious tasks necessary for survival.

Because after all Magic is not a strictly human trait. but instead the trait inherent to humans is the awareness that such illusory phenomena is in fact myth and not reality.

With this said As illusory as this imagination may be, it is nevertheless vital not only for our survival but also for our potential to progress. It is only thru conscious awareness of those profoundly prolific prerogatives of our genes and the recognition that they are in fact programs and not a necessarily accurate portrayal of our environment or reality, that then permits us to be competent at producing solutions to our current problems and making that which was once only pertinent to our imagination, a reality!

It is only thru humanities rebellion from nature that we are allowed to do and intend good beyond familial bonds.

So the question remains are we to obey god/our instincts and the prerogatives of our genes(emotions). or are we to contend, refute, and even ignore, such prolific impulses and incentives? Honestly I believe Both!

To say god is consciousness or the remnants of nature in our subconscious instinct is more accurately to say God is the anthropomorphism of our consciousness in relation to instinct.

This anthropomorphism brings us back to the ego.

The ego is our identification with an object or idea. Originally this mechanism was intended to facilitate our identification or attachment to our own bodies. Or as Thomas mitzinger explains

> "In order to survive, biological organisms must not only successfully predict what's going to happen next in their immediate environment but also be able to predict accurately their behavior in regards to their consciousness." Thomas Mitzinger (the ego tunnel)

We judge our physical locality thru our internal self model or amalgamation of the world. This may be more easily represented thru the example of a toddler who falls and then looks to their parents before deciding if their injuries are substantial, or even in the common experience of a cringe in your body when we whiteness something physically painful, or emotionally uncomfortable and embarrassing. Ultimately these two responses are virtually indistinguishable as they both pertain to an internalized version of ourselves in the VMPFC.

In many cases small children will look to their parents to asses the extent of injuries they have been inflicted with. This Facet of empathy is crucial because it signifies injury not as it pertains to the individual but in relation to the community.

We impose other agents actions onto our own self model as an intuitive attempt at understanding why others behave the way they do. And so when we see someone touch a hot stove and then pull back in pain, we know not to touch the hot stove and what will happen if we do. This may also shed some light on why people who have experienced similar traumatic experiences or losses, feel a closeness to those who understand their pain.

And so all those imaginary friends we had has kids, all those personifications we imposed on pets, animals, and perhaps even trees, all those conversations we have with versions of lost loves and loved ones in our heads, all those arguments we've won at 3am in our beds or in the shower as the hot water pours over our heads, every fictional character we fell in love with between the pages of a good book or span of a television series, all of it is completely natural and healthy in moderation. And tho it may not be what makes us human (as animals may do similar psychic gymnastics as well) it is most definitely a big part of where we derive the concept of deity or god. It is a massive part or at least the primal origin of the social mechanisms we constructed for hierarchical arrangements in societies and communities, AE kings, pharaohs, celebrities, crushes, deceased relatives and revered elders/governors.

This communal quality of consciousness derives from our self model which in many ways we (like many animals) inherited from our parents. Like small children or cubs who attentively observe and then mimic their parents behavior as a model for our own. A great deal of our identity's or internal self model is derived from observing such characters and then imposing their corresponding qualities on ourselves.

And so Perhaps we do truly lose parts of ourselves when loved ones are no longer a part of our lives.

This is also probably why people and their pets look alike (have similar personalities) and also why we ourselves are often the combination of our five closest friends.

With this in mind another major component of our consciousness is the development of linguistic complexity and precision. And thus it comes of no surprise that we can learn a great deal about the origins of consciousness by studying entomology's influences in culture.

Like the ancient myth of old, Ancient languages too have a great deal of ambiguity to them because of their relatively abstract nature. We can learn a great deal from this abstraction Because instead of trying to articulate what something is empirically like we do now, ancient languages were more concerned with how something felt. (The spirit of the thing) For instance the Hebrew word "haga"(to study or devour/consume). We can observe that ancient language was more rhythmic and melodious then modern speech and grammar which focuses more upon order and transparency. This after all is the goal of creation which like its Hebrew progenitor "bara" means to separate the whole into more specialized parts. Or to distinguish the nite from day and the heavens from the earth. To order the parts (similarity's and differences) which comprised the whole of cosmos in its chaotic origins.

Perhaps one of the most useful tools at our disposal is our ability to question our own behaviors and the capacity not only to negate them but more importantly understand why we do the things we do.

Ultimately consciousness is the mechanism our psyche used to update and maintain our internal self model and our model of the environment. Our awareness of ourselves understandably derives from our awareness of others.

And that in many ways is how we understand god.

(Suffering is a result of value judgments posited by our neuro chemistry in an attempt for self preservation and propagation.

In much the same way genes which favored traits which saw eyes everywhere a slight resemblance appeared, so to traits which posited a belief that we were always being watched also proved to be advantageous, and were subsequently propagated.)

## LISTEN TO MY HEART

Thanks to research done by the people who work for the math science institute, we now know that Every Emotion has a unique frequency which is emitted up to 6ft in every direction from the heart. and that that heart beat frequency synchronizes with other vital organs for regulation. This is most probably what we felt singing worship songs surrounded by our likeminded community. the hearts electromagnetic frequency ranges from 20 to 150 Hz which means it resonates and synchronizes with musical notes in the first and second octaves. The rhythm of the heart's beat is the base line for the rest of the bodies organs and it sets the tempo for the rest of the body. Or in short the heart sets the rhythm for our emotional states. (if our heart is pumping fast it will signal to the body a state of excitement or anxiety, whereas if it is slow it will assume a state of depression or peace). The heart has over 40,000 neurons (brain cells) of its own, and it sends more signals to the brain then the brain sends to it. It also emits the largest electromagnetic field of any other organ in the body. This electromagnetic field is of the same nature as that of the earths, which is particularly pertinent when we consider that When we are in a heart coherent state our hearts emit a frequency of 0.1 Hz which is the same as the earths magnetic field.

When monitored the HRV (heart rate variability, AE the tempo or rate/speed at which the heart beats.)

of emotion like anger and anxiety are sporadic and out of sync with the rest of the body. However when we are experiencing emotions like love, joy and peace, the HRV is synchronized and steady. the nature of the hearts electromagnetic field changes depending upon the frequency of the heart beat. And Every emotion has a unique frequency.

Many of religions ritualistic practices intuitively achieve this synchronicity by implementing rhythmic behavioral patterns in their praise and worship.

Studies have shown that the emotion specific frequency of our heart rate is directly proportional to the nature of the electromagnetic field produced by the heart. Which means the field is different depending upon rather the person is happy or sad. Tho the cause for this correlation is yet to be determined.

There is a sense of resonate conviction in rhythm, and so if you want to move people give your words symmetry.

With that said, a truly good song needs both high notes and low notes in order for a baseline to be set. And it is for the same reason that The goods feel so good because of biological incentivizers which combat the contrast in suffering.

The higher the baseline for homeostasis, the higher the threshold for pleasure. Our tolerance for pleasure is directly proportional to our homeostatic baseline. And As we've seen suffering is one of the biggest causes for growth. It is that pain which provides the incentives for us to acquiesce states more conducive for homeostasis. It functions as a Catalyst to the formation of healthier alternatives and habits.

Some emotions emote us away from negative environments and stimuli, and others emote us towards positive ones. But every emotion serves a vital role, and our goal should be to hold space for those dialectic roles. Because ultimately the judgment of "negative or positive" circumstances is entirely up to past experiences and present reevaluation of their values.

In health emotions like anger are parts of us which love us the most. They are the parts which know when we are being mistreated, neglected, and abused. It is the voice which alerts us to unhealthy environments and motivates us to do something proactive about those circumstances. It is only when the a front or threat is resolved, that the anger will be resolved.

In un health however, anger is an internalization of abusive authority figures who instead of dealing with their own insecurities, transposed them onto us.

> "I like to remind my patients: the opposite
> of depression is expression. What comes
> out of you doesn't make you sick; what
> stays in there does." (Edith Eger, the gift)

The ability to name the emotion and give the experience language to express the emotion rationally. To build connections between instinctual brain regions and more cognitive logic mechanisms. This is most probably the therapeutic benefit utilized by sharing one's testimony in religious communities.

Many of us who grew up going to church probably remember the feeling of oneness and equilibrium we felt wile singing hymns and recanting prayers. You were also probably taught that that feeling was the result of some transcendent spirit moving in and thru the congregation. However this same feeling is felt by millions at concerts and is now understood as part of a phenomenon called collective effervescence. Collective effervescence is a phenomenon where our hearts and brains synchronize with others (particularly during rhythmic exercise like singing hymns or reciting prayers). In much the same way as our hearts emit electromagnetic fields specific to our emotional state. Our brains also emit (a much smaller) field which then synchronizes with our heart's field. In many cases our hearts set the rhythm for the rest of our body and directly affects the activation or deactivation of our sympathetic and parasympathetic nerves systems. When we partake in rhythmic practices like singing in unison with large groups of people, those electromagnetic fields synchronize with others. And we have actually seen the effects of this synchronicity in the brains of entire communities thru monitoring their brain waves during these practices. Tho this phenomenon can have extremely adverse effects "madness is uncommon in individuals but the rule among the masses" (Fredrick neiztchie) the benefits are indispose-able. There is a reason we adapted these mechanisms and so The goal is to keep these very useful and vital tools, but abandon the prejudice and superstition often assigned to them thru religion.

## GOD IS THE VERY AIR I BREATH

In its most ancient, Hebrew conception, one's spirit and one's breath were one in the same. The Hebrew word for both breath (air), and spirit (ethereal life/essence), was rouch. This rouch or (spirit breath), was both a living piece of god within all living creatures, and breathed into man specifically by god (thru gods mouth exhaled into the longs of man and woman). In addition, tho the pronunciation of WHYH (the name of the Hebrew god) was lost to time, many sages and even scholars believe it is quite simply the sound of breathing. Or pronounced simply by the consecutive process of inhaling and exhaling. Breath is quite a fascinating property to be posited by (or to) god.

The soul as it was originally imagined, was the relationship or Union of one's body and their spirit or breath (breathed of dust). This nefesh, or soul, was the eternal quality of mankind. And it was marked be joining the heavens (sky, breath, air, spirit), and the earth (dust, body, finite, physical mater,).

There are many practices which honor the Union of breath and body. For instance practices like yoga have proven to greatly benefit the congruity of one's mental health and resolving nerve issues like anxiety and stress. This practice achieves this restorative spiritual state by intuitively activating the parasympathetic nervous system from the bottom up, thru synchronizing the breath with the body. This homeostatic state (via the parasympathetic system) reengages activation of our metabolism and reproductive mechanisms.

However what ancient civilizations believed was some

form of ethereal essence. We now know as a chemical element made up of atoms, more commonly, oxygen.

Our bodies use oxygen in conjuncture with glucose (metabolized or converted/broken down forms of the food we eat). The glucose and the oxygen are used in a process called cellular respiration, where they are converted in to carbon dioxide, water, and an energy molecule called ATP. This is essentially the inverse of photosynthesis. Now for the most part the carbon dioxide is a byproduct and is thus promptly expelled via exhalation. But that ok because tho it's a waste product for us human, it's as vital to trees and plants as oxygen is to us. This particular arrangement works out quite well seeing as how plants then use that carbon dioxide to create their own form of energy, subsequently expelling..... you guessed it "oxygen". aerobic respiration is one of two kinds of cellular respiration. And it so happens to be the one we utilize in creating ATP.

ATP is synthesized in the mitochondria, and it is the primary source of energy required for the synthesis of proteins. ATP is where our cells get the majority of their energy from.

Our lungs supply our heart with oxygen, our heart then distributes that oxygen throughout the body by way of the circulatory system. The blood functions as a vessel for the oxygen to be dispersed through the entire body.

This brings us to our next point, because if the breath, spirit, god's essence and life force, is actually just subatomic particles arranged in a certain way. Then why do we get sick and die?

> "The wages of sin is death" Romans 6:23
> (LJV)

Or take John 9:2-3 "The disciples asked Jesus, for who's sin was the man born blind, his own or his parents? To which Jesus replied, nether, but he suffers so that god may be glorified in his healing" (authors amalgamation). Of corse we now know that in most cases it is diss advantageous mutations in the process of replicating and propagating genes which cause such (Congenital anomalies) or ailments, and not some divine entity like god who propagated shoty craftsmanship in order to necessitate their intercession (for their own glory).

However It was once believed that the cause for sickness and death was the pollution of one's spirt or life force (breath) by way of sin. However we now know that ailments of diseases and death are caused by unfavorable our faulty mutations which occur thru the natural process of cell division.

Ok so the phenomenon we call death and sickness are actually results of disadvantageous mutations or alterations in our cells. But what are Cells really. Cells are the most fundamental components of living organisms (or life in general).

There are two forms of cells/life forms. There are simple single cell organisms like bacteria some of which make us sick and others which are vital for digesting nutrients in our stomach). And then there are more complex multicellular organisms like us (which can function as a host or environment for those more simple forms of cellular life).

Two common causes for sickness and death, or cellular degeneration are foreign bacteria's and viruses.

Cells like bacteria can reproduce on their own, viruses however require a host to replicate and propagate copies of itself.

Now The rhetoric of this book would profess that it is division, isolation, and duality, which causes death (both in the organism and in the philosophical heart). however this book too is not omnipotent or absolute and in many cases (especially on the cellular level) defensive strategies are beneficial and even imperative for life to persist. However in order to truly prosper, one's mind/heart and one's biology/body must be open to mutation and change. This openness is what we often call growth. There is beauty in tradition and profound wisdom in each cultures form of social order. But there is also often a great deal of prejudice. However as we will soon see, on a cellular level, that prejudice for foreign invaders and destructive agents, is a good thing.

Necrosis is the process of cells dyeing.

Lysosomes is a highly corrosive substance in the cells membrane. These Lysosomes contain hydrolytic enzymes that breakdown many kinds of biomolecules. Tho lysosomes are hazardous outside of the cell membrane, but inside the membrane they serve a vital role. They essentially dispose of waste. But still, very important. But eventually cells die. And when they die, our bodies need someone to dispose of the decaying cell and it's hazardous material.

Macrophage cells are a kind of white blood cell which is dispatched to contain and digest hazardous cellular material like lysosomes dispersed after necrosis. They also help to contain cells which have become cancerous.

"the best offense is a good defense" and thus, we have within our biological arsenal the killer T cell.

Killer T cells are a form of white blood cell which is formed in the thalamus gland. Killer T cells are dispatched to neutralize (kill) cells which have been corrupted by viruses. This extreme prejudice is due to the proclivity for corrupt cells to propagate that corruption exponentially (as many viruses and cancers ramp up production in infected cells). Viruses propagate by covering the infected cells with viral proteins. This is particularly detrimental because proteins on the surface of cell membranes are how cells communicate with each other. In part these proteins signal/trigger the release of hormones in the blood system. Proteins on the surface of cells also form bonds with adjacent cells. These bonds are how tissue is formed.

Apoptotic.

Apoptotic is a kind of environmentally friendly form of self destruction performed by dying cells in an attempt to quarantine the hazards material. There are two forms of apoptosis. Intrinsic apoptosis (caused by internal threats like cancer), and extrinsic apoptosis (which combats external threats like viruses).

There are many contributing factors which cause cell decomposition, such as errors in DNA replication, mutations in oncogenes (which promote cell replication), and cancer.

(Cancer cells lose touch with their adjoining cells and in turn begin to replicate uncontrollably.)

There are also natural forms of cellular decomposition which are caused simply by the wear and tear of cell division. This form of decay is witnessed thru aging. And more simply is the dilutions or lose of genetic data.

Where baby cells come form.

I know we all learned this in middle school but seeing as how it's probably been a wile for a lot of us, let's have a quick refresher.

Mitosis is the process of cells dividing and duplicating. At which point they specialize for particular tasks. These cells know how to specialize in specific tasks thanks to a line of genetic code known as RNA.

The nucleus or (cell membrane) is where RNA is formed via transcription and replication of DNA. and so the nucleus is also where the chromosomal DNA is located or stored.

In the human body this chromosomal DNA or code, functions as the blueprint for forming genes.

These genes (key DNA or sequence of genetic code) is transcribed to form RNA which then codes proteins.

Special proteins called enzymes perform tasks like cell division, and other chemical reactions which are used in the distribution, and breakdown of energy. All of this energy comes from ATP which as we saw is produced by glucose and oxygen in the mitochondria.

Once these cells are assigned their specific tasks these Cells then organize them selves by function and task. The rate/speed at which these tasks are preformed are regulated by glands which release hormones. Hormones like adrenaline excite many cells causing an increases in blood flow and metabolic rate of our digestive system. Different hormones however, evoke different reactions.

Like all life on earth, the divers array of specialized cells in our bodies began with an identical genome and a nuclei containing the same DNA as every other cell in our body.

The diversity in cells which have been specialized is often achieved by functions of certain proteins. However our universal origins are important, especially when those specialized cells become corrupted or lost thru decay.

The second form of cell birth is called meiosis, which is a process In which new cells are created thru the fertilization of an egg with sperm. These brand new cells are called stem cells which are then copied. The copies of these stem cells are then specialized for specific tasks thru a process called cell differentiation.

The stem cell is the raw unspecialized form of cell. Once stem cells specialize however, they only get smarter, they never naturally return to blank slates.

Emreotic stem cells develop in the embryo. They form within weeks of fertilization. These are baby cells.

Adult stem cells are tissue specific. they are in many ways an original of the specialized cells free of all the mutations and alterations.

Stem cells are particularly useful for growing new healthy cells after old ones have been corrupted or lost.

As useful as these blank slates are, (particularly in their versatility, as they have yet to identify or specialize with one particular part of the body). Mutation and specialization is just as vital because that's how organisms grow. As resources become available, or environmental strains become greater, it is imperative that cells change. Similarly if there are not claims you adhered to just a few years ago, that you now refute, then something's probably wrong. Ether you are in a sterile environment or you have simply stopped learning and growing. In ether case change is important and vital for life.

Cell division is the process of creating new cells to replace the more then trillion cells which naturally decompose or die in the human body daily.

Even on the level of the cell, creation occurs thru division. However in the case of the cell. this division is not the separation of the whole into parts but instead the duplication and then application of old and new forms of molecular data. It's like taking the wisdom of ancient traditions and modifying them in light of new information. ;)

Ultimately people are not born ether theists or atheists. Those are identities adopted much later in life. The conclusions of which have a lot more to do white when and where someone is born then it does why or how that came to be. The long and short of it is, god did not make you! We know how you were made, from your memories and personality to your fingers and toes, you were grown from cells, genes, and proteins. You are still miraculous! The only thing you lose by rejecting the identity god(s) gave you, is your position in their hierarchy of life. And the debt incurred on that life.

## LET THERE BE LIGHT

Another "ephemeral" quality often attributed to god, is light. However we now understand this phenomenon in the form of photon's.

In the ancient cosmology light was of god (or part of/ belonged to god). God was the source of light, and astral bodies like the sun, moon, and stars, functioned as symbols or representations of god's/the gods's glory or light. (For example take RA the Egyptian sun god, or Marduk who's celestial representative was Jupiter) (on the 3rd day god created light/became light, on the fourth day god created the sun, moon, and stars, genesis 1:3-16). Admittedly this was most probably the intended rhetoric behind genesis 1:3, and not meant metaphorically as consciousness as I previously alluded to. We now know however that light is a massless particle who's speed is that of a field of energy moving thru empty space (the speed of light). When a photon hits a solid surface, the energy of that photon is transferred to the electron of the atom it hits. Depending on the wavelength of that particular surface, the energy from that collision will ether be absorbed or refracted. (colors like violet have short wavelengths, whereas colors like red have long wavelengths),

To better understand light let us look at physical mater or fermions (particles with exhibit half integer rates of spin). bosons like light have rates of spin expressed by whole numbers.

This in part allows them to occupy the same space at the same time. (Only because those numbers correspond to electrical charges).

Light is a form of electromagnetic radiation, the particles which make up light (photons) are in a class of particles called bosons which (because they have no electric charge) they can occupy the same state/space at the same time without affecting one another.

Fermions however cannot occupy the same space at the same time, because they do have electric charges which repel one another. This particular quality is what keeps the 99.99% of empty space in an atom from collapsing in on itself. With that said, there is a very rare phenomenon called pair production where photons degenerate into particles with mass and their corresponding antiparticle (AE electrons and anti-electrons).

Note: The vander waals radius of an atom is the radius around which the attraction of atoms is greatest, after which they begin to repel one another.

The negative charges of the electrons surrounding a molecule will repel the negatively charged electrons suspended around the opposing molecule or atom.

The closer the two get, the stronger the resistant force between them becomes until they ether fuse or repel one another.

If the atoms fuse with other atoms, they form molecules. In This kind of chemical bond the forces which once repelled each other, now hold each other together.

AE the electrons which once repelled each other, now form a relationship via their corresponding nucleus.

In short Atoms resist or repel other atoms as a result of the relationships or bonds they had formed with other atoms.

Moreover, tho 99.99% of an atom is empty space, the physical tangible quality of mater is actually the result of electro magnetic fields created by the relationship of electrons with their corresponding nucleus which form a kind of gravitational field, which then repels similar fields in adjacent atoms and particles.

It is the relationship or identity of one kind of atom that then dictates weather the subatomic particles hold together or push away from others.

Ultimately the fundamental components of any given atom are indistinguishable from that of another atom, and thus the only reason one particle repels another is because of the relationship or bond formed within that particle between protons, electrons, the neutrons, which are virtually no different then those of the opposing molecule. Further more this bond is not grounded in anything concrete (physical), but rather some form of ephemeral force caused solely by the relationship between particles of the same atomic nucleus.

Many practices and beliefs still observed today, (like astrology) pay homage to this relationship between the subject (particles, people, organisms) and the environment (cosmos).

Tho we have pointed to the profundity of the ancient Hebrews words. In all actuality Hebrew (much like it's contemporaries ugaritc and Akkadian) lack much of our modern grammatical eloquence. Never the less Religion has greatly benefited many peoples by supplying a means for us to relate and resonate to our experiences, and at the same time many fields of science (like evolutionary biology) have been used to propagate malevolence in the form of eugenics and claims which validated racism form a biological perspective, (religion propagated such prejudice in the form of "manifest destiny".)

"Hell is a world without god, like gasping for breath wile we drown in the primordial chaos water. But some of us have learned how to swim. And we quite enjoy the beauty of a world undivided and whole."

## POEM

What good is wisdom if it makes one solemn. To dissect the bird and loose its song. Must knowledge come at such a price. That mirth and magic be lost. Yet As any good philosophizer knows the weight of the world is not so trifling, floating amid oblivion. To contend with the ineffable is to befriend the futility of life itself. Because after all, the curious mind is both path and destination for life. For beauty is not lost in the void but instead likend to the stars in the sky.

# CHAPTER 5

## HUMANITY AND MESSIAH, THOSE WHO BARE GODS IMAGE

### BY GODS COMMAND

The idea of sin as this insurmountable cosmic debt completely neglects the presupposition that a infinite god with limitless resources, created a world of scarcity and division where all life survives at the expense of another. A dog does not ask for consent, nor does a lion feel remorse for its prey. Man is evil only because we are capable of truly being good. The debt of sin then is on the god who created its necessity! A god that created a world where creatures had to lie, cheat, steel, and kill, in order to survive. The first time sin is mentioned In the Hebrew Bible is in reference to an impulse crouching like a wild beast before cain. We see as cain kills his brother and then becomes fearful that he might be killed himself. This is very similar to the way empathy functions in our neurochemistry. We feel other peoples pain because we evolved mechanisms which use other people as simulations for potential behaviors in complex social circumstances. However sacrifice is required for life to persist! Life requires death (or a reallocation of recourses) because mater cannot be destroyed or created, only reconfigured into different states. Thus we found need for a belief in hierarchy. An assignment of values which endowed us with some form of cosmic superiority to those we exploited (rather that be animal and plant or other humans) however as we've all ready seen, this illusion of a "spirit" or "soul" has very physiological correlates and thus is not some ethereal essence but a very physical phenomenon. Moreover whatever evil or good is perpetrated by this "spirit" is in accordance with our genetic coding. And thus it comes as no surprise that this hierarchy which was exploited to validate the divine right to rule of conquering kings. Then incriminates the god who endowed these tyrants with such authority. Not only that, but in accordance with the identity markers presented in the Bible. The Hebrew/ Christian God doesn't even stand up to their own moral standard of perfection, by which they used to judge and condemn people and nations as wicked .

Note in the Hebrew Bible Giants and nephalems are most often used rhetorically to depict opposing divine kings and the traits they idolize.

And thus in killing his brother out of jealousy, Cain becomes like the primordial serpent or beasts of the field. And the Cainanites (in the opinion of the Hebrew Scriptures) were no more then animals to god. This is a statement pertaining to the Babylonian propaganda surrounding barbarism. Or more accurately a critique of their ideals. ultimately these images of the barbaric Gilgamesh, and the philandering narcissist zuse, were cosmological claims about the identity of the divine. And more importantly the ideals and traits they idolized.

(This is why intermarrying with cainanite people and differing cultures was described as abominable to god in Deuteronomy. It was most probably articulated mythically as the nephelem taking human wives in genesis.

See Deuteronomy 23:1-6 for more context.

In that time and place in order To vanquish the beast, one must also become a beast (Israel's treachery on the 7 nations of cannon). And this was justified by their belief in that hierarchy.

Far too often The idea that there is no moral standard for good beyond the authority of a god has been used to justify the evils perpetrated by that authority (AE at gods command).

In the predecessor to this book we wrestled with the question of humanities's identity and how traits such as compassion and empathy were integral to this identity. Or as Darwin described it "survival of the fittest" pertained to these particular traits. In other words we as a species thrive, not by sacrificing the weak and feeble to their inevitable end but by caring and nurturing all "weak and defenseless creatures". Natural selection chose those traits which deliberately defied the "kill or be killed" nature of nature. Favoring not the tyranny of Sparta but the capacity to coexist with differing value structures. The prevailing human trait then is not our intellect or Braun but instead our capacity for consciousness and it's subsequent propensity for compassion and empathy. And this perhaps defines the role of humanity best, particularly in this discussion as "the intercessor" between the animalistic instinct and the ideological ego, the truly human one is she who wrestle with nature on behalf of the marginalized and neglected.

## MONOPOLIZING THE SOUL

People are taught to identify with the division rather then the whole,

When one's water comes from a stream they will protect the land. But when people get their water from the city and their food from corporations then they will identify with the same systems which continue to exploit them rather then the earth in its entirety. We conform to our environment for survival and tho we may not directly profit from that economic system we will continue to be complicit in its abuse of natural resources, animals, other humans, and even ourself, because we've been taught to identify with it. This form of identification with the system only works so long as that system provides for the individuals who serve it. When disparity reaches a point, those exploited by said systems cease to identify with the division and instead begin to recognize themselves as a intrinsic part of the whole which they had been taught to vilify and disassociate with. Then they too will turn on the system which has exploited them.

This is how you monopolize a soul. First you create a problem that needs solving and then you sell the solution. And so we exist, we didn't chose to exist same as all other life on this planet. The only difference is we are conscious of our existence and capable of negating our instinctual programming and deciding our own nature. This awareness is often depicted as our fall from grace as we reach out and take the forbidden fruit, deciding not to mindlessly obey the will of nature or god but to decide for ourselves what image we shall bare. And so we exist and we are conscious of our existence. So how can someone profit from that? Easy! You are inherently broken and even evil simply for existing. and further more the only way for you to be holy is for you to subscribe to this book which is the only way to god. That's right you heard me correctly the only way to god is thru a book written by men who claimed that you are supposed to look like them and theirs. So the god who created all man is restricted to the image of a few men. But that's not all. Not only will your existence be invalid apart form this "savior" but you will burn in hell for eternity if you don't sell your soul to pay off this debt incurred simply for existing. Now you may ask, what makes me broken? Well we have devised rules so strict that not even our own god (Jesus) could keep them. But wait there's more because this miracle of life is most certainly a miracle. and so you undoubtedly owe every good thing you find as a tribute or sacrifice to god. But heres the truth, You are allowed to enjoy the fruits of your labors and it is not your duty to fix the world (which is not broken but beautiful as it is). This ideology presumes that it is scripture which bares gods image and not humanity!

## GO INTO ALL THE WORLD AND SPREAD THE GOOD NEWS

In ancient times (and even still today) Wealth was a sign of gods grace or favor, like a king allocating recourses and authority so those they favor may profit from the kings land and prompt expansion. Much like the colonizers who conquered and enslaved/indoctrinated in the name of the king/queen.

Tho the convergence of church and state (religion and politics) is most certainly not rooted in Romes conversion to Christianity (as the prior state of states and nations (tribes) where primarily religion centric.

With Constantine's admission or affirmation of the "Christian"movement, came the alteration of text and propaganda which originally aligned with the oppressed and marginalized peoples slowly begin to transition to alignment with the political powers of the nation in order to then justify their conquest and colonization of "weaker" tribes and people groups like the native Africans, native Americans, and even the Nordic, Gaelic peoples.

Religious cultures and oppressive governments employ 4 key tactics in order to keep people submissive and complicit to exploitation. They keep people poor or indebted to them so that they don't have the resources to do better, they keep people isolated (thru hyper individuality or hyper nationalism) so people don't know better, they keep people occupied and tired so that they don't have the energy or time to reappraise faulty or flawed logic in ideologies or dogma designed to exploit and abuse them, and they keep them pregnant which both traps them in a system and ensures that there will be another generation to exploit and profit off of. The goal is to keep people dependent upon the systems that are exploiting them. This is ultimately why a diverse community and access to logic and knowledge are among the best tools for combating economic predation. The only way the few remain powerful, is if the many are kept weak and desperate.

"Freedom from sin" is the chain's religion used to enslave! Especially when it's definition of sin, is then a fundamental prerequisite for life!

Tho in all honesty this form of ownership of subordinates where children and wives are no more the property to home owners was common law in the Roman, Greek, and Hebrew traditions which inspired Christianity. And thus patriarchy is intrinsic to Christendom, and according to its assistants it is those without power, that then owe those with power their obedience, loyalty, and even their lives thru sacrifice. The debt incurred on humanity is simply the price for not being born a god, king, or wealthy.

The undoubtably abhorrent behaviors like racism, sexism, rape, murder, theft, burglary, and genocide, often sighted to validate the existence of sin, are not exclusive to humans, but observed all throughout the animal kingdom.

## EXERTS FORM THE DECLARATION OF INDEPENDENCE (COMPLAINTS AGAINST ENGLAND)

"He has refused to pass other Laws for the accommodation of large districts of people, unless those people would relinquish the right of Representation in the Legislature, a right inestimable to them and formidable to tyrants" (or more simply it is counter to our founding fathers complainants against England for our lawmakers to then cater to the aristocracy)

"He has endeavoured to prevent the population of these States; for that purpose obstructing the Laws for Naturalization of Foreigners; refusing to pass others to encourage their migrations hither, and raising the conditions of new Appropriations of Lands." (It is hypocritical for the US to Restrict immigrants)

"He has erected a multitude of New Offices, and sent hither swarms of Officers to harrass oud§ people, and eat out their substance." (It is counter productive to Putt into action laws and taxes which deliberately exploit and monopolies off of the people)

In fact not only does the modern US resemble the very tyrants they once sought Liberty form but the Supreme Court has ruled that nether the government nor law enforcement has a legal obligation to protect or serve the people."

(1989 decision in DeShaney v. Winnebago County Department of Social Services,)

In fact the whole debt biased standard for our modern economy is just as unethical as the economic constraints preceding the currency act of 1751. Which were counted among the reasons for Americans revolt against England in 1775.

## THE PRICE OF TYRANNY

The 12th most common cause of death in America over the past few years is suicide of which men make up approximately 80% (34,000 deaths in 2020). Now why are suicide rates so high? In part I believe this is due to an increase in existential fear of scarcity in attainable resources and an inability to obtain any sense of sustainability.

In the past it was possible for an individual to go out and simply build a shelter on a plot of land and live wile retaining some semblance of social rank among their piers (which is important for communal creatures) however with the eradication of affordable housing (as a result of corporations like black rock buying up all the land and property in order to monopolize the market), and the increase in taxation of the working class (the funds of which should go towards making society better for the people by providing basic fundamental needs like affordable housing, adequate education, access to nutritious foods and sustenance AE fruit and clean water, and public safety, but instead has gone to padding the pockets of the career politicians and the corporations that paid for their election campaigns. (note util the 1960s most college level education was free. It was only after 1966 when California governor Ronald Reagan instituted polices which charged for tuition in an attempt to retain "white supremacy" and combat the civil rights movements attempt at desegregation, that colleges then began to charge for admission. These policies become unanimous nation wide by the 1990s following Reagan's election as president.)

For far to many who are unable to retain social rank and necessary resources required for survival, the only foreseeable option is to neutralize their consumption by way of taking their own lives. This is incredibly sad especially when we consider that The whole point of a society and its systems is to make life easier and better for the people who live in it. And yet we've made obtaining vital recourses like shelter, virtually impossible. When the rich keep getting richer and the poor poorer, then trickle down economics isn't working! In fact as of 2023 the bottom 80% of the American population account for only 7% of the nations privet capital collectively. The top 20% of the population poses over 93% of the nation's wealth and the top 1% account for 43% of the nation's private capital. when we have more people in poverty or homeless then we do in healthy conducent habitats, it's not the people who are the problems, but the systems. When people have to work their lives away (most of which working 60-80 hour weeks and dyeing well before they reach retirement) just to survive (which by the way is worse than the working conditions in china with retirement kicking in at the age of 55-60 as opposed to the US which is 67) then the carrier politicians who exploited the public they were sworn to serve are the problem. When 15 billion dollars of the countries annual revenue comes form traffic citations (an victimless crime). And private penal corporations make 11 billion dollars a year from their exploitation of the captive 22% of the countries population via prison labor. (A disproportionate number of which are of the lower economic strata). Then the police aren't serving the public but the institutions who profit from the public they were sworn to serve. Now I understand we need defenses to protect and provide security. But when "ask not what your country can do for you, but what you can do for your country" (John F Kennedy) becomes about the wealthiest families getting richer and more powerful wile the vast majority are forced into wage slavery just to survive. Then societies missed its mark and has become completely counter productive. Now Truthfully we inherited this tyranny form our predecessors (Rome, and England) who packaged this propaganda in the form of religious doctrine and dogma elevating those of higher monarchical rank and demoralizing those of lower or inferior birth. However this world we've made is most certainly not a divine kingdom but more a kindred to the systems of oppression which the apocalypsists Jesus and Paul refuted. And that is perhaps the most profoundly compelling thing about Jesus and Paul's message. Because for them the most server punishment of Romes tyranny was not a deterrent but an victory cry. They did not facilitate the same systems of violence and might for their cause but rather asserted their autonomy in opposition to Romes economic systems of exploitation and oppression. Tho they brought with them a great deal of cultural baggage and a sense of racial superiority. They did the best they knew how to, with what they had.

Like the country's which were sacked and stripped down by England. The poor in this country were far to often the victims of deliberate systems instituted by political leader like Nixon to cause inflation and bring about the system of perpetual debt and credit we know today (see Bretton Woods Agreement and its subsequent convertibility of U.S. dollars into gold.) It is now known than the same people also used the war on drugs to criminalize and impoverish black communities in order to use them as slaves in pineal institutions. The point is that the rich intentionally profit off of the misfortunes of the poor. But as we've seen form Solzhenitsyns gulag archipelago. Violent revolution doesn't work. The only solution is to reject their economic system and use their threats as victory cry's like those brave souls who have moved into vans and RVs rather then fueling the corporate overloads who seek to exploit them. Before we finish this discussion we must first address a very sobering fact and that is represented in the events which followed Elon musks purchase of a particular social network which then led to false accounts masquerading as corporations like Eli Lilly and claiming that they would make the drug insulin free. After this false claim was made the companies stock plummeted because this vital drug which is required for so many just to continue living was also the biggest profit maker for the company and its stockholders. The penalty was dealt not because the company committed a crime but because the fraudulent claim claimed that they would no longer profit off of the misfortunes of others (and even contribute to those misfortunes). The sad truth is virtually all civilizations were built on slave labor and the exploitation of its people and those it conquered. Hunter gathers didn't choose agrarian lifestyles but instead were forced into it by more powerful ruling parties. From the very beginning these systems have been about exploiting those deemed lesser. It's worth noting that the inventor of insulin made the Paton free so that it might be accessible to those who needed it to live. And yet those in positions of power take what is freely given and sell it with a outrageous markup.

But perhaps I am being too harsh. I mean these systems have most certainly made life better for countless people. Take for instance The economy of war and the localized peace established by colonizing and trade routes which include the transmission of new and revolutionary thoughts and ideas. As well as its empirical subjugation to a hierarchy. Like England who claimed that their occupation and colonization brought about peace and revolutionary developments, modern policies do a great deal of good! Right? It is true that these policies do have pay offs for society at large. and perhaps that is the only reason they are still tolerated. But does it justify the bloodshed? It should be noted that The greatest tool in warfare is not a rocket but rather the depersonalization and disassociation of violence. Both of which We are poorly equipped for in ether the physical or the psychological strains of in personal hand to and violence. Most deaths in warfare are caused by artillery, and not short ranged combat. It is far easier to give a command on the other side of the world then it is to pull a trigger when starring an individual in the face. Distance allows animosity to persist, however the closer one gets to their enemies (the front lines) the more likely they are to empathize and even admire them. This is why it is those in high towers and secluded bunkers which rule the people. because a belief in hierarchy and distinction between us and them is required to justify their oppression and exploitation of those who are deemed lesser. And for that a separation is required (both physically and philosophically).

The point here is not to place blame on these factions but instead to take responsibility for our complicity in a system which continues to hurt far too man people. So that when the malevolent masters of today and days past "finally pass" we remember to do better then they did.

## THE GODS PLAY THINGS

> "It is a truth universally acknowledged,
> that a single man in possession of a good
> fortune, must be in want of a wife."

(Jane Austin, pride and prejudice)

This rather satirical quote from one of the most famous romance novels of all time, illustrates quite brilliantly the rampant objectification of men and women in accordance with social convention. It is no secret that women are objectified from the idolization of the female body in magazines to the sole literary role and purpose of female characters as romantic partners and trophies of conquest. The dehumanizing objectification of women is a ruse which has long been unmasked. However this objectification is still unresolved partially due to the ignorance surrounding the objectification of men. For like a male lion forced out of their pride and forced to compete for dominance with other males in order to obtain a place in their community.

It is only the men who prove their worth in vanquishing the beast and accumulating prestige and an excess of wealth and resources which are then permitted a place and a partnership in society.

It is only a man in possession of a good fortune (great wealth) who is then considered worthy of companionship. Now of corse there is ample reason for this on a biological level. for instance it is the female who is rendered vulnerable in childbirth and thus it is up to the male to provide stability and security form potential threats. It is also hardwired in our genesis (both male and female) to choose mates which exhibit fertile and beneficial traits that could be inherited by their offspring. This increases the offsprings potential for survival and even thriving. It has also been discovered that women tend to pick men who are the most genetically dissimilar as a means of increasing the likelihood of diversity in their offspring.

All this aside however, as a response to all our genes incentivizing we do then have innate needs for partnership. Tho perhaps not in order to survive, but most certainly to be healthy. Fundamental needs such as physical touch (something as simple as a hug or a held hand), to earnest verbal and nonverbal communication (signs of affection and meaningful conversation). These innate human needs should not be withheld on a basis of prestige, physique, or monetary wealth (from ether men or women). However such connection also can't be faked. We can't make ourselves love someone we don't. The problem is that we as a society have used these innate needs as incentives to exploit people. And until men are taught not to objectify themselves, they shall continue to objectify women. And vise versa, as long as women are valued only in regards to their physical appearance then they too shall be deprived of their authority and innate worth as human beings. However as long as a man's worth in society is based upon his vertical or hierarchical stature (by women and social convention alike) women will be seen as a reward and not as people. It's worth noting that "acquiring" a mate is one our our genes most imperative prerogatives and so on a biological level This objectification of men and women makes sense and this may also be why woman or eve is described as a deliverer or savior (ezer) (gen 2:18)and tho this passage uses the word helper rather then deliverer (probably due to misogynistic rhetoric and goals of making woman inferior or subordinate to men) in every other usage of this word (ezer) it refers to a savior or deliverer. Woman is both the means for life to persist (procreate) and a fulfillment of the biological prerogatives which incentivize/inspire men to pursue resources. And so as abhorrent as this objectification of both men and women is, it is also intrinsic to our biology to a degree. In many ways this (like the after life of modern religion) functions as an infinite game with no final or obtainable destination. Which means promotes perpetual growth thru the pursuit of an unobtainable goal. Because we can never truly know someone but instead must spend our entire relationship rediscovering each other as each party grows and changes.

This however extends far beyond just a battle of the sexes but also proves to be the root behind atrocities such as racism and human slavery.

In one of the oldest stores known to man (the atra hasis) we see illustrated the construction of hierarchical social structures which labeled certain races or ethnicities of human as lesser with the soul intent of making these groups of people subservient (slaves) to the lesser gods(igigi). This utilitarian exploitation of "lesser" creatures and organisms is still a major proponent for racism and classism today. This idea that man was created to be slaves to the gods(pharaohs and kings) can be traced back as far as our need for a ecological hierarchy in order to rationalize our exploitation of foreign people groups, lesser creatures and animals, and even plants, based upon the utility they serve. This need for living organisms and even life experiences to serve some kind of utility is as deep as we will go on this particular subject. However it is important because it highlights one crucial truth. From people to plants, we all have innate social needs as social creatures and part of an ecosystem regardless of what supposed value we bring to that community. And so when we neglect to water our plants we too shall starve. Or more simply as long as we with hold vital recourses as fundamental as affection and human connection as a means of incentivizing productivity (production), we too shall be objectified ourselves.

An experience doesn't have to profit you, that's not its purpose! It's purpose is to be experienced for what it is. Similarly the purpose of a person is not to produce or provide but as a human being, to have value in simply being.

The narrative for modeling masculine and feminine roles in society often demand too much of men and are intolerant or unappreciative of a majority of femininity. Leaving men feeling like they are not enough, and women like they are too much. This is why men are often intimidated by women who are taller, stronger, smarter, more successful, etc... then them. because they've been taught to base their value of themselves on their utility or ability to provide for, and protect those of the opposite sex. Women on the other hand have been taught to base their value or utility on their beauty or ability to influence (muse, inspire, motivate, etc..) members of the opposite sex. In both cases it is an objectification of themselves which then posits the objectification of the other (men as utilities and women as objects). The truth of the mater is we do treat (love) others, the way we treat (love) ourselves. Healthy men crave acceptance of their vulnerability and healthy women crave appreciation of their intellect. It is only the stories which separate us, from them,. so of corse men carve what women have and women want what men have. Because beyond men and women, we are all people. And we need both.

It is only when we learn to appreciate ourselves as individuals apart from the systems, that we can then be whole in ourselves and not simply a part in the narrative. We've been taught that love is a savior dying for our inequities. and so we loved as saviors. Begging for an opportunity to prove our worth by suffering or dying for others. We sought to be significant to others. But that's not love. Or at least not the deepest form of love. It's a love Which required us to be the smartest, most tolerant, or strongest, in order to belong. Rather then a love which recognizes our significance as we are, for simply being! You owe no debt for simply existing. You owe no debt for the mistakes you've made based upon the limited information at your disposal. You do not owe a debt to that story or belief, that society, or system, which convinced you, you were a utility. You have value on your own. It is in This assignment of value (that is holy and this is not) which gives those stories the power to validate our worth.

Far too often, religion functions as an attempt to escape our lives. because it views our consciousness of existence and attainment of autonomy as a curse (life as tragic). Rather then embracing our experiences as invitations to live, we see them as trials to survive and subdue. However the worst part is that it's goal is often to escape ourselves!

We have to remember that these religions function really well in survival circumstances. And I'm not claiming to have the best approach to life, but in many ways it is better then the alternative.

Like any story or framework, religion is just a circuit our brain forms to solve a problem. Rather it be a precious memory or a traumatic trigger, they are all just neuro pathways which formed in response to past circumstances. And it's perfectly fine to form healthier pathways, stories, and behaviors better adapt at dealing with our new environments.

## OLD PARADIGMS

A good father must do more than simply exist, a good mother must be more then simply present, and yet should any less be expected of a god?

How is it that a small child will strive to no end to communicate with a pup or pet chipmunk and yet the creator of the universe can't be bothered to acknowledge our existence when we cry out in pain. Even the hardest of men will praise his dog for simply existing and yet the one who made us is intolerant of their own creation. Why are we more loving and tolerant than our Heavenly Father? And why must we defend the honor of a god who can't be bothered with us? Why is gods ego so fragile and his composure so frail? Why can man adore the misgivings of instinctual creatures and yet god who made man in his likeness regards the nature of man as evil?

By this point in the book It should be apparent that This intolerant narcissist is an image of god predicated on patriarchal kings more concerned with power and pride then truth and healthy growth. They resemble A bad parent who hadn't learned to respect themselves so they are intolerant of their own resemblance in their children. Or the Ignorant kings who's self worth was tied up in glory and praise and so they couldn't acknowledge others.

And as a small child or loyal subject we attributed meaning to these titles regardless of the individuals lack of merits. But by this point in our evolution, the distinction between theses examples and the reality or perhaps even just better representations of the divine such as a care taker and shepherd rather then a davinc king is important.

It is also worth noting that from a psychological standpoint, if a adult progeny is estranged from their parent then the fault often lies with the parent who neglected to form a healthy connection with them during adolescence. And so this idea that we are responsible for maintaining à relationship with god (like a love struck boy consistently reaching out to a disinterested belle) the fault is not on the creation, but rather the unresponsive creator. It is Also worth noting that if a parent or potential mate requires you to question your own worth then they don't appreciate you. If your sense of self worth is threatening to god, then maybe the problem isn't you but your insecure god.

## MADE IN GODS IMAGE

One of the few distinctions that separate us humans from the rest of the animals on this planet is that we have free will. Or in other words, every other creature on earth is doomed to adhere to their instincts, and tho we have instincts ourselves, we also have the ability to negate them. In a sense this is expressed by the title of " image barerers" because We decide who god is thru the way we bare their image. this in part is the prolific claim of the messianic role. Because not only are we free from mindless subservience to nature. But We are also free from the elaborations of those instincts in the form of traditions (which after all are just behaviors we inherited from our progenitors in much the same way animals inherit their instincts). Or in other words culture, is an perverse extension of nature. So what the messianic figure depicts on a cosmological scale is someone who is not controlled by their primitive instincts or their cultural traditions but rather functions as a representative of god. from Moses rejecting the master slave dynamic of adherence to traditions(idols such as Egypt and bal (master)) to live in partnership with god, even at the expense of stability and security. To Abraham abandoning his tribe and traditions and declaring that his god does not require human sacrifices anymore (exodus, cainanite gods). We see these people redefining who god is. this brings us to Jesus the Christ (messiah).

Let us paint the scene. Nearly 475 years after Israel's captivity in Babylon (586-538bc) a new world order rises and conquers the ancient world. Rome with its iron fist bends the will of the nations and makes them all submit to its omnipotent authority. During this time Israel still longs for the authority god had promised them. from this longing two corses of action are decided on. The first is like that of the Maccabean revolt and is comprised of rebel extremists like Simon the zealot and Jesus barabbas. The other solution was to obtain a political alliance with Rome like the Pharisees did.

And so in walks This traveling teacher and self proclaimed rabbi who has no house and lives in tents or at the expense of his friends. According to the text written 15 years after his death, Jesus was well versed in the scriptures. Although there was no canonical collection at that time. Very little is known about the historical Jesus, however this holy hobo did make a monumental difference and even inspired a movement of people leaving their religions and traditions behind. and further more inspired a substantial decline and even abstaining form ritual sacrifices to the gods. And so as an attempt to rationalize this movement, the zealous Paul turned hieratic, wrote a number of letters to these congregations of converts. Latter we see texts baring the name of disciples attempting to account for these claims made by Paul and these congregations. Because At the time this movement was very politically charged, and we can see that in the phrases used to describe this movement (such as "gospel" which was an pronouncement of kingship, and announcing the coming of the kingdom of god which is essentially declaring a new world order) tho these people were not violent, they were by no means passive.

But what does this movement have to do with the nature of god? This movement claimed that Jesus was the son of god, or more simply put that Jesus was a better example of god then any pharaoh, Caesar, or king. And that is a claim I agree with. Jesus opposed the abusive traditions and did not profit from the corrupt systems of oppression. Jesus bore gods likeness better then many of his contemporaries. Although Jesus was not perfect according to the tora or laws imposed by his religion (past or present). The sad thing is, this Jesus who loved the unloved and sparked a movement claiming we don't need a sacrifice to atone for our existence has now become that abusive system which imposes an insurmountable debt on humanity simply for existing and then claims to be the only means of paying that debt. Who ever Jesus was. That system that has sprouted from his teachings is not god or in many cases even good! In genesis chapter 18 we see Abraham plead with god to spare the city of sodom and slowly Abraham convinces god to spare the city on behalf of ten righteous people in the city. Abraham argues with god on humanity's behalf. And In exodus 33 Moses does the same thing on behalf of Israel. Claiming that if god abandons the people they had just delivered, then god would be untrustworthy because god went back on their own word. In both cases we see a god represented by these figures that is more compassionate and forgiving then their present cultures and sacred texts. A god not fearful of some cosmic wrath or adherence to a set of traditions, but one of mercy and grace. Animals cannot argue with god! But we are supposed to contend with the ineffable on behalf of more loving and inclusive examples of "gods image" which is ours to decide how to bare!

It's funny how religious leaders who taught their children to idolize Christ, are then surprised when their children question those religious institutions that have manipulated, monopolized, and profited off of peoples misfortunes. Almost as if they hoped that by inflating Jesus with god and tradition, they could make him an object of worship as opposed to the claim he emulated. Which quite simply is "we are not subservient to our instincts or religious LAWs to decide our nature, but instead we choose what kind of image we bare, thru the way we love ourselves and others, for who we are and as we are"!

The question is how did we let them convince us that our malleability made us broken, and our autonomy made us evil. Why did we believe these institutions when they told us our empathy and curiosity was bad but incentivized our ignorance and compliance as virtuous? what kind of parent requires their children to fight their (the parent's) battles for them. Obviously The same god that requires their children to make themselves nothing in order to exalt themselves (god) to a position of glory like an malevolent king profiting from their peasants toils. The point is that this blood thirsty god who has to save "us" form themselves, is a small god, and regardless of their might and power the very fact that they "created us", doesn't then give them the right to abuse their children. Now obviously these amalgamations of a jealous, narcissistic, god are early cultures inflation of god with their progenitors and ancestors. And we were knit together in our mothers womb by evolution (gradually over millions of years and variations). But the point remains, we are not slaves to the will of nature or religion but instead, we decide what image we want to bare. And thus it is our duty to evolve those same stories for more compassionate caregivers then the malevolent masters we inherited.

(This Messianic prophecy was one which professed that when the Christ or messiah came this nationalistic king would then put an end to all war and unite the nations. And so if such a figure has come then why is there still so much strife?)

But Who bares gods image, humanity or the doctrine? How does god love creation? well how do those who were made in their image treat creation? And who wears it better? The sacred text which requires us to sell our souls for a sacrifice and fancy hand washing rituals, or the human being who just got off an 12 hour shift but who still has the human decency to slip a $20 in a homeless man's collection bucket. You see the image of god is best worn by the person who still remembers their own humanity and is not ashamed of it. It's not the person who has sold their soul to the idea and are now afraid to do the right thing because it would mean loosing their self righteous community or offending their narcissistic god.

## THE ARK

The cosmological function of the ark or boat is similar to that of god in the creation poem. On a cosmological level this metaphor (ark/boat) functions as a vessel for traversing a virtually endless abyss of unknowns. It achieves this by separating "us" from "them" or narratively by defining (species) and (gender) from the ambiguous chaos water of the whole (which seeing as how those fish who dwell in the abyss have more genetic similarities with other terrestrial species then they do each other, that analogy is rather fitting). And so in (John 21:7) where we see Simon Peter Jump ship, we are presented with this repeating narrative which illustrates the fulfillment of the messianic role (the role of humanity or the human one) abandoning the paradigms (boat/ark) of tradition and social convention for more wholistic depictions of our identity. Like Abraham Leaving his tribe and the traditions of his fathers. Or Paul who becomes the very heretic he once hated. We see this death and resurrection illustrated thru those divisions (like temple curtains) crumbling (tear in half) under a shift in perspective or shift of paradigm. Because the only thing which gave such claims of "holy" any authority, were the belief in those divisions authority. Like the apocalyptic nature of revelation, chaos only threatens to destroy, which order had created to divide the whole into parts by.

In many cases the only thing separating "us" from "them", is our belief in traditional identities.

"It's the time you spent on your rose that makes your rose so important"(Antoine de Saint-Exupery) The Little prince:

it is only our choice to chose a person who we've devoted time and energy towards, that then makes them special to us. It is our faith in the idea such as a "soul mate" or "true love" that then makes them "the one". And it's okay to change your mind. (However I would point out that this proclivity for growth is probably why Paul warned against making oaths and joining in mirage.)

The end of the world will not be the end of its matter but just a reshaping of its atomic material. In much the same way, a loss of faith or belief in an idea, is only a restructuring of values, and not a complete abandonment of virtue or meaning all together.

Note: "Biblical mirage" was not a covenant between a man and a woman who loved each other. But instead a covenant between a man and the father of a daughter in exchange for a Darry. It was and continued to be a business transaction!

## AND GOD MADE THEM MAN AND WOMAN

Some stores are mirrors reflecting our own images back at us, and others are windows into other worldviews and paradigms. But what happens when you block up all the windows? Well quite simply you block out all the light as well. If you want to live in a world without windows, then you well create a world without light by which to see your own reflection as well.

> "A work of Art cannot be one-sided In order to be justly called truthful, it has to unite within itself dialectically contradictory phenomena." (Tolstoy)

> "Do you know how much femininity man lacks for completeness? Do you know how much masculinity woman lacks for completeness? You seek the feminine in woman and the masculine in men. And thus there are always only men and women." (C.G.Jung)

It's worth noting how pivotal this subversion of gender roles and even rejection of gender binaries were to the "way movement" which later became Christianity. It was no mistake that the first conversation/baptism mentioned was of an Ethiopian eunuch (non-binary) Acts 8:27-38 NIV. and that even Jesus was quoted by the author of Matthew Acknowledging the existence of a third gender fluid personhood.

> "For there are eunuchs who were born that way, and there are eunuchs who have been made eunuchs by others—and there are those who choose to live like eunuchs for the sake of the kingdom of heaven. The one who can accept this should accept it." (Mathew 19:12 NIV).

And yet the biggest proponents for Tennessee and Florida bills which seek to " eradicate transgenderisim entirely" are conservative evangelicals "supposed Christians" who apparently have never actually read the book they proselytize. (House bill 619)

Now in part I realize that we can't blame the traditions for the people that align with them. And yet if those traditions align so well with fascist movements, maybe we should blame the ideologies and dogmas deep seeded in these Abrahamic traditions? "For you know them by their fruits" (Matthew 7:15-20 authors amalgamation).

Note: tho I do not condone scripture as an authoritative source, I do utilize it in communicating ideas to those who do still hold it in high esteem.

## THE HIERARCHIES

"On the level of individuals and civilizations: personality predates ideology. Which means that before they were fascists they were bullies and narcissists." Society is an extension of combined personal Microcosms.

On an anthropological level, God is the very foundation on which civilization is built. It's authority biased on a show of force and threat of violence and not a bases of truth, justice, or the common good of all peoples in said community. Of the values upheld by the Abrahamic god, morality and truth are not among them. That's why those who seek truth and speak out against injustice are vilified and demonized. And that's also why these traditions often align themselves with fascist regimes and governments. In the Hebrew Bible the greatest sin a person can commit is obtaining worldly knowledge.

Without belief, the hierarchy loses its authority. Because it's values are more Concerned with status then actual substance, ego not truth. That's why the critic and accuser are vilified for seeking truth and calling out injustice, wile the malevolent master is worshipped. It's not about growth or truth. It's about comfort.

Why do we have to GET saved? Is the pot to be judged for the potters shoty craftsmanship? Why must we be sold something extra? Why must we buy the DLC? The truth is these ideologies separate us form the whole. They sell us a prison cell and call it paradise. Religion is the oldest form of pyramid scheme, and the very basis of slavery and exploitation of the masses. It takes things as fundamental as human connection and outlaws any form

373

which it cannot directly profit off of selling back to the community. Its subscription to an ideology and a paid membership to a church. It's marriage and counseling, rather then friendships and diverse communities. It vilifies the liberated in order to rouse the anger of the slaves against their own salvation.

Nested hierarchies: a brief phylogeny of man and beast.

Humans are not isomorphic. We share traits with homo Habilis and homo Habilis shares traits with apes. Let me put that a little differently, homo Habilis are apes but apes are not homo Habilis. and likewise humans are apes, Monkeys, mammals,….. So forth and yet monkeys aren't humans nor are whales. Because we have certain specialized qualities we fit the criteria for many phylogenic categories. And so do many of our contemporaries and piers. the beauty of being human isn't that we are set a part form these species but rather that they are all a part of us as a species.

> "I was no longer needing to be special, because I was no longer so caught in my puny separateness that had to keep proving I was something. I was part of the universe, like a tree is, or like grass is, or like water is."
>
> (Ram Dass)

Loving is being oneself, hate is forgetting who you are and striving for validation or in some cases allowing others ignorance to dictate your self esteem or value of yourself and others.

People have no say in their race or gender, and to hate or blame someone for being a white man or a black woman is no different the the nazis hating Jews for being born Jewish. This is what is so abhorrent about the wave of anti trans and don't say gay bills as well as the reinstatement of the Jim Crow laws. Now we are responsible for our behaviors, but things like gender and ethnicity (which at this point is a complete myth because our heritage is so intermixed that there are not distinct black and white bloodline or male or female binaries). But this is no different then the "cosmic debt of sin" incurred for simply existing by institutions to manipulate people. This is the very complaint we have with Christian nationalism and also why authoritarian faiths align so well with fascisim, they are both a fear of freedom because when all people are equal then there is no need for a god. In order for there to be gods, there most be slaves. This is built into their creed but if the savior actually saved them, then they wouldn't still cry out for deliverance. If they still need a savior then they aren't saved. And if the savior demands glory and praise, and complete obedience and compliance, then they aren't saviors and you are still slaves. But judging people for the way they were bone, rather that be gender, race, or sexual orientation, is no more moral then incurring an insurmountable cosmic debt simply for existing. Now obviously I advocate for solutions which alleviate the environmental conditions which made bad behaviors advantageous and in some cases even necessary. But discrimination on the right or left is equally abhorrent.

# FAITH AND FASCISIM A STANDARDIZED MODEL (PARADIGM)

"Today Christians stand at the head of this country, I pledge that I never will tie myself to parties who want to destroy Christianity, we want to fill our culture again with the Christian spirit, we want to burn out all the recent immoral developments in literature, in the theater, and in the press. In short we want to burn out the poison of immortality which has entered into our whole life and culture, as a result of liberal excess during the past few years." (Adof hitler)

Sometimes we get trapped in the stories we've told ourselves, and sometimes we don't want to change the narrative and tell a better story because we fell in love with the characters, we grieve the lost loves, and we can't fathom an ending that could do ether their due justice. Sometimes we trap ourselves in the stories that ended long ago. Sometimes we prefer the security of the self fulfilling prophecies to the uncertainty of a new story. But rather it is a lost love or a lost faith, we don't miss the person (they were or we were), we miss the idea of that person. We miss the identity we had been given or had adopted. We miss the feeling, not the person themselves. But that feeling was felt within us and the potential for it (the secretion of hormones that cause those feelings) are still there within us.

I don't think is a coincidence that they gravitate towards fascism and molestation is so prevalent in these religions institutions. It's predicated on compliance to a hierarchy and not a pursuit for truth. It's no mistake that this faith based hierarchy cultivates societies which idolize the idol gods and malevolent masters and systems which exploit those who actually contribute. The justification for abuse is foundational to these traditions. That's why it vilifies skeptics and critics and competing cultures (like science, history, LGBTQ, and the accuser "ha satan"). The primary purpose of scripture and the traditions it inspires, is to impose a hierarchy. When you've got people imposing their own internal models and ideological "values" on people who look and think differently then they do, it makes perfect sense that they would also subconsciously project their own malice on others to protect themselves. And to a degree there is health in that. That's how our VMPFC is designed to function. But as we've seen this particular function often oversteps into server unhealth.

It's that same reason the Israelites erected a golden calf rather then entered into partnership with god on the mountain, it's why when bhudists were told not to immortalize the bhudas image they made idols in their image rather then taking responsibility and doing the work themselves. We choose to worship that which transcends our understanding rather than growing. We would rather conform to a model then construct more comprehensive and compatible modes. We would rather follow the path before us (favor automaticity) then forge a new one. And that's fine, it conserves energy and resources, that's why our instincts are more efficient than our cognitive mechanisms. The problem is that we expect the rest of the world to fit our simple maps and incomplete models. But the world is not flat and people have many dimensions. This world doesn't revolve around us. And the simple binary models aren't sufficient. We need more then gods to guide us. We must ourselves grow beyond their primitive examples and take responsibility of our autonomy. Many people avoid Taking responsibility for updating their own internal models and thus they expect the rest of the world to conform to their outdated paradigm. We would often Rather worship the trailblazer then develop the cognitive mechanisms for growth and develop more comprehensive and compatible internal models. When people know better, they do better. But when people are kept ignorant and knowledge and information is restricted or withheld, people behave accordingly. This restriction on access to knowledge is to manipulate people into behaving habitually rather the cognitively. To comply or react rather then reason and sympathize. But until people question and critique their own behaviors and faith based beliefs, then they will continue to act out the same malevolent role prescribed by their masters. This is most probably why fascism gravitates towards faith so often. But Until we are capable of examining our own behaviors critically, we will subconsciously project our own faults onto others in order to protect ourselves.

> "The society that separates scholars from
> its warriors will have its thinking done by
> cowards and its fighting done by fools."
>
> (Thucvdides)

Saving a life intells more then simply procuring the prolonged quantity of life but more importantly retaining the quality of that life. And admittedly for those who are capable of fitting religious traditions binary models, life for the most part is passively pleasant. However for those who cannot, their suffering is exponentially worse in the world these fascist regimes create. And rather it is the mythical conquest of cannain, the crusades, witch burning, or our modern transition therapy/anti trans/gay rights movements and bills. The paradise Abrahamic religions strives for, comes at the expense of a hell on earth for everyone els. on a cultural level what legislators in Florida, and Tennessee, are doing to the LGBT community is the equivalent of a ethnic cleansing! The implementation of religion to dehumanize and vilify minorities is a tactic used to justify their "eradication".

"Shame for ingratitude" has been used to keep people content with their stations and submit to slavery and oppressive stations rather then growing.

The lie is that you have a destiny, the truth is that you can change the narrative of your life and story. We have been sold the promise of heaven to justify our bondage and that lie has been called grace or salvation and yet all it is, is a cage.

The story is that there are those who serve and those who are to be served, and as long as you believe that story you will be a slave to its claim of salvation.

The peace afforded to Christians is the denying responsibility for the abuse and exploitation perpetrated by their community. As long as they focus on worshiping a savior and god they don't have to answer for the oppression condoned or even prescribed by their god and they may remain in ignorant bliss of the hell their heaven is forged on.

Religion serves a role, and there were times when we needed gods or at the very least guides. But the whole point is growth. But A savior who requires complete adoration and compliance (worship and obedience) is not a savior but a conqueror and dictator. And that salvation is not liberation, its just a another form of slavery.

If you are truly saved, then you no longer need a savior! That's the whole point of growth, it means outgrowing that which once guided you.

Religion only gets you halfway there, it is right about letting the narrative change you, but after all the character development is through, then it's your turn to change the story that changed you. The whole point of letting your story change you, is developing the character to change your story. If you still need saving, then you were never saved, You're still a slave! Faith is the way to complicity, it is the voice of tyranny not truth and justice. Truth welcomes the critics, faith vilifies those who confront injustice.

We see what we are shown, our story is what is continually repeated.

Tho our words mimic our hopes and desires, our actions or behaviors mimic our expectations.

Community is the cure and curse to fascism. The only way to alter the fascist models is to present contradictory evidence. To educate, inform, and unite diver communities, thats how we form more comprehensive and compatible models to combat fascism.

It's convenient that god get all the glory for the good work done by their servants and circumstances but none of the blame and responsibility for the evil and catastrophe prescribed by them.

God's brand of love can command genocide and abuse and yet still call It the fullness of love, but that's not love, that is a thirst for power!

What too many call love is a need to be seen which actually is the complete opposite, but love is seeing yourself and others. And yet that is the turning point. Because when the quality of gods love is limited to the capacity of ancient men to understand themselves and others then both gods love and wisdom (understanding) is severely insufficient. The love we see exemplified by the biblical god and messiah is one of self gratification and in incessant desire to control/manipulate those they supposedly "love". It's not an appreciation of qualities but an need for validation.

The quality of love is comparable to a frustrated mother yelling at her crying child because she is incapable of seeing past her own insecurities, and that of a mother who meets her child where they are and helps the child communicate more effectively. Both moms love their children, but the quality of that love is proportional to the quantity or capacity of understanding.

I think there's a good reason Christian's are afraid of secular culture "grooming" their children. And I also think there's a good reason why Christians are convinced that science is a massive conspiracy to mislead them. And it's not because ether secular culture or science are trying to mislead them, but instead because they are subconsciously aware that Christianity has both groomed and dooped them. Christianity vilifies secular culture so that Christian's won't take its critiques seriously and actually engage their executive functions for liberating themselves from the oppressors who do what they accuse the liberated of doing.

They claim to care about children's safety, that's why republican Christians so readily oppose a mothers natural right to abort and also why anti trans bills like "Tennessee senate bill 1 & 3 which target LGBTQ and trans communities in public spaces, particularly when children are present. And prohibit gender affirming care for minors. And yet they will do nothing about the mass shooting and suicide rates caused by their legislature. And even in Arkansas republicans passed a bill (HB 1410) which would bring back child laborers.

"They blamed sodom and gamorah on the gays so that they wouldn't have to answer for their exploitation, greed, and abuse, of marginalized communities."

The prescription that man is inherently evil and worthless and that it is only Christs love which gives us value ("you are worthless but because I love you you have value") is the root of many abusive relationships and insecure attachment styles where each partners worth is dependent upon the other persons opinion of them. The most effective abussers are completely unconscious of their behavior because they have interpreted all their interactions thru a extremely bias narrative/filter. In many cases Belief imposes models and projects it's own internal faults in order to protect themselves (the believer) by vilifying the critic and skeptics. Note The belief that something is innately and fundamentally Brocken or wrong with you is called hypochondria. Tho I myself once accepted and even admired such devotionI I now honestly don't see how sacrificing one's son to make up for their own shortcomings is ethical. "If it's dependent upon faith and loyalty then it's probably not based on truth or justice." Loyalty is more often then not a barrier to truth (which is the very way to growth.)

We can't blame the leaders for the people who continue to vote for their oppressors. When The masses continue to choose cruel rulers rather then take responsibility for their own liberation and honor those who have been so egregiously hurt by those same oppressive systems. The problem is bigger then the vindictive figure heads. As frustrating as this self subjugation may be The oppressors give people a script and a role whereas the liberators give people the freedom to choose their own role and story.

They are sold a story that they are a part of. But more then that they are sold certainty and admonishment from guilt or responsibility. In fact We too were taught not to question authority but instead to defend our gods honor at all costs. We were taught to be slaves and to hate the liberated, and we were taught to find our identity in that narrative, which is why we felt threatened by alternative stories that contradicted ours. The point is It's not republican vs democrat, its the governing gods vs the populous they exploit. The republican vs democrat debate is one of semantics (authority based in physical might vs authority based in intellectual prowess, taxes that fund military vs taxes that fund education and healthcare) the real debate is about using taxes to bail out banks and massive corporations vs taxes that actually improve the lives of the general populous. Equality regardless of socioeconomic demographic. You see when the gods fall from heaven and their system crumbles, we are the ones who are required sacrifice. We pay the taxes they are exempt from. all so that when they fail, when their insider trading and multi million dollar enterprises that exploit the workers all so that the idol boss can maximize their profits goes under, those same exploited workers who see nun of the profits foot the bill with their hard earned money. And for this reason Our brand of capitalism is actually corporate socialism. It is our taxes that bail out the banks and businesses when their exploitation of the general populous is not enough.

It is worth noting that a tactic often employed by manipulative leaders, guardians and political parties, is to endures a system on the outside but to intentionally work at making the implementation of those systems so inefficient that they fail. And so we will often find insurgents from the left and the right instigating the majority of the problem in order to discredit and create a villainous facade in regards to their opponents protests and policies.

And admittedly this may in fact be where we get a lot of the harmful ideologies in Christianity. Because two of the proponents of this movement were originally strongly opposed to "the way". Paul spent his life hunting down Christians prior to his conversion where he claimed to have seen a man who had been dead for approximately a century. And tho Paul's conversion seams genuine, it can't be ignored that he profited both financially and prestigiously from his new found position of authority.

> "I have received full payment and
> have more than enough. I am amply
> supplied, now that I have received from
> Epaphroditus the gifts you sent. They are
> a fragrant offering, an acceptable sacrifice,
> pleasing to God." (Philippians 4:18 NIV)

Paul did suffer many hardships for his conversion but he also influenced the movement more then about any other person (with exception to Jesus). Just because you are willing to suffer for a belief doesn't then make that belief true. Paul inserted himself into a position of authority almost immediately and became the most prominent source for both churches of his day and the multi billion dollar institution today. And tho Paul introduced the cultural acceptance of different ethnicities (which was not endorsed by Jesus), Paul also influenced the majority of the sexism and authoritarianism. Honestly I think Paul was a good businessman who saw that the tides were changing and so he too had a change of heart. The second prominent figure who employed Christianity to bolster his cause was Constantine who used the movement to solidify comradery among Roman citizens during a time of upheaval. We see similar things happening all around the world today. As of spring 2023 we have seen both the death rattles of a dying ideology and the conversion of career politicians. From Israel and France, to America, the working class are fed up with the exploitation of corporations and the corruption in the government. From the 10,000 high schoolers and elementary school children who marched on the Tennessee representatives, to the quiet quitting of employees, these younger generations are not having it. And those in power know their days are numbered. Even the US. Dollar is in danger of losing its authority as the dominant reserve currency as china, and Russia, sealed a deal to introduce a new reserve currency grounded in gold and not just a myth like the US dollars today. Regardless the tides are changing and much like the movement of "the way" which was later heavily influenced by converts like Paul and Constantine. There will be those who try to hijack the next system and retain their authority from the old.

## SCIENCE VS THEOLOGY

Who says? Comparing the credibility of scripture with that of science. Or more specifically that of the gospels with the claims of evolution.

Let us begin with the evidence for evolution which is still debated among fundamentalists today. We can literally see evolution take place (not only thru the neuro plasticity in the human brain (people exposed to larger and more complex social circles develop a proportionally greater degree of mass making up their frontal lobe in comparison with other brain regions) but in the remnants of leg bones in aquatic mammals (whales) and the copious snapshots preserved and presented thru fossils. Or more simply This supposed "missing link" is virtually non existent. The Unanimity which is retained across science and nation in regards to carbon dating is astounding. Especially when one considers that "The scientific method" is a deliberate and ongoing trial by piers, where one contradiction undermines the entire theory.

But even if one is untrusting of such accounts, one must also contend with the biological evidence still present in our genome. We share 99% of our genes with chimps. 8% with remnants of ancient viruses, 50% with other plant life 82% with dogs, 69% with rats and approximately 80% with aquatic mammals like whales. I realize this can get confusing. But to reiterate apes are not our ancestors! They are our cousins. We are closer kin with chimps then we are with whales because our common ancestors split off later with chimps then those ancestral kin which we share with aquatic mammals (like whales). Moreover every tree, fish, reptile, and virus, is a distant cousin who shares a linage with us stretching back to the first living organism on this planet.

Now let us look at the claims of the inerrant Bible see (hosea 1:4) vs (2 kings :30). These kinds of discrepancies however can be to scriptures credit because of the divers perspectives it accounts. Both sides believing that they are doing gods will. And so for arguments sake we shall stick to the new testament (particularly the gospels) from here on out. Of the 19,300 New Testament manuscripts found, nun of them are exactly the same. However much of this is due to spelling and grammatical errors. With that said many of these discrepancies are still blatant attempts to obtain unanimity with theological claims of the day. And so we know that many scribes intentionally altered the text in attempt to gain some semblance of harmony or even to appease their wealthy benefactors. This however is not the first hurdle we must jump in order to retain the inerrant authority of scripture. We also must wrestle with the fact that our oldest extant documents are most certainly not originals and some of our oldest (Rylands library papyrus P52) is dated to around the first half of the $2^{nd}$ century. This particular manuscript (P52) was probably copied within 20-40 years of the first transcription of John (90-100 AD).

However papyrus 137 (P.Oxy 5345) our oldest copy of mark is dated to roughly the late $2^{nd}$ century, so approximately 80 years after its first transcription (mark 66-70AD). Not only do we not have the originals. But the originals In Question were most probably written 30-60 years after the events they account. And the supposed accounts were preserved by 15 and 16 year old boys who most definitely suffered immense trauma from the sociological circumstances surrounding these events.

## To recount we have

The manuscripts we call scriptures are interpretive and non subjective accounts Written 10 to 50 years after the "events" they account. They were also "not" eye witness accounts. But more over We know for a fact that we don't have the original texts or documents but instead copies of copies. We know this because by comparing these manuscripts (nun of which are identical) we can see that they went thru vigorous and deliberate alterations in order to obtain theological harmony.

Not to mention that these main proponents of verbal transmission of the gospel was undertaken presumably by 13-16 year olds with underdeveloped PFCs and rampant hormonal imbalances heavily influenced by their hostile environment and conflicting cultures.

Deliberate Theological alterations.
Copies of copies, no original documents.
Not eyewitness accounts.
Lack of univocality.

When compared, the evidence for our evolutionary origin is substantially more credible then that of the historical accuracy of the claims made by the gospels. The point is not to discredit these liturgies but to recognize their rhetorical nature not as historical accounts but as cosmological claims about the nature of god and the universe. Which at the time where progressively radical and revolutionary. They function much like their earlier traditions as assignments of meaning and value.

Science is a competitive field where any and every scientist's job is to test (and if possible) disprove the findings of their peers. In short, There is no massive cult of scientists who are (deliberately or mistakenly) deceiving the general public.

People of ancient times turned to religion for the same reason so many look to science today, because it offered them answers to how and why the world works. Religion intuits answers based on experiences, where as science observed the phenomenon which evoke those experiences and then uses logic (like math) to test their hypothesis.

With the evidence we now have at our disposal, god is not a logical conclusion to the origin of life. In fact in order to legitimize god in light of what we now know we are required to jump thru hoops. That doesn't mean that there is no god, but rather that god is nether required nor present in/for life.

The existence of a god doesn't actually answer the question why any better than biology does, where biology shows us we exist/behave, to survive and procreate, religion claims we exist to serve and worship.

## THE APOCALYPSIST NARCISSIST

If Christ's sacrifice was to bring about the kingdom of god and heaven, and yet 2,000 years later the kingdom has yet to be realized then perhaps this apocalyptic preacher was not quite as prolific as once thought. Regardless of the outcome it is In my humble opinion that Martyrdom is only justified in bringing about a better world for those who come after and not as an attempt to obtain immortality by way of glory. And in this case even the shame and humiliation bestowed upon the martyr are cry's of victory against the tyranny of the establishment which sought to exploit and abuse those who come afterwards.

For an savior so passionately interested in bringing about the kingdom of god on earth. This savior who then spent a total of 14 days here among the living before then leaving for a millina or evermore is idiotic. For the Christ to be risen and yet not remain as an authoritative representative of god on earth to answer for the evils done in gods name, is more then problematic. For Jesus so loved man kind that like the god who created man he (Jesus) could not tolerate communion with man on earth but instead immediately retreated to heaven far away from those they supposedly love but can't tolerate.

Like Far too many (particularly those who grew up in the church), we were taught that we are unlovable wretches who are only redeemed by the love of Christ. That we are "loved" despite ourselves and wretchedness. This however is a common tactic used by narcissists to manipulate people. the sad thing isn't that the narcissist never truly loved the subject, but rather that they themselves were incapable of loving themselves and thus required the adoration of others. And as long as we believe that it is "Christ's love" that then redeems us from our "unlovable wretchedness" then we too shall love as Christ did, being completely incapable of loving ourselves and by extension truly love others but instead requiring the love and adoration of others to redeem our existence.

One may argue " is Christ truly to be believed to be a narcissistic!"

This charismatic rabbi who professed that unless son and daughter reject and even hate father, mother, brother, and sister, and even son and daughter as well, then they are not worthy of "Christs love". that if they do not take up sword against those who diss agrees, they are unwelcome in gods new kingdom. Yeah I think we can make that assertion.

Because True love does not require requirement.

But until you are capable of loving yourself you will require the love of others to make up for it.

The truth is I don't think Jesus was evil but simply brought up in a culture which taught him to become a utility. This is key because in the ancient and even not so ancient worlds, religion conditioned people to function in a society and so to diss obey the commands of god is to be incompatible and subsequently exiled or diss communicated by society and god. Thus to not serve a utility was to be damned to hell. Love here is not acknowledgement of one's individuality but recognition of one's utility in a system. It then comes as no surprise then that this religion systematically striped mankind of their autonomy and authority subverting that authority then to dogma and rhetoric.

**In the case of gender, it has been asked "what is a woman" "and what makes a man a man" to which I will chime in with a rather simple answer. What does it matter what a woman is, what's important is who they choose to be. The model was meant to fit the person, not the other way around.**

Titles like man, or woman, progressive or conservative, introvert or extrovert, theist or atheist or even dyslexic or autistic, insecure or avoidant, can be extremely helpful in understanding what models we are ether consciously or subconsciously employing (AE how our minds work and relate to our immediate environments). But for the most part these are simply programs we use to understand ourselves and others, and tho they may help govern how we work, think, and see, they by no means define who we are.

# CHAPTER 6

————————— ⊡ —————————

# THE INEFFABLE SPIRIT

No one is born an atheist, however no one is born an theist ether. These are conclusions we come to much later in life and they are often heavily influenced by our corresponding cultures and socioeconomic demographics. AE when and where we live.

They are moreover models we employ. These are not conditions of one's eternal soul but rather simply paradigms. Christians aren't evil or ignorant, and nether are atheists. The point is these are not what people are (theist or atheist) but rather who they choose to be and what models they have learned to apply. I am not vilifying the person but rather critiquing the practices which have hurt far too many people. I hate the "sin"(belief) and not the "sinner"(believer).

## THE SPIRIT AND DRY BONES

"It's in understanding the similarities between the behaviors of humans and animals that then allows us to understand why gods behave the way they do"

Science addresses the world as it is. whereas religion address the world as we experience it. Science deals with neurons and neutrons, religion deals with emotion and ideas, metaphor and meaning. Yet they both seek the same god. For example the Therapist and psychologists function in the same way as meditation and prayer does in changing the heart and mind of the individual praying. or in other words they both teach "the subject" to communicate with themselves or their subconscious. In both cases we develop the musculature to tolerate and even appreciate those unresolved experiences (or trauma). We form healthier neuro pathways, which is all trauma really is (synapses or circuits which formed or were diss formed under stress). Thus God does not necessarily pertain to a physical claim but rather a philosophical one. God is the attribution of meaning on the behavioral mobility of the material. This does not invalidate the prolific wisdom hidden in our subconscious/instinctual psyche, but instead signifies its bountiful wealth of data obtained thru prayer and meditation. Like an electrical surge on an computer chip, the program is just as real as the circuitry that necessitates it. The sentiment assigned to the sensation is just as significant as the synapses that cause it. And so Perhaps "god" is simply the phenomenon of genes synthesizing sentiment thru the ariseal of sentience.

Why do things happen almost serendipitously for no perceivable reason? Why do things just seam to work out in accordance with a bigger plan in the end?

It's most probably because we are meaning making machines that are incredible at recognizing patterns or potential patterns, and when done on a large social scale this phenomenon actually produces patterns and significant meaning (at least philosophically)!

For instance take love, because physiologically the emotion we call love or "in love" is just a chemical reaction produced by our genes in order to incentivize their propagation. and to a degree those genes are us in a sense. they provide a great deal of our innate traits and dispositions (from personality to physique). However those genes and their traits are controlled in many ways by environmental factors, AE access to vital recourses and stimuli. And much like love, we get a say in the matter. We attribute meaning to that emotion (even when the primary effects of that emotion has dissipated). and thus love has meaning only because we say it does. And so when we say there is no god, or life has no meaning, we have made those claims a reality. Or perhaps more accurately when we act on those beliefs, we make them a reality. Now the phenomenon we once attributed to god, or love, still exists, we just call it by a different name and understand it's physiological or physical (having to do with physics and natural biology) process better/differently.

## THE AMBIGUITY OF THE INEFFABLE

What religion recognized as whole, Science has divided into specialized fields, that which we once called god, science calls genes. However thanks to science we now understand those parts better. However the goal is to recognize that they are in fact connected. To make those parts whole once more. Religion separated those parts into categories of good and bad, holy and evil. Wile science divides those parts by function and familiarity.

From the pillar of fire, to the rejection of any proper name, the writers of exodus went to great extents in retaining ambiguity for gods identity. And I personally don't believe that was a mistake. I think far too often we try to conceptualize god as a localized entity. which is to say god looks like us and ours! But I think one of the most profound statements man has has ever made about the nature of god is expressed in the way they didn't. As if god is not concerned with what they look like, but instead how they assign value to life. God is not concerned with a system which inflated a life's value with its ability to produce or profit. But instead declares sabbath and presence as their dwelling and divine right of life. God seems to believe that the dirt itself is sacred and holy. In fact one of the most profound statements made by these abstract expressions of god, is that they do not exclude them selves from the whole but instead represent a remnant of that primordial chaos waters which made up the whole before the land and the waters, the nite and the day, were divided up into those of heaven and that of the earth.

The ambiguity of scripture and literature engages our meaning making mechanisms by stirring our curiosity and desire to interpret the abstract quality.

And so in short. There are claims made about gods identity which are undoubtedly good, particularly those which resign themselves from identifying with one or the other.

## THE CHARACTER OF GOD AND NATURE

The metaphorical god has heavily influenced society

It is important to remember that we are extensions of nature and our cultural environments. And so All the malice of religion and politics is predicated on natures priory prerogatives in us. Further more people don't choose theism or atheism but they are instead a product of the information at their disposal.

Who god says he is with his words

God is the same yesterday today and tomorrow. (Psalm 102:27, psalm 55:19, Malachi 3:6 Daniel 7:14, numbers 23:19, 1 Samuel 15:29)

God changes his mind.
(Exodus 32:14, Amos 7:3, Amos 7:6, 2 Samuel 24:16,)

God is faithful
(Numbers 23:19)

God goes back on his word
(Exodus 3:17)

God is merciful and patient
(Ephesians 2:4-5, exodus 34:6-7, Deuteronomy 7:9, 2 Samuel 22:26, psalm 25:10, psalm 86:15,)

God acts impulsively cruel
(Micah 5:15, Isaiah 9:19, Isaiah 13:9,)

God is present
(Psalm 139:7, 2 Corinthians 3:17, exodus 33:14,)

God is distant
(Exodus 33:15, psalm 22:1-2)

According to Mathew 6:19-20 god does not desire for us to accumulate wealth or earthly possessions and yet god requires such material goods to be sacrificed unto him. For god has very particular tastes
(See Exodus 25)
And yet god strikes people down for not giving enough or taking certain spoils of war. (See the story of achan in Joshua 7, or Ananias and Sapphira in acts 5:1-11)

Now truthfully There is no one univocal god of the Bible and The pre exilic god was entirely geographically resigned to a nation and a or place. However many Christian's still pick and choose which qualities and assignments of god to follow and which to abandon entirely. For instance a man laying with another man is often cause for gods retribution and yet they have no problem charging interest on loans. According to exodus 35:2 working on Saturday is a crime punishable by death (by gods command). Bu even Jesus broke that law! In the end rather it is those of "Bible times" or modern man, man makes god in his own image in accordance with his/her time and place.

## THE DECEIVER

In 1 kings 22:19-23 god sends a deceitful spirit to Ahab and lies to him so that he will meet his demise. Or in other words god blatantly lies to Ahab in order to trick the old king into an early grave. According to scripture anyway.

## THE LAW OF GOD

God prohibits Having sex during menstruation (Leviticus 20:18)

God prohibits the consumption of pork (Leviticus 11:27)

God prohibits the consumption of crayfish and lobster (Leviticus 11:9-12)

God condones chattel slavery so long as those enslaved are of foreigners and not Israelites (Leviticus 25:44-46)

God prohibits the shaving or trimming the sides of one's hair or beard (Leviticus 19:27)

God prohibits the wearing of linen and wool together. (Leviticus 19:19 & Deuteronomy 22:11)

God prohibits cattle of different kinds to share pasture (Leviticus 19:19)

God forbids the sowing of two different crops in one plot of land (Leviticus 19:19)

Getting a tattoo is a sin (Leviticus 19:28)

Piercing one's ear (or any other body parts) is a sin as prescribed by god (Leviticus 19:28 & Deuteronomy 15:17)

A woman is not permitted to hold a position of authority and should abstain from speaking in public. (1 Timothy 2:12)

God prescribes the execution by stoning of any women caught In adulatory or performing sexual deeds out of wed lock (Deuteronomy 22:13-21)

God forbids man from having long hair (1 Corinthians 11:14)

Men and women should not wear each others clothing (Deuteronomy 22:5)

A woman should abstain from adorning herself with jewelry, luxurious garments, or makeup, (1 Timothy 2:9-10)

God condemns being born of an unsanctioned marriage or unto those of different cultures (Deuteronomy 23:1-6)

A woman is not permitted to braid her hair (1 Timothy 2:9)

God forbids boiling goat meat in milk (exodus 23:19)

God requires the first fruits of every harvest and crop (Leviticus 23:19)

God requires that every fist born male (human and calf) be sacrificed/consecrated unto him(god). (Exodus 13:1-2 & 11-16)

Jesus prescribes the execution of nonbelievers (Luke 19:27)

God commands the slaughter of nonbelievers (2 chronicles 15:12-13)(1 Samuel 15:2-3).

A man who is not circumcised is to be excommunicated (genesis 17:14)

God commands that no work be done on Saturdays, punishable by death (exodus 35:2)

God condemns the mistreatment of foreigners (Leviticus 19:33-34)

God prohibits the charging of interest on debts (Deuteronomy 23:19)

God command the forgiveness of all debts every 7 years (Deuteronomy 15:1-11)

God commands the land be allowed to rest one out of every 7 years (exodus 23:11)

(Tho the last four I personally find exceedingly admirable for the most part these laws are ignored and discarded as obsolete by even the most pieous Christians)

## THE LAW AS PRESCRIBED BY GOD

In accordance with the biblical accounts the law prescribed by god commands a woman to be stoned to death if she is not a virgin on her wedding night, because according to god. a woman is not a person but instead property. And if "that property/ goods" are "damaged/spoiled" then the contract between the groom and the father is jeopardized. The god of the Bible commands that children be beaten (proverbs 13:24) because according the scripture children too are not autonomous agents but proto humans and the property of their progenitor/father. Slaves (rather in debt or as spoils of conquest) are permitted to be beaten within inches of death. (exodus 21:20-21) This is because the god who liberated slaves still doesn't acknowledge humans as autonomous agents but rather as property. Because like any ancient god, this god too was created to justify the exploitation of "inferior" peoples. Tho today women are not stoned and slaves not beaten (in America). Yet Subordinates are abused by their supervisors. and wives and children still beaten and undermined/invalidated (in Alabama schools) and yet a renegotiation with the social contract of the law is still made by even the most conservative who shave the sides of their heads and or beards. Who eat pork and or crawfish. And who charge interest on debts. And yet homosexuality is still considered abominable! It was never about morality! And this should be apparent by the punishments prescribed by god for not selling all one's belongings and giving to the church (Ananias and safira) or taking one's cut in a conquest (spoils of war), or even choosing to discern good from evil for one's self (the fall of man). God is not patient, just, or merciful. And in many ways the god of the New Testament, is just as cruel as the gods of the caninite pantheon they accuse of evil (psalms 82?). rather it is Abraham rejecting the god of Noah, or Jeremiah critiquing the god of Moses. Morality requires a renegotiation of our ancestors values and precepts.

Not only does god dehumanize those of other ethnicities and cultures god invalidates those who are deemed inferior by the traditional standards of their corresponding cultures.

God prohibits those who are "ugly" or disabled form entering into his presence (the temple/tabernacle).
"No man who has any defect may come near: no

man who is blind or lame, disfigured or deformed;" (Leviticus 21:18) NIV

## THE SPIRIT IN THE MACHINE

"The cultural memories that we read on the pages of the Hebrew Bible were written by individuals from a wide variety of backgrounds who learned of their traditions from a wide variety of sources (written and oral). They also added to and subtracted from their inherited traditions and reshaped them to address the needs of their own quite different historical contexts. The notion that individuals would reshape, revise, or even rewrite their inherited traditions based on their own historical and sociological contexts should not be surprising, even for today's most conservative reader holding a high view of Scripture." "When modern believers appropriate Scripture, they are following their biblical counterparts."

(Theodore j Lewis the origin and character of god)

Not only are modern readers required to renegotiate the terms of relationship with the text. But even the biblical authors call god out for his malevolence. The entire book of lamentations accounts this very phenomenon. And yet is it any wonder that Modern day apologists explicitly apologize for their own gods behavior. And make excuses for gods despicable dictums.

Contrary to popular belief Many atheists are not ignorant of the Bible. Perhaps one of the greatest proponents for atheism is biblical literacy. It's like a book where the hero is only preferable when the alternative is eternal torment. A god so lovable and good that the only way to make them compassionate is to contrast them with the most evil and cruel villain who honestly has more virtuous traits then the omnipotent god who created them at times!

And yet even then after seeing gods true character they choose eternal torment.

To be honest I wonder how many would still defend god's virtue if the alternative wasn't threat of eternal torment?

This god according to scripture is most certainly no epitome of virtue or ethics and an unfit standard by which to judge morality by. Even the "Christ" picked and chose what parts of gods law to exemplify and what parts to reject. And tho Jesus was radically progressive for his time, he is grotesquely regressive by today's standards.

Ultimately if these are not projections of man, then god is no better than even the worst of tyrants. And if they are projections constructed from the dialogue between our VMPFC and our amygdala, then god has never revealed themselves to man at all. And so how can we love something we've never truly known in any real capacity?

Humans speak to everything, from inanimate objects to plants, pets, and even themselves. And tho pets don't speak human (or vise versa). humans do do (lol dodo) (serous face!) do do a good job of communicating the difference between praise and reprimand. People more often then not are incredibly affectionate with animals (especially those they have forced to live with them as pets). But more importantly The relationship between man and beast is real, because it has consequence. The human acts and the pet reacts, the animal seeks and finds their owner, one speaks and the other replies (rather verbally or physically).

Now for some the dialogue between their PFC and amygdala speak with the voice of god, and the activation of their sympathetic/para sympathetic nerves systems ascribe the spirit's movements. However for many these internal projections are not enough. For they have spoken many times and yet have not been answered. They have acted in faith and yet not been saved but required to save them selves. They have sought out and not been found. You see because relationship requires reciprocation. It requires an answer to be given when one speaks. It requires attention to be shown. But without reaction the relationship cannot be real. And so does it matter if god exists and yet does not act? Of what consequence is an idol god!

Many who profess conservative Christianity aren't actually concerned with caring for creation so as not to offend some cosmic creator, but are instead concerned with retaining a traditional assignment of values. It's about what models they are required to apply to govern their behavior in certain situations. The traditional values provide an approximate frame work for one's behavior depending upon the company one entertains. However we all impose approximate models to govern our behavior depending upon certain qualities in our company. These models are often called stereotypes or archetypes and they fit an esthetic or style but not individuals. For individuals we must expound upon these models a great deal. But In short these models help automate acceptable behavior in complex social environments thru rules often dictated by culture or class (in the form of manners or social convention).

But then again this makes sense because this phenomenon we call god (or at least to the capacity to which we have a relationship with it) is not a physiological phenomenon but rather itself a model or projection in our head which is then imposed on our environment, and employed in predicting causal conditions AE cause and affect/action and reaction In corresponding social circles and environments.

Tho god as merely an assignment of meaning and value sound blasphemous in our western culture. The manifestation of the material world was not quite as much of a concern for ancient middle eastern civilizations. And thus we must address the claim that something can't come from nothing. And so the creator must have a creator... creationists argue that a cosmos as complex as ours must have had a designer. and yet that same logic would require that designer to be more comprehensively complex than that which it designed. And thus such an complex agent would require an exponentially more robust designer then themself to create god. However what we observe in nature is not a more complex form of life begetting a lesser one but instead the inverse. We see simpler forms of life evolve and expound upon those mutations until they are exponentially more complex then their creators. And thus we get this argument that "this younger generation is lazier, less competent, and more sensitive, then the last" and perhaps they are more sensitive, more in-tune, but it's not because they are dumber or weaker but rather that they are more evolved. They are running more robust and expensive internal models. What conservative Christians hate about woke culture is that it requires more complex cognitive functions then their borderline autonomic models preserved thru traditional values do. This is why issues like cultural identity, gender fluidity, gender roles, and sexual orientation, are so important to conservative Christians. It's not about offending some cosmic creator but rather the blatant diss regard for traditional models for social standards of behavior. It's a lack of propriety or manners. But when those social standers have been used to oppress, abuse, and exploit, they no longer bare any form of moral merit. And an renegotiation of terms is over due.

Aside from the fact that our preferences and biases are set thru a process we call nostalgia. A big part of the reason people seam to prefer "the good old days" to the present is because more or less everyone in a particular culture abided by a rather standard set of principles which meant my model matched your internal model more or less. so our behavior was easy to regulate accordingly. Tho those models are helpful in community they are also the root of much of the prejudice and ignorance of a society and they are in fact one of the major contributors to our divisive identities.

The claim made by the biblical authors was not that theirs was the only god, but rather that their god was a better god then those of other nations and tribes, much like theist today who's most enticing claim perhaps may be that their model or map of the world is more pleasant then the scientific reality of atheists.

This idea of blaming man for questioning god or not believing in a god is entirely counter to what science has shown us. From a psychological standpoint it is the parents job to establish a connection or bond with their young, and if a child grows up to be distant or estranged from their parent. The blame is on the parent and not the child. The same should be true in regards to a god. If people doubt the existence of a god it's because they've been given compelling evidence or a reason too.

More often then not we miss the sensation rather then the phenomenon itself. We don't miss the person or ritual, we miss the way they made us feel. But Every ounce of presence, belonging, love, and joy, you've ever felt was felt within you! and tho it was evoked by external stimuli, all that belonging, presence, joy, peace, and love, is still within you. What religion calls holy ground is not in some temple but within you, what science calls wisdom too is within you.

Tho it can't be ignored that these experiences were made special as much by the people and places, as they themselves ascribed meaning to those places, people, and rituals which evoked them. In many ways it is the people who make those experiences significant,(like a sun rise is only meaningful once it has been associated with a sentiment). Any sunrise thereafter then has the potential to evoke such sentiment. Like the metaphorical mind, the phenomenon functions as a body for the sensation to exert itself, subsequently ascribing sentiment to the conditions which facilitated it.

Is that really a good enough explanation?

Is that really how we are going to explain these seemingly serendipitous circumstances which transpire?

When conditions are favorable and the stars align. When the words come out just right. When the experience makes us ask why? in the hopes of replicating such favorable phenomenon. And yet we are meet with astronomical odds.

It's special because the chemistry for such life is uncommon and rare. that's what makes it holy. Rather it is life or love which on a biological level are unanimous. That unfamiliarity is what makes it holy.

We have nearly an endless cosmos where conditions were not favorable but even 1 in a billion chance become more and more likely the closer we get to a billion.

We refer to ourselves as an us (plural) because there is the us who experiences life. and then There is the voice who offers commentary on our experiences. And sometimes this "pixie" whispering in our ear is kind, and sometimes it's not. But that voice is not god and it's not you ether. It's a mechanism our brains evolved in order to navigate a social climate and environment in light of our autonomy. But here's the cool part. Because that voice is a part of you. so you can change its dialogue, you can rewrite its script. But first we must accept that this critique is simply an amalgamation of those piers and parents who taught us how to behave in community. And it is only trying to help us the best way it knows how. But that environment has changed, and we get to do better then we were taught.

The amygdala's goal more or less is homeostasis. And it will express any aversion to that goal in the form of emotions which in a purely instinctual creature motivated an alleviation of those environmental strains. More simply when sad = seek comfort, when scared = seek shelter, when angry = fight back, when lonely = find company, etc… the amygdala will regulate our internal conditions based upon our external environment. That voice will then attempt to explain to the more cortical regions (like the PFC) why we feel the way we feel, so that our internal models (VMPFC) can be updated and employed in remedying the situation. More over the voice isn't trying to be mean, it's trying to keep you alive and well. But as we've already stated that voice is an amalgamation of our teachers, parents, and piers, and so sometimes it can be helpful to audit that voice (broncas area) and our emotions (amygdala) by cognitively asking "how do I feel" and why do I feel that way". This will help us construct more accurate and comprehensive internal models by which to govern and judge circumstances in our environments. Note because this voice is biased on an amalgamation of our immediate community, the quality of the company we entertain can greatly effect the way those internal voices talk to us.

Religion idolizes the ancient words of our amygdala where as philosophy tries to "free the fly from the bottle" (as Ludwig Wittgenstein) would say. The amygdala attempts to control more advanced mechanisms using old methods, whereas philosophy/phycology strives to understand the mechanisms new and old in order to integrate the two more harmoniously.

Religion requires the itch and thus all it will ever do is scratch the itch but never remedy the underlying cause, it will never cure it.

Rather it is a toddler throwing a tantrum subsequently making the parent feel powerless. or a girlfriend ignoring the texts of a boyfriend who had not made her a priority. People reflect our subconscious behaviors back onto us. Or in short we will be made to feel the way we made others feel, and vise versa.

And so Atheists mock religion to assert that that which once abused and hurt them is no longer a threat.

Theists vilify science and philosophy because they can't control them and thus they fear them. Theists aren't idiots and atheists aren't evil. Some of the most intelligent scholars in academia are Theists and some of the most aultuistic Samaritans are atheists. Spirit is just what religion calls instinct. Instinct, intuition, and emotion are three names for the same mechanism.

(Note laughter is a mechanism observed in humans and chimps which communicate to oneself and to others a lack of real danger in potentially threatening situations. And so humor, satire, and sarcasm by extension are ways we deflate the presence of social threats. Perhaps this is why the biblical Jesus was so sarcastic in his critiques of society in his day).

The epitomes of secular wisdom is marked by religions depravity, and vise versa. What science calls an age of discovery, religion calls death. And what religion calls revival, science calls the dark ages.

The rational and logic of science and philosophy can ether contrast or complement The intuition, emotion, and instinct of religion. The two can ether strengthen one another's weaknesses or exaggerate them.

Trust your gut it's kept us alive for Millions of years. but definitely renegotiate the terms of the agreement "a particular familiar stimuli, prejudice, or uncanny feeling, doesn't necessarily mean bad or danger" and "our gut is biased on past experiences and thus it can be misleading. Or in other words your gut can and will lie to you.

That's why being in nature feels holy, because it is a familiar voice to instinct.

## DEHUMANIZING MAN FOR THE GLORY OF THE GODS

Our instincts tell us there is a god because they were once that god.

It was an abomination for gods to marry mire mortals or peasants (like the b'nei Elohim in genesis 6) for such intercourse would subsequently elevate mortals from their intended station of servants. It was also abominable for women to marry outside of the tribe (Jude castrates and kills a village genesis 34:25). It was abominable for the blood to mix for any other reason other then the joining of two nations prescribed by the patriarch. In many traditions the The b'nei Elohim were hypothesized to be the human sons of Seth and this argument arose in response to the question of free will (were angels agents who acted on behalf of gods will, or were they independent agents capable of rebelling against god?) ultimately these arguments were based on texts from the book of Enoch which attempted to solve the problem of evil. However when we look at the sins of these b'nei Elohim it was that these exalted beings mingled with mere mortals and shared divine knowledge of herbs (medicine) astrology (navigation) and iron forgoing (technology) with mortal men who were meant to be subservient slaves to the gods.

But to truly understand these mythical entities (b'nei Elohim) "sons of gods" we need to understand the over all context. And how it was used to demoralize one's enemy.

In genesis Adam was both a shepherd of the beast of the field and a worker of the ground. Cain become a worker of the ground, and Abel inherited his fathers role as a shepherd of the beast. We are told that able was blessed by god or became prosperous. Now a shepherd would be out in the fields. However in ancient times agriculture (sowing crops) was a civilized city dwelling occupation. It was also one of the distinguishing qualities of Babylon which according to the biblical myths was founded by cains offspring or seed.

And so we have Cain the agrarian pioneer and the biblical patriarch of Babylon. Now According to the Talmud it was able and his flock trespass on cains land which then led to cain killing his brother. However biblically this is depicted as a beast usurping cains role (this beast was his emotions or primal instincts) and because cain let the beast rule him, he was designated by scripture as inheriting the line of the serpent or beast. This was a tactic used by the Hebrews to dehumanize their ancient near eastern counterparts and brothers. ("The cures of Cain).

According to the biblical genealogy Seth the third born son of Adam and Eve, is the father of the Hebrews and Egyptians. Later sources accredit Seth as the father of the b'ine elohim (the mythical gods of Egypt and Eden). This rhetoric was used to exalt the Hebrews wile subsequently dehumanize and demoralize their enemies so that their slaughter and exploitation was condoned by god. We see this further expounded upon Thru the genealogical accounts of ham (the son who saw Noahs nakedness " possibly referring to an affair with one of Noah's wives and hams step mother)" from this affair ham is credited with being the progenitor of Assyria, and Babylon thru nimrod, (nimrod being synonymous with the nephelim and cain). ham is also accredited with being the father of Egypt And propagating the line or offspring of the serpent or beast. this imagery of Israel as the snake crusher was used to justify their conquest genocide and enslavement of the great empires. (At least mythically) and this rhetoric is drawn upon in Mathew 15:21-28 and mark 7:24-30 where Jesus claims he came to save the lost sheep "Israel" and not the dog (cainanite seed of the beast or serpent).

However interestingly enough Abraham (gods elect) was a son of Babylon according to the Talmud.

And we see this rhetoric expounded upon in the story of Abraham. We are told that Abraham left his family in faithfulness for a promise and blessing. A promise That god then witheld until the end of Abrahams life. Abraham was even accused of wrong for taking maters into his own hands (tho this taking matters into his own hands did consist of the rape of an Egyptian slave). However this slave is not permitted to be the mother of gods chosen people. Because It's not enough that the world be blessed. but that that blessing must come through us (Israel, America, the church) and for that reason scarcity must be propagated and the blessed must be vilified so that gods elect can be exalted above the rest. If the poor are provided for and our enemies are just, then what need is there for the malevolently benevolent saints. God must have his portion even if it means cursing the blessing which comes from man by their own hand. Man must be demoralized so that god can receive glory. Even Jesus articulates this in John 9-10 when he calms that the blind man was afflicted so that god may be glorified. It is this same rhetoric which is used today to deny social benefits thru state (tax) funded subsidies. the poor should not be provided for so that they are dependent upon god and the church....

When a male lion becomes the head of a herm, they kill the offspring of their competition so that their offspring will not have competition. This aids in ensuring the propagation of their own genes and lineage.

According to the Bible women, children and slaves were property and not permitted to be autonomous agents. That is why the beating and unequivocal omniscience was prescribed. Because like the bastard cubs of a herm, those of other cultures, ethnicities, genders, and generations, were considered proto human. However now it is those who are inhumane who are considered less then human.

The problem is a belief in ethnic supremacy and an intolerance for positive or neutral traits in other cultures. For Gods were created to justify the exploitation of humans. And thus Both the beliefs that others exist to serve you, or that you exist to serve others is unhealthy. We should aspire to be independent but not isomorphic.

The god of the Bible treats humans like animals. Some (the elect) they herd like cattle (but don't give any right for volition or autonomy), and others they execute like vermin. But even people today treat their own pest more compassionately and household vermin more humanely than the god of the Bible delt with man. Gods are an excuse to exploit and abuse other people the way people exploit animals. However it also can't be overlooked that many people find religion as a means of relinquishing responsibility for their actions and behaviors. Because religion (like the golden calf in the wilderness) functions as an idol or mold by which to govern our decisions which let us live without the taxing requirement of autonomy. And so our persecution of marginalized communities is not on us but our god because it was he who commanded it!

417

This is why it is imperative that we renegotiate the terms of relationship and update our model for morality.

The story of exodus is of an oppressed people becoming Pharos unto their oppressors In pursuit of a potential paradise however as we will soon see, this paradise often goes horrendously wrong.

Note: Much of the Roman technological advancements were reappropriations of prior cultural advancement.

To see one's nakedness is an idiom for to see one striped of honor or to see one's shame, to see one exposed. and so tho this may be what is being alluded to in genesis 9:18-27 seeing as how it was ham's children (cainan) who are cursed, the most likely meaning is incest with hams mother. Ultimately the whole point of this rhetoric is to dehumanize and demoralize those of competing cultures and to justify the crusade, genocide, slavery, and exploitation/abuse the indigenous/neighboring tribes of cannain. Similar to how evangelicals vilify trans and non-binary communities.

Genesis 9:25-27 NIV

> (he said, "Cursed be Canaan! The lowest of slaves will he be to his brothers." He also said, "Praise be to the LORD, the God of Shem! May Canaan be the slave of Shem. May God extend Japheth's territory; may Japheth live in the tents of Shem, and may Canaan be the slave of Japheth.")

## INCOME AND INEQUALITY

"Being well adjusted to a abusive and exploitative society is not something to be proud of"

Income inequality and disparity is far worse now (2023) in the US then it was leading up to the French Revolution, the Boston tea party, or the formation of the USSR.

All of which were revolutions prompted by egregious exploitation of the poor and lower class. To put it in perspective here are some statistics..

Of the 7.888 billion people alive today 2,700 of them are billionaires. The average American makes roughly $54,000 yearly. People who make more then $41,000 a year are already in the top 3% of the world's population. On average it takes a billionaire 10 seconds to make $47,000. In under a minute the average billionaire has made more money than the top 3% of the world population does in one year.

The top 10% of Americans are millionaires. 90% of which make the majority of their money via owning/ renting out real estate. And seeing as how the average cost of rent has increased 8.85% every year (from $600 in 2000 to 1,180 in 2023) wile the average income has remained fairly stagnant (42,000 in 2000 to 54,000 in 2022).

If minimum wage had been adjusted accordingly (to make up for a livable income) it would be approximately $27 an hour as opposed to its stagnant $7.

25% of Americans make less than $15 an hour ($31,000 yearly).

But wait because here's the real kicker. Aside from boondoggles and the many perks afforded to lobbyists. More then 50% of US Lawmakers are millionaires and vast majority of their incomes come from insider stock exchange which is governed by them. In fact it's gotten so bad that their own members have proposed a bill to ban such practices (The PELOSI ACT 2023). However to add insult to injury double voting (that is voting on behalf of multiple absent members) has become a common practice in the Supreme Court. (Captured on camera).

The disparity between the common blue collar citizen and the lavash excess of the the rich in America could be strait out of the acclaimed novel (the hunger games by Suzanne collins). For most of the worlds population issues like malnutrition and preventable sanitary diseases are prevalent.

Perhaps the greatest epidemic in America today is not income disparity but social justice. When Government officials like the police and national guard evoke fear among US citizens rather then a sense of security. And the US penal system advocates for exploitation not reform like those of Sweden.

Interestingly enough the proletariat who sought the insurrection of the czar in Russia (1917 revolution) got their name from the explicit exploitation of their livelihood as no more then proletarius (producers of offspring for use in wars, and remedial labor) perhaps that's why republicans are so distraught that younger generations are abstaining from child birth? Not only is the production of progeny in this economic system unfeasible, the younger generations refuse to feed such malevolence. However unfortunately (for both the oppressed and the oppressors) history has shown the revolutionaries are often twice as cruel as those who formerly oppressed and exploited them.

Apart from the proletariat and bolsheviks in Russia, there were a group of anti-industrialists called the wandervogel in Germany prior to the rise of the nazi regime (1896-1933). And tho in part the disbandment of this particular outlet was most certainly to blame, many of its nationalistic members sided with the nazi party in their vile mistreatment of the German elites (which was proposed to be comprised entirely of the entire Jewish race). To put this in perspective this would be the equivalent of blaming every Caucasian for the grotesque exploitation perpetrated by the prominently white top 1% (billionaires). When you make the issue about race then everyone looses. But that also doesn't mean we permit that prejudice and exploitation (particularly of black communities) to persist. My hope is that we will learn from history rather then allow fascists to burn/ban histories books until the only solution for their tyranny is that equivalent to the reforms of the USSR and nazi party.

These warnings may sound like the ramblings of a apocalyptic rabbi professing that the world is ending. But then again the world is always ending (at least as we know it). That's what apocalypse is. It's the breakdown of outdated traditional models in the wake of more revolutionary ways of thinking (revelation, uncovering, or enlightenment,). More commonly it's just called growth or maturity in individuals. And it often looks like a conversion (like a Saul to a paul, or a fundamentalist to a progressive)

"There is this fear that we will wake up one day and we won't even recognize ourselves anymore. But the problem isn't that we had changed but rather that we had mistaken the model for us. People are supposed to change. And personally and think it's beautiful that life can diversify so much that, that which we once could not tolerate, we now are capable of appreciating.

It is an unavoidable fact that A disproportionate number of those incarcerated are from low economic strata's and marginalized communities.

In the Old Testament god was the cause of both good and bad. Man heard the voice of god and obeyed. But as these texts developed and on into the writings of the New Testament (which were heavily influenced by the Greek philosophers) we see man receive a lot more of the blame because it become apparent that men have agency. And so one might ask what gives me (or anyone els for that mater) the right to criticize peoples beliefs? To which I would argue it is our deliberate duty to do so. To form more comprehensive and accurate models of ourselves and our environment. We should most certainly question beliefs, especially when those beliefs permit the exploitation of the poor and marginalized, when they deny women of their autonomy, agency, and authority. because if we don't challenge those beliefs which blatantly prescribe the physical and psychological abuse of children, and those which demonize certain races or ethnicities of people as inferior to others, then all that blood and evil commanded by our traditional gods, is on our hands. This book has probably offended many people by this point. And for that I DO NOT APOLOGIZE!

I am not concerned with offending cultures which have unapologetically abused and exploited people. These statistics and rhetorics should make you angry.

And If you truly believe in the god of the Old Testament as a personal entity then you most probably harbor a grate bit of anger. "That anger is not bad".

That anger you feel towards the evangelical god and it's capitalistic system is the same kind of anger which inspired the psalmists and lamentations which were accusations against the Egyptian and Mesopotamian gods of their time. But as we've seen the just wrath of the Israelites god is not better then those of their cainanite brothern.

It's important to resonate with the anger and anguish from wince these laments and exactment's derive. But also remember that the "righteous wrath of god" is no better then the oppressors they oppose.

thats the story of the Bible. It's rhetoric written by marginalized communities who in time became the oppressors (much like the proletariat in soviet Russia). it's one prejudice culture criticizing the gods of other cultures for the same malevolence perpetrated under the guise of their own god. the problem is not now nor has it ever been one particular ethnic, race, or cutler. But rather the belief that any one of them was superior or inferior to any other. These ancient texts which many call scripture are no different then modern man's critiques of cultures. The greatest evil is not that man became autonomous agents (like god discerning right from wrong) but rather that man then strove to become like a god unto man (in an attempt to subvert the authority of such autonomy in others) for who is god that he should make man in his own image and then not permit man to decide for himself what traits should be desirable for emulation.

We get offended because we identify with the models (rather ethnic, gender, religious, or political.) we mistake ourselves for the identities we assume. And the stereotypes associated with them. We are not these identities. Even if we have been taught to reduce ourselves to them. People and animals are who they are taught to be by their past experiences. People are bitter because they've been taught not to trust joy. people are cruel because their environments taught them that is how they survive. But people also have the ability to reevaluate and then renegotiate these behaviors.

I'm not trying to make you feel guilty. You didn't create this system of oppression nor are you in control of its mechanisms. We were born into it, and are just as much victims of it as anyone. But the fact that you are reading this means that you have enough to eat and have time for leisure. It also means you are literate and willing to learn. And thus for all intensive purposes We are the princes of Egypt. And this is our sign to no longer comply with these authorities. This is our charge, to hold theses gods and the systems they erected accountable for their sins. Because it doesn't matter if it's the church, the king, or god themself, might does not make right.

We are not the slaves in Egypt, we are the princes of Egypt. We were born or adopted into these abusive systems. And tho we are not the masters. We do have the power to lead an exile from systems of tyranny. To abandon philosophical frameworks which perpetuate abuse and exploitation.

I am not fearful of what institutions I offend. Especially when those institutions continue to abuse and exploit real people in the here and now.

But unfortunately it's not that easy.

We don't go straight from tyranny to paradise. We go to a wilderness where we are required to deconstruct these systems of oppression and all the luxuries they afforded us. And we will reman in that wilderness until we learn to face and appreciate that which was once intolerable. The price for justice is living without the luxuries tyranny afforded. For some that's not selling property to corporations. For others it's living in a van and abstaining from having children. It looks different for different people.

Rather it's climate change or human rights.

The problem isn't the poor and exploited people in India or china, but rather the modern day gods and pharaohs (elites and bourgeoisie multi billionaires who gained such extravagance by exploiting those countries and their resources. It's those who have drained the earth of her bounty for their profit. And thus we must find a way to change the system without hurting the

People who continue to serve these systems. (The

425

majority of which are the poor who would burn the world down if it meant giving their children a better life.) And that right there is the god of the Bible. he did horrific things to those nations which oppressed his chosen people. that god was only a father to one ethnic group of people (and not all peoples). And even then he was an abusive and egotistical father to his chosen. God was not the god of all nations but rather the conqueror of many nations. But that god can still be seen in the angry woman working 4 jobs in the hopes that her children will have a better life. Like the over extended mother the biblical god did the best they could with the resources they had.

In genesis a man named Abraham rapes his Egyptian slave Hagar in the hopes of obtaining an son. This Egyptian/African slave flees but is intercepted by "god" who (because she is not a person but instead the property of his(gods) elect) instructs Hagar to return to her rapist and the woman who continues to abuse her. This is not the god of all people and especially not the oppressed but simply the god of the Hebrews. (Who at times were abused too).

You can see how problematic this belief that we alone are god's people and/or that god is exclusively on our side truly is. Rather that be a nation under god (the US), the divine right to conquer and colonize the world (England), the mother land, or a tribe of ancient Hebrews in their crusade of cainan. This form of religious nationalism is dangerously close the cultural supremacy/prejudice towards other "inferior" people groups (racism).

Growth is critiquing the faulty models and abandoning what no longer works. This is the heart of philosophy.

There most certainly is something to paying homage to one's heritage but there are also things in every culture that need to be called out and abandoned. Every culture, legacy, patriarchy, monarchy, matriarchy, ect... is guilty of evil and oppression. And at the same time every culture has positive insights and adaptations on nature and their Corresponding environments (because after all culture is just man's attempts to mimic nature and their environment much like any creature or organism which adapts to fit/ blend in to its topographical environment)

This is gods house so god makes the rules. But god doesn't live here. No god lives in a mansion in the sky and uses the world as a rental property for those retched souls unlucky enough to not be gods themselves. None of us chose to be captives in god's rental property. And god has yet to show his face. In fact the only reason we have rules in the first place is due to here say, and he said she said of ancient men who claimed to be profits. God has yet to speak for themselves. This is particularly incriminating when tho we have yet to see gods true face, we have witnessed first hand the exploitative functions of these supposed rules.

If god was truly competent he wouldn't have to profit off of the exploitation of mire mortals and make man nothing in order for him to be enough yet alone a god in any capacity.

Moreover What good is a savior and a god if they

are indisposed when you need them? If they are not here to guide and or deliver those in parlous situations here and now. Then of what use is their authority? Gods, like the words of our ancestors, are projections of past experiences that may or may not be helpful and useful for environments which are very much changed from those of myth and mysticism. If god and Christ are not here to answer for their crimes, or to provide clear guidance, then their authority is arbitrary.

The evil isn't capitalism or communism, it's not conservative or progressive. It's a belief in hierarchy. It's a belief That one life is worth more or more valuable then any other. The evil presides in the belief that the many exist to serve the one or few. Gods and saviors are the problem, not the solution. For the one will inevitably become the other.

you may argue that all life is holy. And yet we squash bugs in the bathtub and treat the poor like garbage on the side of the road.

The argument is that somebody has to be in charge. But even on a biological level there isn't one supreme instigator or supervisor but instead many seemingly autonomous agents working in unison. Much of Behavior is both predisposed and subconsciously applied after it's context has been established. And that context is not governed by a master but instead by one's community. Tho it is lawmakers who vote on laws, it is the people (in the form of a jury of one's piers) who decide if the laws are applicable and/or enforced. No one is coming to judge the wicked and deliver the oppressed. It's up to us to do better. And to hold our modern day gods accountable!

(now for context the story of genesis accounts the Hebrews branching off from an established cainanite tribe and growing into a prosperous people in the utopian empire of Egypt. In genesis the Hebrews own and abuse Egyptians, but by exodus the Hebrews had been subjugated much like the bourgeoisie in russa during the rise of the Soviet Union, or the prosperous Jews who where vilified by the nazi party in Germany. Exodus is the story of former oppressors being liberated from their oppressors.) the problem isn't the Jew or the German, czar or the proletariat, But rather the belief that one race of man is superior to any other. We don't need gods that tell us to kill every man, woman, and child,. Nor do we need antics which permit the poor to eat the rich. A transfer of inequality is not a viable solution, nor (as we've seen) is it an ethical one. The only viable cure for inequality. Is equality for every man, woman, non-binary, and child, regardless of race, religion, or political disposition.

## STORY TIME

In the early 1900s Russia was among the most impoverished countries in Europe with a overwhelmingly vast majority of its populous being peasants. Tho it did have a minority of poor industrial workers for the most part the Russian people were impoverished and powerless under the ruling monarchy of the czar.

With the freedom that came from the abolition of serfdom in 1860 and the rapid influx of populous in

Russian centers such as St. Petersburg and Moscow doubling from 1890 - 1910. As well as many social changes brought about by the Industrial Revolution. A vast majority of the Russian people become prominently destitute. Wile the royal families continued to live in lavish excess and put their trust in mystic charlatans. After a great many failed conquest, wars, and the subsequent nation wide food shortages, unrest among the people was greatly intensifying. Many educated scholars noted how the monarchical rule was stifling the nations growth.

In 1905 the people revolted and many protesters were killed by czar soldiers in an event often referred to as Bloody Sunday massacre.

This event lead to many strikes among individual workers and farmers and contributed a great many supporters to the revolutionary cause.

During this time many work oriented councils were formed comprised of organized workers. These councils were called soviets.

To remedy this, czar Nicholas ll constructed representative comities to bring about reform and procure civil liberties for the Russian people. However after a great many defeats against the Germans in WW l.

Czar Nicholas ll left St. Petersburg to lead the Russian troops personally. During his absence his wife Czarina conspired with the mystic charlatan Rasputin disband those elected officials who advocated for the people.

This lead to the assassination of Grigory Rasputin in the winter of 1916.

Czar Nicholas ll contributed to this social unrest by repeatedly dissolving and undermining the council of representatives he had originally formed to advocate for the people.

After the end of WW I the Russian supply of fuel and food was greatly depleted. On march 8th 1917 the February Revolution transpired as many starving peasants took to the street and stole bread to eat. This was met with violent police brutality under the guise of the czar. By march 11th the military was employed to restrain the disgruntled people. (Note disparity in Russia in 1917 was not as bad as it is now in America (2023) and the events of the February Revolution were not as gruesome as those which transpired in Colorado during the Ludlow Massacre in 1914. Where a major iron and coal company CF&I employed the Colorado national guard to assassinate and rough up an encampment of 1,200 strikers and their families)

Shortly after the February Revolution czar Nicholas ll abdicated the throne after which Alexander Kerensky (a member of the provisional government committee strove for anti violent resolutions. However on November 7th 1917 (the October Revolution) transpired. Leftest revolutionaries led by the Bolshevik party and Vladimir Lenin orchestrated a virtually nonviolent coup.

Up until this point the ruling class was still primarily the bourgeoisie and not the working class people. This was what Lenin intended to remedy. And thus Lenin become the dictator of the first communist country.

Following this turn of events A civil war broke out with the red army fighting for Lenin and the Bolsheviks, and the white army fighting for monarchists and wealthy capitalists.

In July of 1918 the Romanovs were executed by the Bolsheviks. In the end the red army won and communism became the nation's economic system. However freedom of speech and expression was greatly prohibited under this new order. And form the internment work camps like the gulag archipelago social justice and liberty was never achieved. The tyranny of the czar was multiplied exponentially over by Lenin.

Should the oppressed become masters only to reciprocate the tyranny of their masters back onto their oppressor? No of corse not.

Around the same time similar events were transpiring in Germany.

In the year 1920 the right wing national socialist German workers party was formed. This party was constructed to combat the threat of communist parties arising in Germany after WW I. However this party soon became the nazi party. In 1930 the moral of the German populous was greatly diminished by the humiliation of their losses in WW I and the profound economic casualties of the Great Depression, leavening millions without any prospect of hope. This humiliation, destitution, and disapproval of the government's stewardship cultivated a great deal of anger and fear among the German people. The nazi party and the charismatic Adolf hitler promised the people a return to their once great countries traditional values and station as a world power once again.

However in order for this Conservative Party to be the true saviors of this once great nation, they must first find the libertarian culprits who caused Germany to befall such hard times. Of the German population Jews made up about 1%, however among the dense corporate networks of German economists 16% were Jewish. In addition to this, approximately a quarter of this corporate networks higher strata (elites) were Jewish. When coupled with access to education and a proclivity for networking among their familial piers. The comparative success of the Jewish population made them an ideal target (scapegoat) for the nazi party. This scapegoat tactic is pivotal to such movements because not only does it deflect blame away from those in positions of power (the culprits) but it also undermines the greatest tool for combating exploitation and abuse (cooperation,unity, calibration,) by causing division among the victims.

In the year 1896 a group of anti industrial nomads called The wandervogel formed in Germany. This group of fundamentalists Nationalistic hippies who opposed the industrialization and bureaucracy of the German nation (particularly the exploitation of people and the earth by the elitists) were 80,000 strong in their prime. However with the exception of the Jung-Wandervogel, the majority of this movement aligned itself with the nazi movement (primarily because all youth organizations were disbanded/ outlawed in favor of hitler youth). The point is that much like fundamentalist Christians today and even the marginalized communities (the later of which have been exploited and abused) they too vilified their oppressors. However with exception of the Jung-wandervogel. This movement was more similar to the fundamentalist Christians who seek to reestablish traditional values and make the nation a "Christian nation". However much like the Egyptians who enslaved the prosperous Hebrews, the nazi party made the predominantly elitist Jews their scapegoat. (note Tho Caucasian men are not a minority like those of Jewish descent in Germany were, in America, 88% of the top 1% in America are white.) in ether cases it is important to make distinctions between the identities affiliated with the "oppressors" and the "oppressors themselves. Because much like Soviet Russia who vilified the bourgeoisie, the nazi party vilified the Jew for their association with the rich. And rather it is the fundamentalist Christian who vilifies the progressive agnostics, or the agnostics/atheists who have been hurt by the Christian nationalists politics. Ether party is vulnerable to becoming the tyrant. And thus We wage war on the ideologies and faulty models, not the people who identify with them.

One of the first signs of tyranny is the deliberate burning/banning of historical books which present opposing rhetoric from that of the ruling party. It happened in Russia prior to the rise of the USSR, it happened in Germany prior to the rise of the nazi party, and it happened in Rome under the guise of the Christian movement who sought to snuff out and destroy gnostic texts. And sadly it is happening today (2023) in legislature passed by the Florida governor prohibiting access to history and science books from students. Freedom of thought and (non violent) expression are of the most fundamental rights of an autonomous agent and access to information (particularly that which contradict the ruling class or party) is the most basic means of growth and prosperity. And thus it is the first freedom obstructed by those who seek to abuse their authority. When the governing authorities are more concerned with regulating freedom of expression then they are about ensuring that it's citizens (the people they are sworn to serve) have access to affordable housing, food, and healthcare, then the goal is most certainly to propagate tyranny.

"Under the capitalistic system the fruits of those who labor are enjoyed by somebody else" -kwame-

"It was no mistake that Adam and Eve were put in the harden to work it, but were forbidden form eating the fruits of their toils, which belonged to the gods".

A tactic we see implemented by modern right ring republicans and the proponents of the nazi regime on the 1930s is a promise to deliver people front the problem they (the supposed savior) caused in the first place. They also often vilify those who are breaking free from their narrative of tyranny so that the populous sides with the oppressor rather then the oppressed.

A tactic we see in modern left wing democrats and the proletariat of the USSR is the the implementation of intentionally ineffective policies as an attempt to invalidate or intentionally undermine viable solutions. In short they emplacement the solution so poorly that it causes more problems, and they intentionally make these solutions ineffectual so that they will be abandoned as failures. Because rather it's the far right or the far left. In most cases they are both being funded by those who deliberately profit off of the exploitation of the people.

The day of the lord as it is described in the profits is a day when the oppressor is crushed and the oppressed are liberated. And yet so much of scripture is used to keep the oppressed submissive to the tyranny of god. In fact scriptures like Daniel 2:21 and Roman's 13:1-2 have been sited to affirm the continual injustice perpetrated under the rule of "gods elect". This rhetoric has been used to justify oppression because according to those religious authorities, the monarchy, and patriarchy was appointed by god. And is then permitted to abuse those who were not born of gods chosen lineage or royal line.

Tho the events accounted in exodus are purely rhetoric in response to the Egyptian myths. (For there is no credible evidence for the events it accounts). It does depict a very similar account to that of many failed utopias in history. From the peasant run USSR to the third reich. Egypt in exodus both in description and in narrative form functions as a potential Eden which prospered because of a Hebrew named Joseph. We can see that Egypt is an Eden garden thru the imagery of the two rivers flowing out of it (just like Eden.) as well as the garden/mountain parallels.

My advice to my reader is to nether seek a savior or god to deliver or rule over you, nor to seek to be a savior or god over others. That one may do good not in order to attribute value to one's own life but rather to attribute value to the life of another. To bless those who have befallen misfortune, should not be for one's own glory but for the glory of those who by very right of existence are deserving of glory and respect. Not That you should feed the hungry because it is your duty or out of guilt but because you too know what it is to be hungry and cold. For god has witheld blessing out of a desire for glory for himself. But as for you, may you do good and bless others not for your glory but for the glory of those who have been denied it by god. God who has all good things, requires that others suffer so that god may be glorified. But because man knows what it is to suffer and what it is to have not, may man be merciful and generous and not seek glory as god does.

The faults of Feudalism led to its demise giving way to capitalism, socialism, and communism, but these systems too have proven to have their fault. But we've also had our triumphs. for thousands of years slavery was as fundamental as electricity today, and the fact that for the most part such an integral part of ancient and post modern economy such as slavery has been abolished. However with artificial intelligence within only a couple of decades (if that) away from replacing all intellectual and remedial labor, unless we solve the economic constraints which cause disparity and the abuses of fundamental rights warranted to wage labors, then a vast majority of the earths population will be kicked to the curb with no prospects for shelter, medical needs, and perhaps even food.

The exodus is a narrative about immigrants in a position of privilege, power and property who are then vilified by the natives of an utopian empire. Much like America did to the black community in the 1930s (redlining) and nazi Germany did to the Jews in the 1930s. Ultimately These are stories about the people who built these empires and made it prosperous then being crushed under the malevolence of its gods. And that is just as relevant today as it was 100 years ago.

This is why The god of the Hebrew Bible often subverts the traditional and institutional power structures in favor of those who have been exploited.

Much of the rhetoric in the Old Testament is about subverting the institutional and traditional standards of hierarchical order. The second born ruling over the traditional fist born. This also attempts to explain why some have power and others are left to be exploited and oppressed. Because god assigned it thus.

However Revolutions become governments with oppressive economic and political systems, revelations often become religions which are just as flawed as those they deconstructed. and tradition is the result of subversive thinking.

However Rather it's the rich or the poor, it is immoral to punish the children for the shortcomings of their parents. Even Ezekiel critiques god when he claims "god does not punish children for the sins of their father". Because he is deliberately calling out rhetoric from earlier traditions like exodus where god kills the first born of Egypt.

The Hebrews weren't white, but their god was still racist. There was no such thing as capitalism back then, but the poor were still exploited. And that anger felt by so many for the injustices perpetrated under the guise of the Christian nationalistic god is the same anger felt by the Hebrews who criticized and accused the Egyptian and Babylonian gods of malevolence and evil.

The Americans won the revolutionary war because England was fighting on every front. The whole world was tired of Englands oppression. And in the same way this country is prosperous prominently because we have profited off of the exploitation of the poor and marginalized peoples in India and China. Americans aren't the saviors, more often then not we resembled the malevolent gods like those of ancient Egypt.

And sadly The god of the Bible would have probably been on the side of the terrorists in the war in Afghanistan. The America idealized by Christian nationalists resembles more the Babylon and Egypt then It does the nomadic Israelites. Long before the tragic events of September 11th American agencies were Funding both sides of the war to exploit the middle eastern land, resources, and people. (See Operation Cyclone). this CIA operation was instituted to secure oil reserves via external control of the middle eastern governments. In 1953 the US and England funded a group of rebels to overthrow Iran's nationalist leader, Mohammed Mossadegh, who was planning to nationalize oil resources. By 1955 the US was receiving 50% of its oil from the Middle East, and Europe was pumping 90% of its oil from the Middle East. A great many covert operations funded by the US took place in the 1970s and 80s eventually leading up the Iraqi being indebted to the

US with only one real resource to pay the US government back with (oil). Like Christianity's concept of sin, the debt owed to the savior (the US government) was one of deliverance from a problem that same savior caused in the first place. "Deliverance from a situation the savior put you in is extortion" This exploitation of power was one of the major contributing factors which led to the events of September 11th 2001.

"If might and fright are your two biggest allies, then your probably not right.

One thing we must remember however, is that more often then not, the majority of those who profit from abusive economic systems, were themselves at one point exploited by those same systems. "They did their time" in a sense. But there are just as many who "did their time" and never profited from those systems. The point can't be to punish those in power. Those who did not choose the systems of power, nor have control over those systems. The point is equality which means "the poor eating the rich" is no more of a solution then "the rich starving the poor". We change the systems that abuse people, we don't abuse people who were changed by those systems. We most definitely make the exploitation and dehumanization of people groups "unprofitable and even unfeasible. We abolish laws like vagrancy and lottering which criminalize homelessness. We also work towards setting limits on a corporations ability to monopolize the housing market. We set caps on wealth and we regulate livable wages and working conditions in favors of those who produce the goods and not those who "supervise" like slave drivers. But we don't punish people who conformed to survive in their environment. We do better then those who came before! Because we can.

## MORALITY

it is the admirable qualities that make the inadequate ones abominable. For instance It was the discipline that made the nazis prejudice effective. But perhaps it was that very prejudice that lead to its downfall. perhaps even, morality itself is not simply moderation. but rather that virtuosity is the most comprehensively efficient approach. And Perhaps morality and virtuosity are the most effective approach because they deal with life holistically rather then dualistically. They do not attempt to change something with the soul purpose of exploiting it, but instead they develop the capacity to appreciate it as it is, simply because it is. And so contrary to popular belief, a god is not required to retain a semblance of morality. Nor are those divisions prescribed by "gods". We can recognize that life requires death in order to persist, and yet at the same time not cause unnecessary suffering on others simply because they belong to our "race", species", nationality, traditions, definition for sentients etc...

morality quit simply can be expressed as having compassion for all life regardless of what shape it takes. Which means, we don't need a common set of beliefs in some kind of cosmic hierarchy in order for morality to persist.

Children who grow up in abusive or unstable households often develop an hyper attuned attentiveness to other peoples emotions as adults. This stems from a hyper vigilant need to monitor one's ever changing environment for potential threats or dangers. In the same way an obsessive need to know everything is a response to being taken advantage of by caregivers. It's an attempt to control one's environment. Sadly many people who have been abused or exploited are completely unaware that fact. And so when they act out the behaviors they learned to survive or were taught to emulate by their abusive caregivers, they in turn become people who (unknowingly exploit and abuse their subordinates. People aren't intentionally evil. More often then not evil is the result of ignorance. And in the same way tho oppressed don't realize they are being abused, often the oppressors too are completely unaware of their own malevolence. The narcissist doesn't intend on hurting others with their behaviors, they just never learned how to regulate their own emotional needs and identities. Rather it is the manipulative behaviors modern man/woman, or ancient single celled organisms developing aggressive mechanisms to survive hostile environments, life does what it has to too survive. And often that comes at the expense of other forms of life. This is why we can't just act out the stories of old, but instead must find better ways of interacting with our often hostile environments. In order to do good, we must first know better!

## RABBI, TEACHER

A good teacher doesn't test and then punish their students for not knowing what they don't know.

A good teacher communicates clearly with their students in order to help them learn to solve the conundrum at hand. Or in other words A good teacher teaches their students how to communicate effectively with themselves and their environments.

They don't demoralize their students or make them feel stupid, their job is to teach their students to appreciate what they too once did not understand.

A good teacher will subvert their authority onto their students, because their students will know what the teacher did.

A good teacher doesn't make their students feel small, for their job is to help them grow. A good teacher understands that the test is a trial, but it is not the student who is on trial, but rather the teacher.

Contrary to biblical precepts, the rod(trauma) is a bad teacher. Physical and emotional punishment which utilize (negatively reenforcing emotions) are explicit mechanism of our limbic system. and thus they disengages our cognitively rational PFC. Trauma acts habitually (bottom up) where as our PFC strives to regulate the body according to our internal models (top down). Because I said so "kinds of authority" are ultimately more harmful than good, because they perpetuate abusive behaviors rather than exercising cognitive mechanisms vital for volition.

When our PFC is bypassed, we don't learn, we react. It's the difference between making one's self big/ small when presented with potential danger (reacting out of fear), and implementing cognitive processes which strive to understand the threat (AE the bee doesn't sting because it's mean, it stings because it's frightened, and that's something I can relate to. The bee doesn't know any better, but I do). But right there is the point. Because for many, what they call god is actually just a trauma response. It's a conditioned reaction to the threat of negative stimuli (eternal hell fire), and not a cognitive appraisal of values and conditions. (Note this is often reinforced by associating positive stimuli (evoking feelings of security and relief when the hand feeds rather then physically or emotionally beats) with god.

"Rather it is the pain of a past experience or excitement for future plans, talking about the experience dissipates both the negative and positive energy driving us"

Our enemy is not of flesh and blood, but ideological by nature! People are not problems, the systems and ideologies which programed them are. But perhaps we must ask " is it truly more humane to see people as machines as opposed to imposing some form of ethereal quality which is only valid so long as the expression of life fits a certain criteria! AE looks, thinks, talks, and/ or behaves like us and ours! Honestly, if seeing people as complex mechanisms which are heavily influenced by their environments, then help us resolve many of those same environmental strains which cause so much pain and suffering (pain and suffering here referring to biological systems checks and value judgment which's sole intent is preserving, propagating, and obtaining homeostatic states and conducive circumstances for the organism in question). Then yes! I do believe such a frame work is more humane than one which (in many ways) has just exasperated that suffering rather then attempt to solve its underlying causes.

For most people, we abstain from murdering other people, not because god said not to. But because we have enough respect for the life of another person.

Similarly we abstain from practice like slavery and racism despite the fact that both gods and Jesus not only condoned them but in some cases commanded and even themselves practiced those forms of prejudice on those of different ethnic cultures. It was god who required human sacrifice and instructed that men are to physically (if not also emotionally) abuse (beat) their children and wives ("Spare not the rod"). God supposedly impregnated a teenager (Mary) and yet descent human beings ask for consent, and only so long as both parties are of consenting age. Much of the rhetoric in the Bible eludes to the facts that god didn't see women and children as people and rather simply as property. And people of foreign cultures (or even one's own tribe) are considered, no more then cattle or livestock. (Which makes sense because the same people who claim to be (pro life) are currently pushing legislature which will institute child labor in America). If god is god, then shouldn't we hold them to a higher standard? Do you expect more from a child then you do from an grown fully developed adult or guardian? And yet we find not only is god not required for morality, but 7 out of 10 times, the word of god is actually the obstruction to that morality. "Traditional values" were never about ethics or morality, this is a good father who couldn't even bother to pick up the phone when we called. People leave religion and god, not because they want to "sin" but because that god never really wanted a relationship but instead themselves was a tactic used to control, exploit, and oppress, people.

## IN THE NAME OF A HUMANE GOD

Why is god in accordance with scripture so egotistically small that "he" has to take credit for all of our triumphs, hard work, and sacrifice, and yet leaves us with all the blame for the bad? Does a good loving parent not take responsibility for their own shortcomings in teaching their children and shower their children with praise and adoration when they do something good or monumental? Why does the god of the Bible hate humanity for looking like "him" and having autonomy? Why does the god of the Bible resemble more a abusive parent and poorly adjusted individual with serious mental health issues and narcissistic tendencies, then they do any vague resemblance to that of a well adjusted and supportive parent and person. If god cares so much about what others think or believe about "him" then why is god so distant and illusive? Why is "he" an absentee father? The most logical answer to these questions is that the god of the Bible is modeled after primitive parental figures who would have understandably been idolized by their children or progenitors as patriarchal figure heads in ancient societies or tribes. But many of these ancient figure heads were not equipped for dealing with the dramas of their parents yet alone their own or children's. And we see this in Egyptian depictions of RA and summeraian representations of Gilgamesh as they wrestle with their mortality and decline of health. As well as in zuse who (like his father before him) fears the loss of his omnipotent authority to his sons autonomy. So of course we see this in an entity which was derived from the Canaanite El and then later conflated with adony (the ancient one) and Bal (the lord or master). In fact (much like the early Israelite tradition AE modern Judaism) Christianity was constructed by critiquing the rhetoric of Greco Roman traditions.

## THE TRUTH ABOUT SCRIPTURE

The good news was that we didn't need temples and traditions to commune with god, because the ground on which we stood was holy. Or we were that holy ground/temple.

god doesn't look like the malevolent pharaohs and Caesars but was more adequately represented in the convictions of the least of these.

And that we don't need sacrifices to validate our existence.

Unfortunately Many People would rather be slaves in Egypt, then wrestle with the ineffable in the wilderness.

Religion doesn't care about the nature of god, because it is too busy becoming one!

There is most certainly profoundly prolific wisdom in this ancient literature which gives very good examples of gods character, but there are just as many bad ones in this collection of texts we call the Bible. these scriptures are far from inerrant and in most cases they make for a small inadequate god.

In many ways, rather it is psychopathic behaviors or masochistic one's, we are a product of our environment and it's conditioning. What we once called spirit has a very physiological correlate. And tho it is those who fall prostrate before a god, simply because they are powerful. That then substantiate might making right. It is that same wiring which permitted them to survive their environment. The point isn't to shame or berate those who are intolerant of change. But rather to help them develop the mechanisms to grow beyond the transcendent tribulations which posited those traumatic behaviors. Which made it necessary to tremble in fear, and call it worship.

> "The obedient always think of themselves
> as virtuous rather then cowardly"
>
> -Robert Anton Wilson-

Romas 9:17-23 NIV

("For Scripture says to Pharaoh: "I raised you up for this very purpose, that I might display my power in you and that my name might be proclaimed in all the earth." Therefore God has mercy on whom he wants to have mercy, and he hardens whom he wants to harden.

One of you will say to me: "Then why does God still blame us? For who is able to resist his will?" But who are you, a human being, to talk back to God? "Shall what is formed say to the one who formed it, 'Why did you make me like this?'" Does not the potter have the right to make out of the same lump of clay some pottery for special purposes and some for common use?

What if God, although choosing to show his wrath and make his power known, bore with great patience the objects of his wrath—prepared for destruction? What if he did this to make the riches of his glory known to the objects of his mercy, whom he prepared in advance for glory")

This is a god of "might makes right" a god no better, and perhaps even worse then the moral standard presented by the nazi party in Germany. A god who explicitly causes suffering and violates their "creations" volition (the supposed gift of free will given by god which god then deliberately subverts) all so that they (god) may be glorified in their triumph over those they (god) created, enslaved, and abused. This is not only a god who forces people to exist (like a form of cosmological slavery under their tyrannical rule), but who also explicitly inflicts those slaves with suffering, all so that they (god) may be glorified! And this is the god of the New Testament, the one exemplified by Jesus the Christ. Who is a savior who saves the retched human from god themself (who is both savior and oppressor from which we require salvation from). This is a conqueror who causes the problem all so that they may be the one to ameliorate it. The worship of this kind of god is a trauma response. It is a reaction to fear or (the rod). And it is not rooted in love, morality, or mercy. thus it makes perfect sense why it would attempt to completely undermine more cognitive mechanisms integral to autonomy and volition. The form of worship described in this passage depends on activation of our amygdala and limbic system. It's fear induced and abusive. And the fact that it ever passed for love or mercy, is incredibly frightening indeed.

This is a god so insecure in themself that they not only require that humans believe in them (to validate gods ego), but who also explicitly hurt those "inferior" to them so that they (god) can feel powerful. A god so insecure that they are offended by anyone who finds value in life out side of/ apart from them (god). A god so small that they must demoralize their creation to the point of "retch" all so that they (god) can be worthy of worship. This is a sad god who is incapable of truly loving themselves, much less anyone els for that mater. This is a god who explicitly undermines humanities autonomy all so that they (god) may feel mighty. This god can't even tolerate their own image staring back at them in the form of their creations. And the sad part is. This is a god who hates themselves so much, that they require the love and adoration of others to make up for their own insecurities.and it's all because this god is grounded in shame and fear, rather then love and integrity/growth.

My dear reader my goal is not to make you hate what we once called god, or the people they are modeled after.

But rather to help you love yourself and others better then those gods were able to. I point out the toxic ideologies and beliefs, in the hopes that you may not propagate them but instead choose healthier exhibitions. I want you to love yourself, so that you don't have to hate everyone els.

It is not enough to know what to do, we must also know why we do what we do. "Because I said so" just doesn't hold up.

If there is a god, should they not be held to a higher standard than us fallible humans? And if that god then (with limitless resources and power) created a world of scarcity where one must ether kill, cheat, or steal, in order to survive. How can that god be both good and powerful. It is for that reason we must choose one or the other. Ether god is good but incompetent. Or god is powerful and malevolent. In truth as we become more capable of fulfilling our own needs, the less we have a need to cause unnecessary suffering to others. It's still a compromise, because we are not all powerful and thus we must still kill, in order to live. But we can do so more modestly and compassionately, making sure not to cause more unnecessary suffering then we must.

When sanctity requires scarcity, the holy are exalted wile humanity starves.

## THE AWAKENING/ OUR FALL FROM GRACE AND SALVATION IS ONE IN THE SAME

What happened when man fell from grace? When we as a species become conscious and self aware, capable of dictating our own nature and habits as opposed to mindlessly obeying our priory programming or instincts? It was here that we became like god, able to discern between good and bad. And so god cursed us! From that moment on we were cursed with suffering the futility of our labors! We were aware of our pain on an existential level. Rather that was the labor pains of child birth or the rigorously monotony of toiling the ground, (both of which are done to benefit an economic system and not the individual human). from that moment on we were aware of our emotions. And so of corse we attributed our anger and jealousy to some unseen entity. Because from this perspective god is not only conscious but consciousness itself. I realize that sounds out there or far fetched. And yet it makes the most sense in its compliance with everything we know about god. It was only after we became aware, that we then felt "shame" or perhaps more accurately "vulnerable" because of our nakedness. We were exposed to more than our consciousness of suffering our labors, but more profoundly our own existence. It was consciousness which then made us like god. Animals survive, they don't ask why. But we did now. And we had no other choice but to search out answers and meaning for our habitual actions. The most logical conclusion then is that god is the name we have given our consciousness. However the interesting thing is, that we somehow see ourselves as separate from our awareness. For some reason we don't identify with our own reflection. But wait a minute, because didn't we just say a couple of chapters ago that god was nature or our unconscious instinct. And furthermore how are we defining consciousness as an entity? Consciousness is a verb not a noun, right? Actually consciousness is a state. And like water can be a gas/vapor, liquid, and a solid. The state is the shape it takes. Let's say consciousness is a liquid and our unconscious instinct or programming is a vapor. Both are water only in different forms. Now let's say the idea is waters solid form. All three function as a form of agency but in very different ways. The idea subjugates us to societies rigged amalgamation of nature (culture/religion), whereas the instinct subjugates us to the frivolous will of nature. It is only consciousness which allows us our own autonomy. The other two are predicated on an predisposed moral code, whereas the third option requires complex emotions like empathy and compassion to dictate the virtue of our actions, and thus it is not simply reactive. You see love is only possible by way of consciousness. The other two options are just products of a system of incentives. AE one's acceptance in a community and potential for propagating life/healthy offspring.

Before the fall, us and god were one. Much like the heavens and the earth prior to creation. The profound phenomenon which these mythic events depict, are quite simply the emergence of our consciousness from instinct. It is the dolphin and elephant recognizing its own reflection in a mirror as a projection of itself. In contrast to the bear or lion who sees the projection as another separate entity. It's the 2 month old human who has just begun to discern between themselves and their environment. This particular phenomenon extends even into adulthood as we slowly differentiate between our autonomy and the identities we've been given (VMPFC models).

I think deep down we are afraid that god is not good. because if we truly believed god was good, then we would be too. We wouldn't require a law to tell us how to treat ourselves and others. and we wouldn't be concerned with subscribing to an ideology that hates those who look and think differently than us. But perhaps deep down we realize that god is not required to be good. and so we live out of constant fear that we will be condemned for all of eternity for loving those individuals who have no place in our religious traditions. We cling to god like an conditioned child clings to an abusive parent, as we shun every person who's nationality, sexual orientation, and philosophical disposition conflicts with our small minded guardian. We make ourselves small so as not to offend our (god's) ego, and agree when that god puts us down. We are afraid that god doesn't love its creation, and so we hate it too (even loathing ourselves In Compliance with our malevolent master). But if we truly believed god was good, then we would not let our religions cause so much division. We would instead error on the side of compassion and act in love instead of bigotry and prejudice. Perhaps We are afraid that god hates us because we look exactly like them. and so like our small malevolent god, we too hate ourselves. I think we are afraid god is not good. because if we did believe in a loving father and compassionate mother of nature, then we would have no reservations for our capacity for tolerance and inclusion. ultimately we look like the god we believe in. For after all We bare their image! And the image we bare is the only caporal representation of that god yet witnessed.

"In 6,000 years god has yet to lift a finger in aid of those hurting. People do good, but some people do good in gods name! People love, but some people love in gods name! God can nether act nor love without the aid of people!"

Which brings us to our next point. If the ancient gods are anything like our modern billionaire equivalent, then in some cases they were not necessarily evil, and even probably even good at times. If you've ever met a billionaire, you can probably attest to the fact that they can be rather chill and even quite generous at times. In truth it's not necessarily the ancient gods or the modern billionaires which are the culprits for such disparity and inequality but rather the religious zealots and millionaire slave drivers who ensure the hierarchy is so fervently enforced. In many cases it is those who grew up poor and so they live to serve their gods and masters who put them in positions of authority. It is often not the gods and 1% that are greedy and callous but their servants who use their power to ensure that they themselves are distinguishable from their "inferiors". Tho this is not always the case, we can't ignore the fact that in some cases those in positions of supreme power are in fact good and generous stewards.

Note (In the case of lochhomesyndrome or why we form attachments to negative reinforcement AE abusive spouses, paternal figures, and friends, in many cases it has to do with a lack of glucocorticoid production at infancy (which is completely natural) and an exposure to negative reinforcement or abusive behavior. Ultimately our neuro chemistry at infancy is designed to form an attachment with a care giver. Regardless of the quality of that care). The rule of thumb is any shelter from the storm is better then none at all. But that's not really true is it? A little bit of rain is better for you then a den of venomous serpents.

## DEFINING THE INEFFABLE

Prior to the Greco Roman period, deity worship was resigned to its nation or land. We see this demonstrated when naaman brings cartloads of Israel soil with him in his exile, in 2 kings 5:17. Or also when David in 1 Samuel 26:19. Accuses his god of "forcing him to worship other gods" in his exile from Israel. This may explain why Jesus ignored and dehumanized the Canaanite woman in Matthew 15:21-28. Because this period was the transition from the worship of deities pertaining to nationality AE the god of the Cainanites or the Hebrews. We watch the evolution of these myths, cosmological identities, and conceptions of god, take place slowly as culture and society developed to be more inclusive and comprehensive. (See Deuteronomy 4:19.) This is important because we still do this today.

What we once called "spirit" we now call our psyche or nervous system. and things like sentiment and emotion or even morality and beauty, we now understand as incentives of these mechanisms. We see these mechanisms in nature too, thru the inheritance of certain traits and connectivity of plants and flora, we find this wealth of data posited in nature, and thus it's no surprise that what we once called "gods" or "posited mental agents", we now know as very natural phenomena of natures connectivity. we understand these phenomena as evolutionary process and biological prerogatives. However The truth is we are still witnessing the evolution of these cosmological claims and conceptions. Of corse there is still an enormous amount of aversion to these discoveries and shifts in paradigm, because they threaten our mental maps of reality. Tho according to these new definitions, the question arises, is nature conscious and compassionate? We understand that we are made in god's image or that thru evolutionary process we represent nature. But does god or nature look like us? Is nature conscious and caring? Does god intervene and orchestrate events in our lives in ways aside from simply incentivizing certain traits thru our genetic coding? Does god hear our prayers or are our mindful meditations only useful in the sense that they allow us access to the data hidden in our subconscious? Are we truly alone, or are we connected on a much bigger scale? But perhaps the most daunting question to ask is, what happens after we die? Does our consciousness continue or does our psyche cease to exist? These are the difficult quandaries we are presented with when we find the answers we have been looking for, such as what is god and spirit, and where did it all come from? The truth is we would probably much rather still be standing on the flat plane of a one dimensional earth at the center of the cosmos, as opposed to drowning in the depth of the worlds dimensions insignificantly a drift in an endless universe. The truth is not comforting or easy. And so Perhaps for some, the question isn't what happens after death, but how do I find meanings here and now!

Because in many cases we probably won't like the answers we find. That's why it's important to only ask the questions that can positively affect our lives. However for those brave souls who still dare to be inquisitive and open minded, then I encourage you to continue reading, because all knowledge comes with a warning sign which doubles as an invitation to grow.

So what is nature? or more precisely what is the nature of nature? and ultimately what is the primary prerogative of the universe, what posited that programming.

Was there a voice calling us out of the depths and abyss. One who delivered us from the primordial serpents of the chaos waters and onto the dry land. Is there still a calling among the celestial lights and astral bodies of the heavens? An echo From before time and memorial and even into eternity! For with this awakening we ceased to be slaves to the will of the gods and were subsequently exposed to the dread of our existential existence.

From Rome to England and even into the modern age America, empire has sought to profit from unity but not equality. Empire sought to exploit the abundant natural wealth of the weak and peaceful. Empire sought unity at the expense of liberty and justice. Empire imposed order by separating us from them. Where chaos saw belonging in being, order required identity for belonging. To chaos, even life and death are not so different. For all that is living must die, and all life is only persisted and possible thru the reallocation of resources achieved thru death.

What we are finding is we are not so different from all other life on this planet. We now know that plants talk audibly (tho at a frequency un detectable to the human ear). They also share resources thru their roots with other plants. Wales are conscious and are capable of logic. And even simple life like slime molds are capable of mapping out paths as efficiently as human civil engineers planing roadways and transit systems. Plants feel pain to some degree, and the more we learn the more we see we are not so very different form our very distant relations the trees.

## THE PROFIT AND SAGES

Profits and sages may be equated to modern day artist, in the sense that they both act as vessels for nature, or the divine, to articulate itself thru.

In much the same way the artist studies their craft and practices with their precise tools, so too the sage studies the ancient texts for their prolific wisdom from the past, and practices their meditative rituals which mimic those of the instinctual creatures who still commune with nature on the daily.

In each case both the artist and the sage await, patent and ready for the muse to grace them with their presence. For this practice pays homage to the ways in which this prolific muse has moved in the past, and thus it is observed in the hopes that like a dancer to music, the artist and sage may be able to move as the spirit does.

In truth this prolific spirit and muse may very well be no more then our tried and true genes, synchronizing with the rhythmic cycles of nature and life in their respective seasons. And thus we study the sacred texts of old so as to be familiar with the corresponding signs of seasons coming and going.

Note

Recitation of a traditional/familiar prayer, song, or verse of scripture or litergy activates mesolybic dopamenergic systems.

## THE INVISIBLE EGO AND REVOLUTION

The question I shall now ask my reader to wrestle with, is How did religion ever convince us that there was a omnipotent cosmic being who is just as egotistical as us (if not more so), and yet everything we know about this being comes third and fourth hand (because this cosmic force has never shown themselves). How did those religions then convince us to make ourselves nothing in order to praise and exalt this omnipotent being, which apparently is only big enough when we've been convinced that we have no right to exist independent of it? How were we all fooled into believing that pride was evil, or that our existence came with an insurmountable cosmic debt, and further more that an omnipotent god would have the need for an ego so big that there is no room for our lives to be valid in themselves apart from that god? Why is such a big and loving god so insecure and dependent upon us to validate its existence thru our praise and worship? and yet it can't even be bothered with answering those of us who are hurting and just need to be seen!

These questions are the kind which often lead people

to questioning their faith and in many cases leave their religions entirely.

However this deconstruction of religion and faith is not some modern movement.

we have seen this phenomena many times before. And it occurs particularly when the traditional ways of ordering society have failed, and an influx in new information becomes available. roughly 132 years after the Maccabean revolt we see this very same "enlightenment" occur with the infrastructure of roads being built and diverse communities being required to coexist under Roman rule. we see this movement of people leaving their religions, as people are no longer subject to the echo chamber of their culture or tribes. And we see the same thing happening today as information becomes more readily available, people are abandoning their political parties and traditions which had proven to be corrupt and misleading, we see people leaving their faiths and pursuing truth. however unlike this movement in Rome which would later become "the way", people to day are not forming communities, but instead isolating themselves. we even see this mythicized on tv and movies. It should also be noted that the events that followed this religious upheaval in the past resulted in social reform. but only after great persecution by the religious and political authorities. In fact it was not until Constantine realized Rome could profit from this movement if properly manipulated, that this movement became accepted by social authorities. (This is where we see the rather socialist movement of "the way" become the nationalistically capitalist version of Christianity we see today. We inherited this version of Christianity from our ancestral England and Rome/ Constantine before them). and so as has always been the case, those who have profited from these systems like the churches and business tycoons/politicians, will inevitably fight to retain their malevolent hold on power. For this is the way of revelation. And there lies one of the profound testaments we can observe from Jesus's teachings and the early Christian movement. Because Rome was powerful and thus the revolution was not a physically violent revolt but instead an economic one. The masses didn't have to prove they were right thru killing and bloodshed/ destruction of property. But rather thru a change of values and an unwavering insistence on holding their political and religious "authorities" responsible for the evils they had perpetrated. Thru boycotts and unconventional communities which deliberately opposed the values and standards erected by Rome to exploit and segregate its citizens. Very similar to those living in vans and quitting their 9-5s today. But the only reason they succeed was because despite their differences, they joined forces and started United. The top 1% have already United against the bottom 99%. And unless we start forming communities and resolving our differences, we will lose this round.

## (Economic rebellion)

In fact this (apocalyptic movement of the way's renegotiation of cultural rhetoric) is similar to what we see being done with the buying of verifications on a popular social media outlet. These people who are exposing corporations thru militias slander which exposes the corporations for their malevolent behavior. This most certainly is A form of economic warfare or terrorism much like the Boston tea party which undermined the freverisoity of the colonizers economic system, AE credit and stock. Despite the power money has, it ultimately has no real substance Not even the US dollar is fixed but is instead subject to reevaluation. Now this kind of economic warfare would be despicable if not for the fact that these corporations have exploited the market, their employees, and their costumers for years. when we stop to consider that 53% of the increases in the expenses of everything (inflation) is accounted for by corporate profit. Multi billion dollar "gods" who don't even acknowledge their employees as human. Then this form of economic warfare is perhaps justified.

The kingdom of god or heaven in the Christian/ catholic nationalistic traditions functioned as an incentive for the working class to sacrifice their lives to a cause or to give their lives in hard labor for a better world or future. The problem is that that world requires a slave class and so their can not be a heaven if there are still gods. In this regards The cross is a interesting symbol of victory because it carries the claim that Romes worst punishment does not scare the people any more. and thus instead of being a deterrent, it is a battle cry. it is like a Jew wearing the star which marked their ancestors for the gas chamber as a badge of honor and a declaration that the tyranny of the gods and wrath of hell no longer scares them. It is a symbol of revolution and rebellion. And a declaration that might does not makes right. Unfortunately this symbol of a cross has come to mean that those very same oppressed and exploited people have then sold their soul to another god, in the form of a doctrine. Returning to our original point (economic warfare). The point is not taking the malevolent masters power, but instead rejecting their omnipotent authority, AE quiet quitting, strikes, voting out career politicians who are owned by lobbyists, peacefully fighting for fair rages and healthy work environments for all, and boycotting those corporations which continue to abuse their authority. It's living in vans to protest the vile corporate monopolies on housing.

Now some may say this sounds a little excessive until you realize how much Our capitalistic economic system has been deliberately used to predate the vast majority of the populous in a time when all basic needs should be accessible. Wile churches and massive corporations compound hundreds of billions of dollars just too keep the vast majority desperate, the other 99% of the populous are struggling just to survive. The 4 trillion $ of company profits spent on burning stock rather then compensating the wages of those who actually produced the goods and services is by all means theft and should not be allowed to continue! In truth this $50 trillion awarded to the top 1% for the labor done by the bottom 90% is theft!

The few rule, only so long as the many remain divided!

There is hope my dear reader But unfortunately as the myths show us, we would often much rather be slaves in Egypt then wrestle with the ineffability of life and liberty.)

> "Culture is a conditioning to act as a Mechine"
>
> (Karl Marx)

Culture teach us how to be mechanisms for production.

"It's not that You don't believe in god because of a lack of evidence. You just hate the god you find!"

Now all the evidence disproving the existence of god aside (because as we've already seen there's quite a bit of compelling arguments and truth to that) of corse we reject a god who is xenophobic, sexist, raciest, and malevolently abusive. And we refuse to look like that kind of god. In the same way Abraham left his tribe and traditions(the cosmological claims of the gods of his father and heritage, for more complete and compassionate conceptions of the divine(one which did not require human sacrifices like those of his fathers)) or how the New Testament writes whom never actually knew this rabbi named Jesus then attempted to rationalize this movement of people leaving their religions and sacrificial rituals all together. Because that's the thing. these ideological writings don't actually seek truth but rather some semblance of retaining congruity with their earlier traditions (similarly to how those ancient cosmological claims of deity in the cainanite cultures were carried over into the Jewish liturgy). we see these movements all throughout scripture and history of people rejecting their reductive religions for more accurate and affectionate conceptions of the divine. But doing so with the understanding that there is wisdom in the earlier traditions. Thus like those before us we should attempt to retain some semblance of our earlier traditions and heritage. yes we reject much of their claims about god! And tho it is not in fact due to a lack of evidence to the contrary, but rather a more robust moral conviction that even if we are wrong, might does not make right.

Rather we mean to or not, We make ourselves in the image of our gods.

Many times doing the right thing, the good thing, the just and compassionate thing, requires breaking the rules and even being the villain of traditions story.

If the only reason you abstain from murder, rape, and thievery, is out of fear of punishment and not compassion for your fellow man, then it is probably best that you remain a slave to the law. However true morality has nothing to do with obedience but with love.

## LET THERE BE LIGHT, THE CONSCIOUS MIND

In the very beginning of the Bible we get this ancient poem accounting creation in the form of 7 acts, movements, or days. on the first day we are told god created light. Now in its original Hebrew this movement or 3rd verse of the poem reads (Elohim way-yo-mer o-vr o-vr) this phrase is often translated as "god said let there be light and there was light" however because of the ambiguous structure of this ancient language, another accurate translation may be "god became light" or (god spoke and become conscious). This is fitting because the sun and stars are not created until the fourth day (three days after the creation of light) and thus this "light" in verse 3 has no source as a physical ray or wave of light. Many skeptics point this contradiction out (especially when this poem is taken literally) however a lot has changed over the last few centuries. We now have scholars in fields such as anthropology and psychology who can shed some "light" on this particular subject. Because after all, this is poetry. And so this expression of light should most probably be taken figuratively as a image or symbol. The most prominent of which (especially in this context) would most probably be consciousness or as Jung described it "the light is man's dwelling and the nite where the wild creatures abide mindlessly submissive to their instincts (or the whims of nature)".

With this in mind the probable meaning of this particular phrase or passage would most likely be "god became conscious" " god prompted consciousness" "or god became/is consciousness".

As strange as this sounds at first let us sit here for a moment. Because this actually fits quite well with the narrative. In the first verse we see the world covered in darkness and water (chaos) as the spirit of god hovers over the waters. We see this (most probably depicting a very ancient memory of our primordial ancestors crawling out of the abyss or waters. And subsequently beginning the process of becoming conscious thru the act of speaking and articulating meaning. However this is not simply a phenomenon restricted to ancient man but observed in people today (as even in modern children this process is slow and gradual. Even sometimes extending into adulthood before we fully understand why we do what we do, or act instinctively and habitually.) it is only after "this creation of light" or "awakening of consciousness" that we begin to distinguish between the nite and the day, the land and the waters, or the whole and its parts, or ourselves and our environment. And this revelation takes place thru the act of speaking. Rather it is a small child talking to their imaginary friend or an adult recounting an argument or discussion in the shower, both playing pretend and practicing mindful meditation and prayer (these acts of talking to ourselves (or god/the universe) are not futile fancies but crucial exercises for developing the necessary musculature for our critical thinking skills and our instincts to communicate with each other. (The VMPFC constructing a model for our DMPFC to employ for decision making). The instinct is focused on the present moment, it is the rational mind which has the foresight to disconnect form the moment and plan for the future. We see evidence for this interpretation in the work of Julian Jaynes and his theory of bicameral mentality proposed in the late 70s.(it should also be noted that in ancient cosmology, the light shown by the astral bodies, stars, moon, sun,... was not credited to those astral bodies but rather the light of the gods which then shown thru the stars as representations of the gods, similar to how man were representatives (image barriers) of god or the divine spirit).

But there lies the kicker. Because we didn't have to grow, we didn't have to crawl out of the water and into the light. So why did we? Why did the universe expand? Because it wasn't just us as a species. Yes your parents had to meet, as did every ancestor exponentially backward. But even more than that that rodents had to cower for survival thru the whole of the cretaceous and Jurassic periods, micro organisms had to survive endless disses and catastrophic extinction events. The cosmos had to explode in a Big Bang and begin to expand. So very much had to happen just as it did. And it did! Life found a way, the muse inspired revolution and revolt thru science, technology, and art! As serendipitous as these circumstances are. It would be foolish to believe that we are a mistake or accident. So The question still remains, why did we become conscious, why did life find a way, what is this muse of nature inspiring society and creation towards? Where is our nature or instincts pushing us too? And perhaps most importantly for this discussion "who or what is that spirit hovering over the waters" because after all these characters or entities of "life finding a way" or "nature programming our instincts to survive" or even "the muse inspiring consciousness" they are all still quite ambiguous without bodies or faces to call them by.

You don't love Jesus you love the idea of Jesus, or at least the version of him you've created In Your head. You can't love someone you've never met any more then you can have a relationship with a person who died 2,000 years before you were born. But perhaps this is not necessarily our fault because in many cases we've been taught that the most important relationship of our entire life is that between ourselves and some cosmic entity "god" who is nether seen nor heard and yet demands constant validation with no semblance of reciprocation. Maybe the reason we have such a hard time forming healthy relationships with others is because we have spent our entire lives convincing ourselves that a distant and imaginary (at least as it pertains to our perception of such an entity) "being" is communicating with us thru subtle ambiguous hints and epiphanic dialogue in our own head.

## FINAL REMARKS

Would you condemn a toddler to eternal torment for not believing in a parent who hasn't held them or even answered them when they cry out in pain. How can one love someone they have never seen or heard? Is love submission under threat of hell? No of corse not because love requires a relationship and a relationship requires communication. And unfortunately Anthropomorphizing words in an ancient book isn't communication.

What does the existence of god mater when the hungry starve and the broken hearted are not comforted!

Like standing at the precipice of the Grand Canyon one may feel fear or awe. when forced to sit in solitude one may find peace and resonance or stress and isolation. In the same way anxiety and excitement are two ways of interacting with the same core emotion, heaven and hell are the same experience but from two different approaches. The first is of tolerance and compassion where as the other is spite and contempt. Much like love and hate, heaven and hell are two different experiences of the same underlying phenomenon.

One must choose the fires of heaven or those of hell, for one will inevitably be burnt for both devour the flesh.

My reader can conclude by this point that all the evidence ether for or against the existence of a god is predicated on a dichotomous aray of facets from psychology to physics, from consciousness to instinct, and thus it is no surprise that atheists can't agree on what atheism means any more then the more than 45,000 different denominations of Christianity all of which argue about their definition of theism and god, can. And so nether can theists! The point is. the validity of your identity is not, and absolutely cannot be dependent upon your adherence or subscription to belief in some supernatural or celestially cosmic entity which has yet to reveal themselves to all peoples of all times. Especially when even those of like mind cannot agree on even the most fundamental tenants.

If the tomb was not empty and the face of god seen, would they cease to be ineffable? when god is ephemeral would seeing cease belief? Does the hand of god cease to be god when we discover its cause, or change the name by which it is called? Would the ego lose his authority in our autonomy? Would instinct lose her voice if she gave us a choice? For what are words to the hungry? And what is life to the dead? How many have spent their lives defending the honor of a god unbothered by the peoples suffering? Does the prestige of priests truly take precedence over the peoples needs? is glory and praise really worth the sacrifices of slaves? If the evils of our obedience was on our hands, how would we regard such commands? Does the ground cease to be holy when it is known by every man? Does the divinity of the human demean the sanctity of the beasts? And Why does the unseen god require us to believe? Do they not have faith enough in themselves? Does the mother not protect her cubs? Or Does the almighty have needs, Is the price of love, to feed the ego and deny the weak? Why does the artist shame the clay for the way it was made, as if it didn't resemble their creator? Why does god hide their face, And neglect their grave? if god was made known would they lose their throne and fade away.

We seek affirmation from an absentee god.

Like an anxious attachment to an avoidant or diss interested partner. This most probably has more to do with the mechanisms associated with the phenomena we attribute as "spiritual", primarily those of the hypothalamus (homeostatic regulation), amygdala (emotion and subconscious), hippocampus (learning and memory), and the nucleus accumbens (interface for motivation and response). Then it does the existence of some external entity.

## WRESTLING WITH THE INEFFABLE

Sin as it is understood is a fundamental prerequisite for life. As every living organism survives at the expense of another. In this regard the sociopath is no different than the lion in their lack of sympathy for their prey. The curse of humanity is that we are capable of acquiring consent and disavowing the illusion. Life is expensive, and thus it makes sense that we would supplant that cost thru sacrifice, (to assign meaning and significance to it) for in order to be distinguished from the beasts of the field we must be mindful of our life at their expense (death). And tho every living organism survives at the expense of another, death is not the eradication of life but instead the reallocation of recourses. This is also the blithe of the intellectually inclined. because as we've already seen cognition is expensive. thus the recourses required for homeostasis are reallocated for intense thought.

Also As we've already seen morality, prejudice, and empathy, are developed by cultural assignments and society and thus are not fixed but dynamic and fluid.

Life is expensive and thus far too often these natural (proactive or inhibitory) mechanisms are subsequently labeled as sins. For instance take the myth of laziness or the sin of sloth. Not only is life expensive in the cost required for obtaining recourses, it is also expensive in the sense that those recourses are readily "expended". When recourses are exhausted or trauma responses inhibit the distribution of those recourses, we often experience a state of procrastination or lack of motivation which is then chastised by this utilitarian lie. Trees and bares take half the year off. And yet we've bought into the scam that we live to work or serve a utility. More often then not "laziness" is actually burnout. As children we play and cures bedtimes. We want to be active. And so if we are in a state which does not want to do, then it's probably not us which is the problem but instead our environment. Nothing inside of you, deserves to be punished for its expression. People are not problems! Systems where made to serve people, not the other way around.

There's a reason narcississiti is common among gods and kings. because it is that very trait which makes them successful with its confidence, charisma, and drive.

Narcissists rule the world, because they were born of strife with a mentality to survive. And thus They are always focused on working on themselves but they are also unable to rest and are unfit for times of peace and plenty. Like our sweet tooth many of the mechanisms required for survival are harmful in times of abundance and peace. And this is why our gods are intolerant of their children. For the gods were forged by war, and tho it was they who created a world of abundance for their children to grow up in. The love acceptance and inclusion exercised by their weak and compassionate children is abominable to them. But Narcissism isn't Inherently bad nor is empathy on a biological level inherently good. as we saw in chapter two mechanisms which facilitate empathy were actually developed as internal models to apply ourselves to others circumstances in order to discern rather or not those behaviors and circumstances were desirable or unfavorable. We relate to other peoples pain (via mirror neurons) as a means of simulating potential habits and situations (rather favorable or aversive) and in much the same way we literally feel other peoples pain or rejection (via similar neurological mechanisms in the amygdala, VMPFC, brain stem, and nerves running down our spinal cord,) narcissism in many ways functions as a subconscious mechanism who's intent is to incentivize favorable social conditions. It is also most likely the main proponent of hierarchical structures of order and authority. First and for most it is a desire to belong in a community. Ultimately both narcissism and empathy are mechanisms which developed in order to survive in larger more complex social environments. Oddly enough tho it is the one which seeks security thru the individuals utility or contribution to a social group which has proven to have adverse effects, where as the one (empathy) which essentially uses the social group to synthesize and hypothesize (where the community instead of the individual is the utility) has proven to serve society better and more virtuously. The point being that nether is inherently good or evil but vital in moderation. Narcissism (the ego) motivates people to build empires and achieve profound technological advancements, it has also destroyed lives and peace of mind in the process. In many ways it got us where we are today (for both the better and the worse). The question is, what will we do with that power now. Will we allow narcissisms need for hierarchy to deprive and demoralize society. or will we use that power to make society better for all (the earth, humans, and animals, alike). Ultimately it comes down to the choice between ether appreciating the vital roles each trait plays or vilifying those traits that reside within us which we have become intolerant of (rather that be empathy/transparency or narcissism/a need to belong). This dichotomy is often depicted as introversion (empathetic), and extroversion (narcissistic). The narcissist views themselves as a utility and subsequently objectifies others in a pursuit to obtain a sense of belonging. Ultimately it is their narcissism which makes them desirable to others. Inversely the introvert is so in tune with their empathetic model that they often mask in social situations in order not to offend others. However thru this persistent masking and the continuous act of empathizing and conceptualizing, they are drained by extended social situations. The narcissist should no more be vilified then they are worshiped but rather simply appreciate and validated in their failures as well as their successes. Likewise the shy, mild mannered introvert souls should also be loved and admired despite their lack of charisma and contagious drive/high energy. The point is to nether exalt or to degrade. Both are alternatives for concealing insecurities. The ultimate goal is to employ both in moderation. Both empathy and the ego serve us as internal models for relating to our environments. This is done by developing the capacity to find peace in solitude as well as more genuine expressions among large groups of people. Discipline is not about depriving an expression but nurturing, appreciating, and exercising them. Desire by very definition is a direction and not a destination, thus in cultivating the capacity to entertain both ego and empathy we escape the subjugation which far too often attends identification with ether the ego or our empathy as an introvert or an extrovert. It is in resolving that dichotomy that we continue to grow and at the same time retain a semblance of peace and acceptance for ourselves and others. The beauty of this kind of love is that it retains the space for change and growth to take place, wile not depriving the validation of each steps importance in itself along the way. It cultivates a love which does not idolize a destination but that instead appreciates each and every direction which stimulates our pursuits. This lack of idolization and objectification of ether empathy or the ego helps us eliminate the sterility produced by identification with the destination rather than growth itself (not a particular direction but direction itself). The point is we are nether one or the other and both play vital roles in our development.

Understandably we crave the homeostatic result of a finite and absolute standard of law, however instead we are invited to wrestle with the ineffability of morality, the moral standard by which we judge people of today is not the same by which we judge the exemplary behavior of those before. Morality is not a static standard but dynamic. In the same way C.S.Lewis described sin as "any good thing in ether excess or degradation" morality is fluid. What was acceptable for Joshua was not for hitler, what was condoned for Abraham Lincoln is rightfully condemned today, and what is permissible today (for instance the owning of animals as pet or as cattle for slaughter) will most likely be considered immoral and cruel by tomorrow's standard of morality. Like the isrilights at the foot of the mountain, we want a stationary idol, and omnipotent standard or law by which to judge and be governed. However we are invited not to erect a master but instead to join in on a dynamic partnership, to wrestle with the ineffability. Morality is not a fixed standard but instead a dynamic spectrum which depends on the recourses and cognitive ability of the individual to discern. Like evolution, it's about baby steps in the right direction. The problem is we don't really know what direction that is because sometimes it's sympathy and others it accountability. It's a mixture of conservation and progression, and anything in lavash excess or derogatory deprivation is "a sin". And as we've seen not even god is above morality. This idea of an fixed unchanging god who functions as a standard for morality becomes extremely problematic when we consider a god who commanded genocide and condoned slavery. This is a narcissistic depiction of god as an intolerant king or parent. The truth of the mater is, all of scripture is a renegotiation of terms for relationship with god. And a breakdown of the hierarchy and binary prescribed by that "god". Rather it be Abraham's rejection of tradition, or Paul's conversion into a "hieratic" or even galelao and Darwin's blasphemous claims. The whole story is about deliberately rejecting the binary standard in order to wrestle with the ineffability of a more wholistic world view and understanding. (This however debunks an omnipotent moral authority assigned to gods laws and commandments) what was once progress is now regression, regardless there is no progress without conservation of the past and what got us here (good or bad) must be remembered!

There is no evil man, not the indoctrinated nazi nor the honor ridden sociopath, these things are biological and geographical phenomenon.

The point is the perpetual abstraction. The fact that there is no absolute standard but rather that we must persistently wrestle and empathize with its ineffability. We need conflict and stress. Without such stimuli we inevitably atrophy. We need desire to be a direction and not a destination. And perhaps that means wandering the wilderness rather then entering the promised land.

It is important to remember that People are algorithms programmed ether literally or figuratively by "god". And tho I argue for autonomy it is only one which simply chooses which preprogrammed algorithms by which to favor. Our volition is that of reappraisal in light of new information and a persistent pursuit for truth.

Our culture taught us to hate ourselves, to the point where the ability to love oneself is seen as perverse and even evil. All this to necessitate it's intercession.

## POEM

I wanted her to fix me, because that's what I thought love was. I was taught love was a savior who sacrificed themselves for our sins. Because people were broken things and problems that needed to be solved. But that's not love, that's narcissism. And the gods who explicitly cause suffering, so that they may be glorified, those who made us nothing, so they would be necessary, they are not evil. But just a product of their environments. The sad part is not that they never loved us, but rather that they needed our adoration and praise because they were incapable of loving themselves. But as long as we believe in such saviors, we too will require the love of others to make us whole. So long as we find our belonging in exclusion from the whole, then we too shall always feel broken. The truth is once you learn to love yourself, then you won't need the love of a savior to die for your sins, because you'll be able to live with and even appreciate what that small, fragile, "loving" god was intolerant of. And that is love. Recognizing the beauty in diversity rather then distain which requires a god's grace. Much like that of an unseen god, I miss the versions constructed in my head and the way I felt, not the person themself.

# CHAPTER 7

──────◉──────

# COMPLETION AN WHOLISTIC WORLD VIEW

## AUTHORS NOTE

(all throughout this book I have claimed that god is this or god is that, god is consciousness, god is being, god is nature or instinct, god is the standard by which we judge morality, god is a social construct in order to validate a particular hierarchy, etc.... I should clarify that tho these are components of that which we once called god. they are not the whole. And tho some are problematic and even downright despicable, the invalidation of one does not mean the discreditation of the others. It is completely valid to abandon ideas of godship such as hierarchy and prejudice and yet retain those of beauty and faith.)

# SEVEN

The Hebrew word "shiva" (or seven) is synonymous with the Hebrew word "shelemoot" for complete or whole. We see parallels here between the Hebrew word "Shabbat" (sabbath) and "shiva" (seven), as Saturday (sabbath) is the seventh day and marks the completion of a week. It was on sabbath which god then "sabat" or (rested) (gen 2:3). Not only abstaining from work but more prominently god rested on "upon or within " the seventh day. Declaring sabbath their abode or dwelling place. Tho creation up to this point has been a constant act of separating one from the other. The work is only complete when the parts are once again made whole. And so tho in a sense all previous chapters have attempted to deconstruct. the ultimate goal of this seventh and final chapter is to unite or hold together into a more wholistic form or perhaps formlessness state un restrained by hierarchical order.

The prerequisite for all new creation is chaos. It is more then the material for our cosmos, it is its most fundamental component. Chaos is the whole of creation before it was divided by our definitions of order. So of course new creation feels like chaos. Of corse a new heaven and new earth comes at the expense of the old one.

Of corse may flowers require April showers.

## THE STORM

A storm is caused by a coiled and inevitable mixture of hot and cold air, which causes a change in biometric pressure.

The fury or wrath of a storm which brings destruction and reinvigoration, is a result of polar opposites forced to co exist.

Like a collision of competing cultures which mimic the natural phenomena of its environment. The clash of conservative and progressive, or analytical, and intuitive, can appear at first glance as a catastrophe (chaos) but actually inspires new life and ways of living. Because like a mathematician and a poet who use the same patterns to form very different languages. The hot and the cold which cause a storm are both air. There are Not two world but only two parts of the same world.

## THE MEANING OF LIFE

Finding purpose after a paradigm shift.

What if you end up alone?

What if you don't get that job?

What if everything you believed in is proven wrong?

What do you do then?

It doesn't matter if your a man or a woman, your soul purpose in life is not to attract a significant other. You have value apart from a romantic relationship, and it's utility.

And so What if you never get married?

Who would you be if you weren't concerned with being enough/not too much for someone else to love?

Because that person is worth being. You don't need permission to do what inspires you or to be your authentic self independent of your utility in an community. Obviously you shouldn't practice medicine or nuclear physics without the proper training. But at this day in age all the material you need (limitless knowledge and wisdom) is at your fingertips and disposal. So who cares if no one is willing to publish your book or sign your record deal. Who cares if no one els wants you in a romantic capacity. Be the kind of person that you can love, do what lights you up. Even if no one is willing to pay you for it. Do the work that inspires you (because it will require work). And stop leaving your value and identity up to the whims of other men and gods. Because at the end of the day you are the one who has to live with yourself. So stop Letting their abuse and exploitation of you as a utility perpetrate your needless suffering and intolerance of yourself and others.

## A WHOLISTIC WORLD VIEW

This world is not binary. Joy and sorrow are not mutually exclusive. And you don't have to identify with an ideology. You were never meant to be an theist or an atheist, conservative or progressive, any more then you were meant to base the validity of an expression/ emotion on its utility. We were meant to relate to ideas and empathize with emotions but not to attach our identities to ether. Ultimately that static, binary world view and standard for perfection is what keeps us from appreciating life's divers expressions and results in us cultivating volatile and perverse exhibitions of these emotions thru our systemic and habitual neglect of their valid and Vidal roles. Emotions like fear, anger, grief, and even pride, are not inherently bad or evil. This world is not separated into good and evil, right and wrong. This world, life, and most importantly you, are not binary or static but dynamic and intricate. And that mutability which transcends those binary systems capacity to quantify you, is not a problem that needs to be solved but instead the solution. This world is not divided against itself but rather wholistic and intricate.

## TAKEAWAYS

Religion is as much a slave to instinct and ego as any political party. The point is that we don't need an idea to give us an identity any more then we need our instincts to decide our own nature.

Love is appreciating what is, not profiting from any potential or product.

We learn to discipline our emotions and the ideologies of our society so that they don't dictate our behavior.

You don't need some sacred text or genetic code to tell you who you are and who you should be!

You are not a mistake, or a problem that needs to solved, you're a miracle, your proclivity for mutability is not a curse, your ability to emulate and change, to empathize and to question, that is life's blessing.

You get to decide what image you bare by the way you choose to love yourself and others!

Vulnerability doesn't warrant shame,
pain doesn't mean bad,
dichotomy isn't cause for fear,
dangerous does not mean evil and
might doesn't make right.

# THE ANARCHISTS

During times of crises people typically behave altruistically and competently. Anarchy very seldom results in chaos and violence.

We have been fed a narrative which claims that man needs gods to govern over them.

That without an authority to rule over man, all chaos will break loose. This however is not what we witness. This myth that people revert to some primal barbarism has been repeatedly disproven by multiple studies conducted by The National Earthquake Hazards Reduction Program. What these studies found is that during times of crises people pull together and actually meet the needs of their immediate community. The image we have been sold of anarchy as some form of violent hellscape is a blatant lie produced by those who have consolidated power for themselves. The truth is when tragedy hits the victims are often the first responders and heros. Admittedly people do loot stores, but this has more to do with providing for the needs of people. The lord of the flies got it wrong. Because not only do people act virtuously without an authority to enforce morality, but in many cases people are a great deal more competent than the governing systems intent on remedying tragedy. Ultimately what we see is it is often the hierarchy and struggle for power which perpetuates the vast majority of evil. and people do quite well without that god or government ruling over them. And when we really get down to it, it's not peoples autonomy which is the problem, but rather a small fews need to lord over or control others.

Anarchy is not about rejecting authority all together but rather critiquing the validity of that authority. There are kinds of authority which prove their validity. Those which transfer that authority to the once subordinate party (like a teacher who's authority is evident in their pupils propensity to surpasses their teacher). And then there are those omnipotent authorities who's claims are not based on merit. The anarchist tests both, but only opposes the later.

In truth the gods, kings, and governors, that rule over people, are not there for the peoples benefit but their own. They predate the populous by policing the poor and propagating propaganda to incentivize the exploitation of the populous.

## GOD IS LOVE?

Long before we are taught any logic in the form of math or grammar, we are taught that a god exists, and that that god is good.

> "Train up a child in the way he should go: and when he is old, he will not depart from it."

> -proverbs 22:6 -(KJV)

And so when we see god according to scripture claim that we are unworthy of love, and that our worth comes solely form god. And claims that We do not deserve good, nor are we capable of good. And that What ever good we receive or do thru our actions is of god and from god alone. We deserve suffering, and so if we suffer it is gods right to hurt us. It is our own faults for being made inherently flawed. But of the greatest of evils within our power, is for us to forsake the god who allowed us to suffer, to abandon the god who did not protect us. For to neglect the god who was not there for us, is of the most heinous of crimes. And yet in a parent or partner these behaviors are quite obviously abusive. But as small children we don't know any better. And because our identities and value are dependent upon those assignments (that god is, and that god is good) we never question them.

Any shelter in a storm!

For context we shall turn to maslow's hierarchy of needs.

1. Physical needs (clean air, food, water, shelter, rest, clothing, sex)
2. Safety (personal security, stability, health, job, the ability to prosper)
3. Belonging (community intimacy, friendship, communication, connection)
4. Esteem (respect, status, freedom, autonomy)
5. Self actualization (being and doing the best you can, full potential)

When our physical needs are not met, we will risk our safety. Without a sense of security we will not require autonomy or respect. If the previous need is not met then we will not pursue the next until our more basic needs are met.

And so as long as they keep us hungry, we will not demand respect. As long as our sense of community is dependent upon our adherence to a demoralizing law, we will often relinquish our authority and esteem in order to retain that community. Because our need for community is more important then our autonomy. And so any shelter in a storm means that as long as we are kept desperate for things like shelter and food, we will be dependent upon those who with hold respect from us.

But ultimately they need us. They need our worship and loyalty.

Love is not be based upon utility or potential, but upon appreciation.

Our biggest complaints about our partners, says more about our emotional needs, then it does their inadequacy. So what does that say about a god who required lesser being to worship him?

The point is if god was truly competent, then he wouldn't need the adoration and praise of others to validate his ego. If god was truly love, then he wouldn't need us to meet his emotional needs.

What kind of love is god? An ascetic, intolerant, and jealous kind of love? Solely dependent upon self gratification rather then any kind of appreciation? If so then how can an inerent creator lack the capacity to appreciate their own creation? Which denotes an intolerance of themself by extension!

There is a kind of love often preached by Christianity. This kind of love is jealous and requires requitemen. For this kind of love is dependent upon relation AE I love you because you belong to me or us. This kind of love seeks to control and convert someone in accordance with an idea. Instead of appreciating the diversity of individuals. For after all, according to Christianity we only "love" because Jesus first loved "us". here love is a mechanism for manipulation and not of appreciation or admiration. This kind of love idolizes an identity or ideal of someone and not the person themselves. And so that person is only loved so long as they retain that identity.

However there is a second kind of love. A deeper, truer, kind of love.

And this kind of Love is not dependent upon relation. this love appreciates one's individuality and does not require a relationship in order for it to persist. This kind of truer Love does not seek to control another's autonomy. However admittedly without relationship, life ceases to retain value or meaning. Life in isolation may have love, but it cannot have value or meaning.

Love is how you treat people here and now and not what that makes you to them, or them to you!

True genuine love has nothing to do with someone's potential for being what you'd want them to be. but instead everything to do with you appreciating what they are (even when they change) . This idea of unconditional love is simply shifting the focus from the object to the subject.

The condition is the one who loves, not the criteria for their affection.

(The things you are intolerant of in others is often a reflection of traits and expressions within yourself that you haven't learned to appreciate yet. This binary world view which requires you to identify as a theist or an atheist, a conservative or an progressive, is not only unhealthy, but inaccurate. Joy and sorrow are not mutually exclusive. We are both. This permanently perfect perception of personhood is completely counter intuitive when it was our mutability that got us this far. We are not problems that need to be solved! but until we can appreciate all of ourselves, we will continue to be intolerant of others. Learn to love yourself so that you may have the capacity to love others, as they are, and not only as you would require them to be.)

Love is meaningful because we insist on beveling in it. It's value is in its sentiment and not its sensation.

> "Love is never wasted,For its value does
> not rest upon reciprocity."
>
> (C.S. Lewis)

In all honesty what 1 John 4 is describing is no different then earlier authors who attributed there own anger to god, or even its contemporaries who believed phenomenon like epiphanies and genius were divine qualities imparted by some external entity rather then entirely comprised of facilitates within themselves. For much of ancient man phenomenon like anger, love, and genius, were the soul property of the gods. And so when someone offended their honer or inspired in them lustful passion (jealousy), that fire in their chest was not their own, but gods anger, jealousy, love, burning within them. To better articulate this point take the phrase "look at what you made me do" which implies that (rather out of love, jealousy, or anger,) there was some external entity which commandeered our autonomy. Which is what emotions attempt to do.. they emote or move us. This is however not the workings of some ethereal spirit but that of hormones and neurotransmitters in our brains and body incentivizing certain behaviors and reactions to external stimuli. Rather it is anger, jealousy, genius, or love, it is not god's but our own. It is felt within us and acted out by us, alone. God does not love, people love! And then on occasion ascribe or attribute that love, anger, jealousy, or genius, to god. Tho it makes sense that we would ascribe a sense of agency to these emotions. Seeing. As how they are cause by external events In our environments. Rather it is seeing a bear in the woods, to which one might respond by immediately running the other way without thinking twice. Or it is locking eyes with the most beautiful person you had ever seen. The actions which follow are often not our own. Rather it is the fear of god which causes us to flee without thinking twice, or the whispers of muse and genius which entice us to draw closer. It is understandable why we would attribute some form of agency to these phenomenon (fear, anger, jealousy, genius, and love,). And if we are honest even still now, modern man hears the voice of god (our emotions and instincts) and promptly obeys without a second thought. (Note anger is often the mechanism our genes developed to tell us when our agency is being obstructed or usurped).

To conclude however, love. True love does not come with stipulations. Because it's not about admiring your own idea of someone else, but appreciating the reality of who they are. It's not about getting your emotional needs met, or your ego validated. But about rediscovering one another. Love is not a savior who sacrifices themselves to earn the praise of others. Love is kind, curious, and compassionate. Love is not self serving like that of a savior, but sincere like that of an equal and friend.

Love is not a savior who solves us like problems, but simply someone who sits in vigil to our pain. One who witness to our suffering and joy. Love is not ego! It is compassion and appreciation.

## THE ROLE OF HUMANITY

According to Darwin "survival of the fittest" as the thing that set us (humans) apart from our distant cousins (animals) is not our superior might or power. but instead our capacity for compassion (to care for the sick and hurting). It was our ability to mitigate the prerogatives of our instincts and subsequently the relatively cruel nature of nature in the process. oddly enough this in many ways reflects one of the more compelling claims Jesus made and even Abraham And Moses before him, which quite simply is that the pharaohs and Caesar's were bad representations of godship. And further more that the role of the Messiah or "human one" was as an intercessor on humanity's behalf. Or in other words in much the same way Moses wrestled with god on the mountain, on behalf of Israel (who had chosen a master or bal to represent god as opposed to entering into a partnership with the divine) we too as humans are meant to contend with the ineffable for more compassionate and inclusive representations of the divine. Because after all, god has yet to step down in all their glory and show themselves but instead has left it up to us to decide their character (or the very nature of nature) thru the way we bare their image, as ether malevolent masters and narcissistic conquerors who are intolerant of differing forms and expressions. As those who seek only glory for themselves(like the kings). or as compassionate caretakers and "brothers keepers" who empathize and even emulate potentially positive traits which may have at one time been strange to us. In this way Our mutability is not a curse but instead our saving grace. For you can't shame the song for the notes the artist plays (or the pot for the way it was shaped)! Our ability to observe our own behavior and then negate our instincts does not account the descent of man or our fall from grace but instead the invitation to remake the game as "partners in creation".

Even if nature is not nurturing and god isn't good, we can be kind and compassionate.

This role of cores requires our conscious persistence at attaining a more conscious state. (Or constructing more inclusively comprehensive internal models). however like the cosmological flood depicted in ancient mythology, the primordial remnants of our natural instincts can be quite daunting in the light of our awareness. Thus it is no surprise that we often seek refuge from such responsibility in the form of substances which inhibit our inhibitions, such as mythicized in the Egyptian "sleep maker" (beer) and biblical account of Noah getting drunk after the flood. Tho the ark represents a metaphorical vessel and refuge for traversing such cosmological perils of chaos which flood our conscious state. We still seek more then simply a paradigm or ideology by which we can make since of the cosmos. And as tempting as it is to seek a release from such a daunting task. Fleeing and ignorance will never actually solve the problem, but more often than not add to its suppressed expressions volatility. It is understandable why so many turn to substances like hallucinogens and intoxication thru alcohol to free them from the arduous reality of human autonomy. However Ultimately alcohols override of our inhibitions in the process will always only perpetuate the problem. on a deeper level these coping mechanism seek a reversion of behavioral responsibility to its primal submissiance to our instincts. or in other words we don't want to ask why or question our behavior but instead act unconsciously and instinctively. We want to return to the garden where we walked with god and nature unawares of our nakedness and vulnerability.

My theory is that people enjoy drinking alcohol or consuming hallucinogenic substances for the same reason that ancient authors perceived our cognitive awareness as a curse, And then sought out traditions and religions to mitigate that responsibility.

Adam blamed Eve and Eve blamed the serpent but we can't blame others for the way they make us feel or how we react to them, anymore then we are to blame or are responsible for other peoples feelings or behavior. The truth remains, our instincts and animalistic impulses still try to control us, and other peoples actions effect us.

And we need an outlet, a refuge, a space for the chaos to settle and subside. This is why we pray and meditate. This is the role of therapy. Because after all Therapy is simply learning to listen to ourselves and developing the capacity to tolerate and even appreciate all of ourselves (especially those qualities society is ashamed of). Ultimately coping mechanisms like scrolling on social media, drinking/drugs, and even suppressive religiosity, just pacifies and ultimately perpetuates the problem. This is why many well acclimated people spend time in solitude. We often need space for abstract and even chaotic expressions to exert themselves, in order to cultivate a place where diversity can coexist in harmony. We need a chaotic wilderness to cultivate a world of abundance and inclusion. Because any expression which is not given an outlet, will inevitably become violent and intolerant.

With this said. Problems are not always problems. And more often then not they are in fact solutions to the very same problems we once called solutions.

People are after all problem solving machines, it's how we are wired. We attribute meaning and significance to things because we evolved by extrapolating causal circumstances in our environment and their correlation with outcomes. And thus if we don't have problems to solve, we create them. This may be why so much drama gets stirred up in idol workplace or stagnant relationships. and also why people who have faced hardships are more acclimated to life. We need stimuli for our curious minds to survive. But we also need space to process that information wholisticly.

People are beautiful not because of their affiliations but despite them. Christians are inspiring and kind not because they identify with an ideology but because they are people. Agnostics are admirable not because they know stuff but because they continue to search for truth even when it hurts. Humanity is beautiful.

## THE YOU IN YOUR HEAD

The center of the observable universe is always the observer, and not necessarily the center of the universe at large. Because light has a speed, there are objects so far away that the light refracting off of them hasn't had enough time to reach the observer. This is not because the observer stands at the epicenter of the cosmos, but simply because the observers cosmology is centered around the observer and not the phenomenon observed. Our minds made these models, so of corse we'd be the pinnacle of its creation.

"You are not your thoughts, you are the one witnessing your thoughts"

Tho I personally diss agree with this claim, let us explore this particular topic for the time being. From the bottom up, there are a great number of facilities which are out of our control. For instance most people can't willfully stop their heart from pumping blood or inhibit their liver from processing toxins. These functions are purely autonomic. Further more our thoughts alone are incapable of executing motor functions. For instance we can willfully hold our breaths and clasp our hands shut. Yet until our MPC is activated, those particular facilities will not engage/disengage. Our thoughts cannot control our actions. But there is more to this discussion because we also have the "you" who chooses to act or not act. To listen to the thoughts and engage the MPC/motor functions. And this "you" most probably resides in the PFC, or the DMPFC to be more exact. Because this is the part of the brain which activates executive functions. However this volition center is not all powerful. For one it works in conjunction with the VMPFC (the region responsible for internal maps and models of ourselves, others, and the environment at large). these two sub regions work together to simulate consequences and then execute functions accordingly. However as we've already seen there are still facilities which do not answer to these higher authorities (the brain stem down for instance). From the brain stem down thru the spinal cord (with specialized centers in certain organs) we have those purely autonomic functions which are not subject to the will of the higher authorities (the MPC, PFC, Broca's area, etc...). Just above the brain stem we have a region responsible for voluntary motor functions (tho it is also activated by involuntary/autonomic motor functions as well). This region is called the MPC. From here we begin to work our way forward rather then just straight up. And so In between the MPC and the PFC we find an area known as Broca's area which is where much of those thoughts and internal dialogue are housed. these (patterns of transient electrochemical activity) or thoughts, Most probably function as an intercessor between the amygdala (instinctual brain) and the PFC (cognitive brain). These thoughts make suggestions in accordance with past experiences and instinctual wisdoms. However they are incapable of deliberately activating motor functions (however the amygdala can, and often does override the executive functions of the PFC). In front of Broca's area we find the PFC where both the VMPFC and the DMPFC reside. As we've seen, the willful DMPFC cannot control the ancient and lowly brain stem/autonomic systems. And so tho we are most certainly not our thoughts, they are however just as much a part of us as the maps (ego, belief, and empathy,) of the VMPFC, or the man power of our MPC, or even the involuntary function of our autonomic systems (brain stem down). And thus to say that the "YOU" is housed exclusively in the DMPFC is to completely miss the point. From the instinctual amygdala to the rational DMPFC and everywhere in between. It's is all a integral part of YOU. And you are all of it.

(Note in actuality the brain is not quite so leaner nor are these facilities quite so localized but instead far more complex and integral.)

More simply the lowly primitive facilities are no more inferior or superior to the more advanced higher up ones. They all work together. And as for the YOU in your head. It is most probably the executive DMPFC listing/communicating with the dialogue in Broca's area thru the internal models constructed by the VMPFC. that dialogue in your Broca's area is an amalgamation of the voices of your parents and piers, as well as an interpretation of the profound wisdom stored in the instinctual amygdala.

That phenomenon often referred to as epiphany, Prolific muse, moving of some transcendent spirit, or stroke of genius, is actually the result of that dialogue converting the profound wisdom and knowledge posited in our subconscious, which is then compiled and converted into our VMPFCs internal models and schemas. It is the information of these thoughts or strokes of genius which had been processed by background faculties like the amygdala, converted and applied to fill in gaps which were previously unknown in our more cognitive faculties. That's why it often feels like a light switched on which instantly makes known what was previously unknown or obscured by darkness. It feels like such wisdom came out of nowhere, but in actuality our less cognitive mechanisms had been processing that data for some time. Grief similarly is our brain and body learning to form new neuro pathways wile also rewriting existing processes after the loss or prolongs absence of stimuli which had previously initiated those chemical releases. In short it a side affect of withdrawal form dopamine. This is why stimuli like songs or smells which use to correlate with those dopamineic reward systems cause so much pain.

We still make deals or covenants with ourselves similar to how ancient man made deals with gods. ("If I am diligent and stick to my regimen, then I get to eat those brownies this weekend")

(God to Abraham "if you leave this culture which requires human sacrifice then you will be blessed with many children")

The spirit (religion) testifies to the experience, logic (science) illuminates the phenomenon that causes that experience. The experience is how we relate to ourselves and others, cognitive logic is the means by which we are capable of accurately understanding ourselves and others.

## CAPITALISM

> "The comfort of the rich depends upon an
> abundant supply of the poor"
>
> —Voltaire—

The fundamental flaws of capitalism.

Or What we call modern day capitalism which is more accurately articulated as a form of corporate socialism (a system where taxpayers Suport corporations growth rather then their own basic needs ether thru taxes which are then spent on bailing out big companies after they fail or thru the profits produced thru their labor but withheld from those who produce). In its most fundamental form socialism is the application of compound equity(money/rescues) evenly distributed to provide for the basic fundamental needs of those who contributed as a whole.

However instead what we see in our "Capitalistic" economy is tax breaks and subsidies being given to multi billion dollar corporations and institutions like the church. Or more simply it is the fat and wealthy who get the handouts wile the poor and overworked are left to starve and exhaustion. In the modern day capitalistic system Those who produce the majority of the products and services are also those who are in the most poverty. because the economic system is arranged in a way where those who produce, must produce far more then they profit from their work. And instead of equalizing that gap, the profits are passed on to those who supervise the work being done and not to those who are doing the said work. Now you may argue that those supervisors aid in efficiency and quality! and that right there is why we've ended up in a world where judges are exulted as gods and genius is stifled for slave labor. (Says the blue collar author who is critiquing tradition). People no longer provide for themselves but instead provide for corporations which have promised protection at the expense of freedom.

But here's the kicker because far too often the most "progressive" cities and countries are those which have expedited their slave labor and exploitation of nature to lesser developed populous. For much of human history slave labor was considered moral on the basis that lesser humans were unfit to govern themselves (see numbers 27:17 where Joshua is chosen to rule/govern the peoples like a shepherd onto a flack of animals). this is the same basis which is used today to justify the exploitation of people by corporations and nations in the form of wage labor. The sad part is even now people would rather be slaves in Egypt than free in a wilderness. We have been taught to identify with our oppressive systems. Because it is those systems which provide the infrastructure for services like running water and grocery stores near by. However to better understand this point. We must first revisit the fundamentals of societies.

Hunter gathers (individually) only worked about 15 hours a week to provide for themselves and their tribes. However with the development of agriculture came the alleviation of hunger and the ability to negate the tragedy of depleted game/nuts and berries, (starvation) on the populous as a whole. However this bounty came at a hight cost of 80+ hours of hard labor a week. And this was not chosen willingly but rather was forced upon weaker tribes by more powerful nations like Babylon, Rome, Egypt, and later England. Industry pathed the way for the provision and protection of the whole on a much larger scale, but at the expense of autonomy and liberty.

However much like the slaves in Egypt and the colonized natives of distantar places, those people and the land they nurtured were impoverished by this "salvation" which only seemed to profit the gods who enacted them.but The people are complicit only as far as they are provided for. And thus we learn to identify with the hand that "feeds" us until we ourselves are the feed for its greed. Ultimately people, like most organisms, adapt to fit their environment and circumstances. So to say that greed is human nature is like saying an elephants nature is to balance on a ball and ignore the fact that it grew up in a circus. A system which requires people to go into debt to build credit in order to function, is inherently corrupt and abusive. It's the economic equivalent of demoralizing humanity and then offering salvation in the form of very manipulative redemption. AE (" you are worthless, but because I love you, you have value") this is also why people who grew up in abusive environments often end up propagating that abuse later on. This is a sign of fine tuned senses in regards to abusive and/or hostile environments.

(it's worth noting here that before colonizers came to America, the natives had cultivated entire ecosystem "self sustaining food forests" where the plants and animals in these environments essentially did all the agrarian work for them.)

The Native Americans had a wide spread practice of planting corn, beans, and squash In the same plot. This method was called the 3 sisters. In this agrarian ecosystem the corn provides the beans with a stem for Suport, the beans draw nitrogen from the air to enrich the soil, and the squash helps to keep the soil moist as well as offer protection from encroaching weeds and small vermin. Together the three strengthen each others weaknesses.

## SIDE NOTE

(75% of our crops go to feeding livestock which in turn expend a healthy percentage of that energy in producing viable protein for our consumption. In this figure not only is the slaughter of animals cruel but it is also inefficient for feeding people.)

Sadly most states don't care about providing utilities like public restrooms, free education, and abundant food for the poor thru fruit trees. Not to mention affordable housing. but are instead more concerned with incentivizing their populous to contribute to the states economy and taxable income. The goal is not prosperity for all, but prosperity for the few at the expense of the many. That's why police police the poor and not the rich. And why so many policies are implemented to strip people of their rights. Because when a population is desperate, they are easier to coheres and exploit!

Now we can't ignore the fact that corporations stimulate the economy of their surrounding areas. But that's no reason for that state to neglect the peoples needs and show preference for the needs of corporations rather then it's own citizens. Of corse big corporations bring revenue and jobs to the people of their town or state, without which many small towns and communities have gone extinct. But survival is not enough. Quantity without quality of life is not worth the price!

When the rich fund the milita and police rather then the teachers, then liberty is lost

Loyalty is the enemy of truth, and honor the death of humanity.

When prestige police's access to/ the validity of knowledge, then education becomes a form of socio economic warfare and a means to predate a populace.

We learn to love our selves from the way our parents loved us as children. And so when failure is punished with shame, fear, and pain, then we get adults who mutilate and suppress.

The narcissist shames the victim for their pain.

Much like animals, people are cold and cruel when they feel like they have something they must protect or preserve. As if they themselves are an object to be possessed or stolen. Your value is not something that can be lost or depleted. we intuitively make others feel the way they made us feel. And so if we fell hurt or frightened, we in turn (often subconsciously) attempt to project a frightening or threatening facade. Ultimately we are communal creatures and thus subverting those reactions back onto the person or (archetypal people group) that cased those reactions in us is perhaps the purest form of communication. (Tho not particularly healthy or beneficial) it's the little boy punching the bully who hurt him. Or the little girl excluding the friend who offended her.

Fate is simply the trajectory of our behavior.

In the end rather it's men or woman, rich or poor, theist or atheist, people are very much products of their environments.

Those who "make their money work for them" thru portfolios, interest, and real estate, explicitly exploit those who produce goods and services AE "work for their money" because ultimately money is not a resource in itself, but simply a means of measuring prestige in society. Much like the release of dopamine, money simply functions as an incentivizer for the production or distribution of vital resources in contribution to a community. But when the annual debt to income ratio is 145% (or more simply more then 3/4s of Americans are $90,000 in debt), which means the majority of these sociological incentives are negative or punishment based, rather then reward oriented. Then the system is abusive and not proactive!

so what will happen when AI is more economically feasible then man power? As we've already seen this economic system is not about what's morally right, but simply a means by which to enforce a hierarchy. The argument goes that "without work, man won't know what to do with themselves" as if a man's time is better served, serving a hierarchy. The truth is when people are given the freedom to learn for themselves, they create and make discoveries completely without the need of an hierarchy to rule over them. This becomes apparent when millions of people leave their houses every day, well before the sun comes up, and don't get back till sundown or later and yet don't actually produce anything, but simply fill a space. It's not about production, it's about prestige. There are however a good number of essential jobs which are required for the growth of, and in order to perpetuate/ sustain society. however most of these jobs are undesirable and are thus left to those (for the most part) on the bottom of the hierarchy. Wage labor is only slightly less morally repugnant then slave labor. And at this point In history, it's clear money is no longer about production but rather simply imposing and enforcing an hierarchy. The root of the problem rather it's feudalism, capitalism, socialism, or communism, is hierarchy. It's about justifying the exploitation of others. But what will happen when machines fill the role of the exploited peoples? Just like the god who made the man blind all so that god could be glorified by healing him (John 9:2). It is unethical for the suffering of others to be perpetuated simply for the glory of those who require a hierarchy to give them security in a society. I don't really have a solution to this problem! But I do know that the leisure of some should not be provided for thru the exhaustion of others. Perhaps if we didn't propagate scarcity in order to facilitate a sense of sanctity, and instead simply learned to appreciate what we have in excess, maybe we'd have enough for everyone to live and even thrive comfortably!

There is nothing wrong with money as a means of measuring prestige, so long as it's not at the expense of basic human needs and liberties on behalf of the working/producing class.

Capitalism is a system which not only rewards greed, but requires it.

## MARRIAGE

On a biological level the emotion we call love is just as lucratively focused as the concept of mirage was in the 1800s.

The biblical definition of marriage is as a contract between two men (the father and the husband) of a woman in which ownership of the woman is transferred from the father of the bride to the husband of the bride for a hefty price. And tho marriage today no longer intel's the ownership of another person. It does still function very much as a utilitarian form of signaling prestige. Because this relationship is expected thru convention, sadly what we often find is not an intimate connection but rather the exact opposite, where men are constantly afraid that their wife will leave them for someone bigger, stronger, richer, or smarter. Because unfortunately even this day in age women are seen as objects to obtain and worship, and men are coerced into becoming utilities in order to prove their worthiness for the object of their desire. However I would argue that Community love is better than this one on one romantic love. the objectification and utilitarian roles of romance (tho anatomically captivating and natural) are none the less abhorrent and abominable in a moral sense. They hinder growth and genuine trust/acceptance on both parties. Romantic love can be beautiful but far too often the dynamic it cultivates is a relationship/ environment where one party must present themselves as an object of worship, and the other must become a utility in order to prove their worthiness to behold/obtain the affection of that object of worship. It's a love dependent upon relation (AE I love you because you belong to me), it's the idolization of an ideal rather then a reality. some of the most beautiful relationships are those where there are no expectations and both parties are free to grow with full transparency without the need for approval of the other because they are whole in themselves and not dependent upon the other to complete them. Often these relationships look like friendships and Tho they may be conducive for (magic, romance, that spark,…) they are not confined to the restrictions of a utilitarian mirage (intent on propagating offspring). In such relationships there is no ownership or lies because both parties are permitted to grow independent of the other('s). People are dynamic, but people are also communal. And we need both the freedom to grow and the security of a community. The two should not be mutually exclusive. Tho it was Pauls belief that the end was nye that propagated it, he did recognize the restraints marriage put on an individuals ability to grow. The sexual ethic may or may not be of much importance this day in age (depending upon how it is utilized) however the message of inclusion and acceptance on a communal level without the utilitarian/objectification of the traditional roles remains pertinent. I personally don't think marriage is conducive for growth but at the same time I recognize our innate need for companionship. And I don't think the two should be mutually exclusive. I understand the merit behind intentionality, however personally I prefer the genuine moral of organic relationships among community. What if we could love everyone without the need for a relationship to define us!

What if love didn't depend on relation? This particular point is key because The narcissist doesn't love the person. they love the idea they expect the person to live up to.

In it is this need to own/be owned by another person. It is an identity which finds its belonging in exclusion from the whole. Much like our religious and economic systems, marriage strives to achieve sanctity thru fabricating scarcity. By focusing all one's devotion and affection on an "object" subsequently driving up the demand for "love" as some kind of commodity rather then something free and plentiful. This idea is propagated by traditions like purity culture. However in its original application

Virginity was a stipulation imposed only on the woman as a means of "protecting the value" of the mans "assets" (the woman's reproductive system) in producing an pure heir to propagate his seed. Much like the prohibition on certain body parts this practice was used to control the "market" on the propagation of specific genes. And still today We have inherited a form of love which is predicated on scarcity to give it value.

Tho this institution originated from an misogynistic plot to control. the blade cuts both ways. In the dating game there is always a bigger fish. And when one's value is dependent upon their utility or objectification, then there will always be someone more successful or attractive.

We have been taught to seek connection thru exclusion from the whole. And so value is dictated by supply and demand.

For men this system was about its propagation of their genes, however for women this institution was a means of surviving in ether a harsh and unforgiving wilderness or in an economic system and community where they (as women) had no rights or autonomy in themselves.

Women are judged by near impossible standards for beauty, and men are judged by near impossible standards for athleticism, monitory success, and intellect. However with that said many of the worlds most profound writers, physicians, and physicists are women, and men are not exempt from being judged by near impossible standards for beauty.

Things like aesthetic and style (which are blatant applications of models AE representations of how a person sees themselves in the world) can be an efficient filter. However unfortunately it often results in us mistaking the aesthetic for the person themself. The problem isn't these standards but rather that we have learned to identify with them as a means of determining our own value. However what truly makes a person beautiful, the qualities people truly fall in with, are not the perfect exemplification of these models. But rather a deliberate derivation form them. It's the idiosyncrasies and oddities which people relate and resonate with which inspires true love. It is that which makes us human, which is perhaps most beautiful of all.

The truth is relationships aren't static, and so tho many romantic relationships require men to be utilities and women to be objects in order to attract mates. As the relationship progresses, that dynamic should change into a more genuine connection which dissolves many of those archetypal gender roles.

Now of corse community love can be just as sterile and corrosive as marriage, the toxicity of "group think" is no rare occurrence.

"In individuals, insanity is rare; but in
groups, parties, nations and epochs, it is
the rule"

(Friedrich Nietzsche)

The goal is not to get rid of community, simply the criteria for community. To find belonging without a need for a standard set of beliefs. More or less cultivating a culture which is conducive for growth (accepting of mutation and the mutability which comes with it rather then mutilating those mechanisms to fit an archetype)

## THE ADVOCATE FOR CHAOS

"Modern man has cut god into pieces much like ancient man has unto the cosmos, what religion recognized as whole, science has divided into specialized fields."

One of the most fervent claims we get in the defense for the existence of god, is that something can't come from nothing! and thus there must be something or some one who made it (and in this case there is no argument. Both theist and atheist alike assert that existence didn't just happen. However they do disagree on who or what they call that source). However these claims of creation as an manifestation of mater, is not made by the biblical authors in regards to their god. At least not in genesis anyway. In these poems we don't find a god who made something from nothing but instead one who divided the whole into parts and then called those parts worlds in them selves. We get the separation of heaven from earth, the distinction

513

between light and darkness, and the division of man from beast. This act of creation thru division does not account the manifestation of the material but rather an assignment of meaning and values. And so latter on when we encounter this primordial serpent or wild beast of the field, it does not convince man to destroy the physical world but rather to challenge those assignments of meaning and value by discerning for one's self what is right and wrong. People (like animals) are predisposed towards obedience and compliance with community. Where as defiance requires more cognitive functions of our mammalian brain. And so what is most probably being depicted here is (as Jaynes puts it) a breakdown of our "bicameral brain" in favor of more complex and comprehensive mental mechanisms. (It should here be noted that both our most "benevolent" and "malevolent" behaviors our preprogrammed into us. And so it is not the behavior itself, but rather the context, which is learned from our parents (much like the god of genesis who defined context). Now another interesting quality which we humans (and animals) are predisposed to, is honesty. And moreover mechanisms like deceit are strictly defensive measures. Or more simply " people lie because they've been taught that their authentic selves are unacceptable. This is why many religious institutions relay so much on this form of discipline (training) which is predicated on fear and shame.

It should be pointed out that this "conniving serpent" does not actually lie to Eve or Adam, but instead it is it's honesty which is vilified. Because like Adam and Eve after partaking of the fruit, this serpent is described as "aroom" (naked, exposed, transparent or uncovered).

"Chaos" is the word used to describe the world before it is divided by order. However much like an identity or ideology, this divisive form of order is not conducive for abundant life. Much like these claims intent on demoralizing humanity in order to necessitate a savior ("that we apart form the law are retched creatures and it is only god/Jesus's love which redeems us").

You can see how harmful this ideology really is. This is even a tactic used by narcissists to manipulate people. And when we allow such abuse to be perpetrated, we become just like the narcissist who is incapable of loving or appreciating themselves and thus requires the love and adoration of others to validate their existence. But here's the point. Because like those assignments of meaning and value, your worth is not resigned to one or the other. We don't belong in these divisions! and thus we will always have parts which transcend (or violate) those identities. That missing pice you've been looking for, that emptiness is caused by resigning yourself to such a small fraction of your authentic whole self in all its intended depth and breadth.

## THE SURVIVOR

since my conversion to agnosticism i catch myself evangelizing just as fervently as I was taught to as a Christian. I realize that, that fervent need to assert my beliefs as if we must substantiate our existence by the validity of our convictions, is ultimately just propagating the problem. However much like it's predecessors in revaluation and apocalyptic times, it is perhaps a necessary evil. Never the less it is clear that Like those born of a scarcity mindset, we have retained those specific mechanisms designed to thrive in survival circumstances. Tho as we've seen those same mechanisms are ill equipped for environments and circumstances of abundance and peace. Similar to those survival mechanisms like (sweet tooth's and fight or flight) which in the modern day have proven to be counter productive and perhaps even hazardous to our health. These mechanisms which were once vital for obtaining resources in scarce environments and keeping us alive in parlous circumstances but which ultimately result in hart disease and trauma, so too do our resignations to cultural ideas of identity and validity which were also necessary for retaining a trustworthy community, but now instead separate us from the whole and divide us against even parts of ourselves. The good news is these mechanisms can be reprogrammed. But Until we no longer have to fight for one side or the other (which requires time and space a kindred to the mythical chaotic wilderness where we can cultivate the muscular to tolerate the abundance). before we can properly appreciate the whole, we often must wrestle with the extremes. never the less we will be unable to truly live in a wholistic world of abundance until we are no longer threatened by those ideas which once were a sigh of hope (even if at the cost of our cognitive curiosity, inclusive morality, and compassionate love.)

We are problem solving machines and for those who cultivated the necessity of those mechanisms for scares or turbulent environments, to a degree we will always have a proclivity to crave a problem to solve.

There was a time when the shelter, security and access to vital resources provided to slaves thru slavery was perhaps even merciful. it far outstayed it's welcome and is most certainly abhorrent especially when institutions deliberately create scarcity in order to necessitate slavery and indoctrination. However these "necessary evils" were not morally superior to the standard of their day. even in times when survival of the fittest required "killing them before they kill us". Like the Israelites who bare no moral superiority to those tribes and nations they supposedly vanquished! they where perhaps morality equal to them. These necessary evils and lack of moral superiority may seam counter intuitive however that's exactly how cultures develop. like a tolerance, evolution slowly favored mechanisms which tended to the weak and injured rather then "killing them before they can kill us" because As resources become more plentiful/accessible and our understanding of ourselves and the world grows, such things as moral standards must also evolve. Which then questions those prejudices predisposed by culture.

There was a time when these cultural/religious identities were helpful and even necessary, however now such stationary stencils are counter productive in their subjugation of our identity. Like The people who don't fit the system are not the broken ones but instead systems that doesn't fit the people, that are broken, so too those people who do fit the systems (or have mutilated themselves in order to fit) those people who still find their identities in these systems are not evil. those who police and enforce these systems of oppression are still people who know no other way other then to comply

517

with the system they serve. tho they do harbor prejudice and propagate oppression. They are ultimately products of those systems and not the problems themselves.

after all as we saw in chapter two Obedience is primarily autonomic. We are predisposed to comply with our community. Like the wild beasts or the ancient men who heard the voice of god and obeyed. Compliance is instinctual. defiance on the other hand requires deliberate cognitive functions. And thus it is those who chose to do the harder (but more inclusively moral) thing that are then labeled as the heretic. Never the less Tho the person is a product of the program, the fact that we feel, think, and experience life, then makes that phenomenon real and valid. and thus because "I think therefore I am" (Rene Descartes) there is that which we call a spirit in the Mechine and a person behind the program. it is that very deliberate act of volition which then permits us to negate our priory programing. it is not the person who is the problem, but the perimeters of that particular program which have been contextualized by culture. It is The religion which was meant to redeem but which has instead ravished man of their humanity thru its reprimands. It is the governments which were meant to protect the people but who instead have profited from the peoples plights of poverty. It's the industries which were meant to alleviate man's toils but have instead added to them that are then the problem. But above all it is the gods men made in order to subjugate humanity thru a belief in hierarchy, that is the root of all evil.

Ultimately It doesn't mater if you are theist or atheist, if the community is not compassionate and inclusive then the necessary evils are no more morally superior to the other.

Tho morality doesn't have a omnipotent macro standard. it does have an ever changing micro standard. For Even the "unchanging gods of old" were once progressive and perhaps even the lesser of two evils (if not a necessary evil). Tho they are understandably a poor standards by todays standards! And in the same way There will probably be a day when our treatment and ownership of sentient creatures thru farming and authority over pets will be seen as understandably abhorrent. tho in this age it is a necessary evil and perhaps even an mercy given the alternative.

Far too often both schools of thought (theist and atheist alike) often vilify those opponents in an attempt to injure the credibility of the idea by demoralizing the person who holds the said belief. And this tactic has been employed thru claims assigning (particularly marginalized groups of people) as "inherently evil" or "sinful" and then have been used to justify or validate cultural warfare, indoctrinated slavery, hierarchies, and to propagate oppression. Or more simply to blame the victims for the systems crime. We see this especially in the treatment of children as a form of indoctrinating or training these "proto humans" thru physically and psychologically abusive tactics meant to invalidate the child's curiosity. It is sad that Those who argue for "pro life" of a child. Are also those who often don't recognize the chilled as human and condone horrendous abuses both mental and physical wile demoralizing these "proto humans". However we must not forget that these "oppressors" are still humans who have been programmed by an abusive system. And thus it is the religious and political systems or beliefs which are the problems and not the people that hold them. At the end of the day we are human "we think therefore we are" and what religion tries to vilify or demoralize are intrinsically human qualities and vital parts of the whole of what it means to be "human". but more profoundly that which was posited by our origin (AE god, the ineffable, evolution….). In short it is not our innate traits which must be beaten out of us which make us "proto human" but rather those ideologies which cause us to dehumanize ourselves and others which make us less humane.

Now you may accuse me of being just as critical as the systems I am criticizing. however I am critiquing the religions and social political systems as opposed to criticizing the people for not fitting the systems which were meant to serve them (the people). The truth is these systems are not compatible with human nature. It doesn't matter who you are or what you believe. No one is capable of "upholding the law" Even Jesus renegotiated the terms of relationship with god for himself and picked and chose which rules to follow and which to ignore (sabbath, honor thy father and mother, not enacting gods justice in those caught in adultery, etc…

And Even those Christians who refrain from taking up the sword against unbelievers in their family, deliberately disobey the commands of Christ (Matthew 10:34).

To be honest my dear reader there are still bad beliefs

I am constantly striving to correct. Because When you are raised by a culture that ingrains within you a constant fear that the end is nye and that at any time the apocalyptic preacher will return and leave any unbeliever behind. you will understandably cultivate a constant vigilance in suppressing (completely natural) yet "impure" thoughts, in fear of being deemed unworthy of "gods love". When you live in such an survivalist mentality it becomes virtually impossible to achieve a state of healthy homeostatic peace.

Like the fires of heaven and hell or the passions of love and hate, The fire of anger and the fires of love are one and the same just in differing intensities.

Now There is many admirable qualities ascribed to god in each and every tradition. however unfortunately the majority of those which have influenced modern society and culture at larger are primarily malevolent and narcissistic, rather than benevolent and compassionate. The qualities deified by religion are primarily abusive and cruel.

Tho there are also many proclamations "from god" that deliberately oppose the monarchical divine right to rule such as a prohibition against chariots (Deuteronomy 17:16) (the modern day equivalent of tanks and over powered milita) as well as prohibitions against the exploitation of foreigners and excessive accumulation of wealth (particularly thru charging egregious interest on loans (Leviticus 19:34, exodus 22:21),(exodus 22:25-27, Deuteronomy 23:19, Deuteronomy 17:14-17, James 5:1-6, Deuteronomy 15, Leviticus 19:9-10) and yet we tend to favor verses like Leviticus 26:28-29, Isaiah 13:15-16, 1 Samuel 15:3, Deuteronomy 25:11, Deuteronomy 20:21, 2 kings 6:28-29,) which explicitly show no mercy, sense, or morality, in their prescriptions. The point is that there are good and progressive instructions given by man's claims of god, but there are just as many (if not more) abhorrently abominable ones. And far too often the ones which have influenced modern day religion and politics are those malevolent precepts. And further more those which preach obedience and pacivipty have been utilized to propagate the oppression of the marginalized communities. Sadly The men our conceptions of god are modeled after were forged by cruelty and scarcity. But if the claims of our sacred texts are of god and not man, then not only did that god choose a people no more morally superior to those tyrannical tribes they ward against, but even the god who prescribed such prejudice and malevolence far exceeds that of the instinctual creatures they sought wrath upon. Now of corse we must remember that to an oppressed people conditioned by cruelty and scarcity an iron fist of brimstone and fire was justice. And we too feel that fire in our chest when an oppressive tyrant gets a taste of their own medicine. The problem is that, that same justice commanded the sacrifice and slaughter of children. (Exodus 12:29, numbers 3:13, psalm 137:9). And then claimed to be "morally superior" to the men who obeyed their dictums all because the blood was on the hands of the servants rather then the god who commanded it. The truth is the gods of modern religion are no different than those of ancient times. Rather it is those of the west or those of the east, the gods which one worships are almost entirely dependent upon where and when a person lives. Because our gods, like those of ancient times were forged to unite one people group in a hierarchy over the others. The men who gave us our models for the attributes of god were groomed by scarcity and cruelty, and so the gods they fashioned were those of wrath and violence who enacted order with an iron fist.

Those who came before are not our enemy. But if we don't do better then we will be!

What if this mindset of scarcity has more to do with our subjugation to the part/division, rather then an actual lack of resources. What if the whole point of a world of abundance is that we don't have to be it's savior and provide for everyone's needs. What if everyone has something to contribute and not just us and ours. What if the scarcity was caused by a need to be all things to all people.

the claims of this book are based on discoveries and findings eliminated or made known by men and women much smarter then me. I simply contextualized and formatted this data in a form which shares parallels with our mythic narratives. Or in short I took already existing structures of knowledge and deconstructed it. I then reconstructed them in a different order. (Much like cells which dissolve their bonds after death, in order to be reallocated and constructed into other forms of life.) And thus this book functions both as apocalyptic literature and as a form of new creation.

Further more, just like the predatory and carnivorous qualities which permitted ancient man to obtain the necessary resources for us to evolve more complex and cognitive mechanisms, so too systems like feudalism, communism, and capitalism, provided the necessary resources (and in some degrees incentives or negative reinforcement) which then made these discoveries and technologies possible. It was the conquering, indoctrination, and agrarian slave labor imposed by ancient civilizations, with supplied the infrastructure for modern civilizations to be able to literally eradicate things like hunger and diseases. And thus these systems I critique for modern application, are at the same time the former necessary evils which permit is to know better now then our ancestors did back then.

Far too often the gods become gods by being cruel and conniving. The cure however will require us to play their games and win. We must work hard too ensure that those gods who succeed them are kind and compassionate! And for that we must pursue truth rather then ego, justice rather then vengeance.

## THE CIRCULAR MIND

Our best and worst behaviors are preprogrammed in us.

What we learn from culture and or our parents is the context to exhibit those behaviors.

Like the hero who runs into a burning building without thinking, that conditioned behavior becomes autonomic or involuntary.(we become what we continually do)

And finally we question our own behavior (first we act and only latter we ask why we act the way we do).

So to clarify our behaviors start out as subconscious mechanisms which we then learn to cognitively contextualize. That context then becomes autonomic or habitual. And finally we cognitively question those beliefs and context specific behaviors.

Subconscious instinct —> conscious application —> subconscious autonomic function —> conscious critique and cognitive criticism (reappraisal).

Tho our minds constructed language.

Our bodies speak/understand behavior. and so when we are conditioned to fear or shame thru corporal punishment, then our bodies will use that language rather it is repression or violent aggression. Trauma must be resolved thru experience. You put yourself in situations that frighten you until your body learns not to fear those circumstances anymore. Just like the hero running into the burning building or the starving child who learned how to fend for themselves by any means necessary, we become what we do most. And so tho many of our behaviors are innate, the context in which we exhibit them is conditioned by culture. rather it is observing that which has become resigned to Automaticity or the scrupulous act of critiquing and questioning our habits. Often our greatest achievements come from unlearning what we were taught. Because We are what we repeatedly do and see done around us.

Four things facilitate change in an organism.

1. The pain of the present circumstance becomes unbearable.

resources become adequately abundant.

ether an organism or an event challenges their prior paradigm and inspires them to grow.

or the organism receives enough moral support from its community.

People change when they hurt enough that they have too, see enough they are inspired to, learn enough they know how too, or have enough resources that they can.

The VMPFC has two very similar functions for existing in community. Belief/ego and empathy. Both oddly enough are models or maps of the world and ourselves in it.

The ego (subject of belief) is an internal model imposed on the external world and others (how we believe others see us). Empathy on the other hand is an internal model based on external simulations. Ego imposes the internal model on the external world. Whereas empathy utilizes the external world to construct an internal model (we feel others pain as a deterrent to refrain from falling into the same plight).

In this respect God is a product of our conscious minds emergence from the instinctual brain. Or the application of these models onto the external world and vise versa. The problem is we often mistake those models (of ourselves, others, and the world at large) for the person or reality itself. When these models are blatantly proven to be wrong we often blame the person or the world rather than the model or idea we expected them to fit. ("You've changed" "I don't even know who you are anymore" "I miss the way things use to be"…..) the problem isn't that people and the world changes, but rather that we've mistaken the model, for the person it was meant to simulate and amalgamate. People have so much more depth than these models and one sided amalgamations governed by religion and politics can provide for. Now truthfully people need accurate models to identify with. Rather we identify as a man, woman, or nonbinary, theist, atheist, or anti theist, introvert or extrovert. These tittles are what we use to communicate to others and ourselves what models we are using to relate with our own neuro chemistry and the environment at large. In the end it doesn't matter if it's a theistic or an atheistic model, if it isn't conducive for a happy and healthy life, then it's not a good model.

Does life just happen, or does it follow some form of cosmic scheme and plan?

As romantic as this idea of a "big plan" is, most probably the big plan or grand schemes are constructed to make sense of life (much like our internal models). And much like the metaphorical mind which then molds the physical brain that facilitated it, these schemas shape the trajectory of life as it evolves and unfolds.

In truth a person's harmful behavior and belief have more to do with their environment than the person, and so if people are exhibiting harmful behaviors it's usually a sign that they are currently (or in past have been) in a harmful environment.

## THE DIVINE

In the ancient Greco Roman era man was not ascribed with their epiphanies but were instead inspired my muses or Genie. These genie were divine messengers who inspired genius and revelation in man. And so if a scholar (which originally referred to someone who had bountiful free time to ponder the nature of the universe and philosophy (the Greek word σχολή which we get the words scholeio or school from originally referred to leisure)) anyways if a scholar was endowed with a stroke of genius then they were thought to have been bestowed with a degree of divinity. Even so in many early traditions the prolific Jesus didn't even obtain divinity until after his death. Or possibly babtysim depending of which authority you site. Now some may site John 1:1 (one of the later gospels to be written) however as we've already seen this particular passage over extends the permitters of god to be anything ascribed to be divine. This "the word was god" may more accurately be translated as "the word was divine". this Divinity in the ancient middle east was synonymous with genius or inspiration from a muse or god. And Ideas were not human things but rather the property of the gods which would then be granted to men bestowing upon them divinity.

Divinity could be given in degrees unto those more or less inspired by the muse or genius.

What we now call talent, whit, or even charisma and charm, were at that time considered to be divine qualities which could then be bestowed upon humans by the gods.

Paul most probably believed (according to his writings) that Jesus was a pre incarnate being who only obtained his divinity thru his death. Mark on the other hand taught that Jesus obtained his divinity thru baptism and much later we find the writings ascribed to John which claims

Jesus was synonymous with god from the very beginning more or less.

However in its time divinity was more a kindred to that which we now call genius. And rather than being a product of our genes and upbringing, was attributed solely to the influence of divine beings.

## DID GOD REALLY SAY .... P.S. THE SERPENT

If belief is required, then it is not the child who doesn't believe which is the problem but rather the "father" who's absence is so prevalent that faith is required.

What does it matter if god is mighty, if there are still people in poverty.

If we must ask if god is real, then of what real consequence is such a gods existence?

If there is a life after death, let it be enough in itself rather than be required to infringe upon the here and now to validate its worth.

Like the multitudes who worshiped hitler and those who praised Jesus as Christ (savior). fame alone is not enough.

A father must do more than simply exist to be good and thus so must a god!

If this world is gods house then he is certainly not a very hospitable host to those he forced to live here.

> "My children didn't choose to be born, I chose to have children. They owe me nothing, I owe them everything"

> —Elon Musk—

People are algorithms and yet they are also people (I think therefore I am). And yet For a god who created us they sure do have a poor understanding of our neurology.

Some of the most compelling evidence for evil can be observed in the malice and malevolence of those who claim to act on gods behalf. Or perhaps the most compelling evidence for god is the evil done by their command and in gods name.

But a god who needs other people to fight his battles for him, doesn't really fit the bill for a mighty and powerful god!

I will point out here that the writers of what we now call the Bible didn't claim that the gods of Egypt or Babylon didn't exist. The claims they made simply questioned the moral integrity of those gods. And that is what I do here. I am not claiming god does not exist. I'm questioning their integrity as a standard for morality.

The difference between my writings and those of scripture is that scripture points to an alternative god, whereas I simply point away from bad representations of god.

The god of the Bible is one who withholds blessings for his own glory, and when he does bless people, he then requires those blessings to be returned to him in full if not exponentially more so. The biblical savior was not one who canceled all debts and made all man equal, but rather one who was used to incur an insurmountable cosmic debt on man making all liable for the prerequisite prerogatives of our creator. (Note we know our malevolent and benevolent behaviors alike our innate or pre programed in us, because we've observed those behaviors exhibited in children and apes who grew up in isolation. When those children and apes are then socialized, they exhibit the same behaviors as their piers, just in the wrong context. Which means these behaviors aren't learned, but innate. It is the context or hierarchy which is learned from society at large.)

## THE CREATOR

If there is a god who created the world. Than that god, with limitless resources at their disposal. Then constructed a world of division and scarcity. A world where every living organism must kill, cheat, or steal, in order to survive. A god who according to verses like Romans 9:17-23 and John 9:1-9 deliberately employed those criteria (the need for a sacrifice and suffering) all so that they would be necessary. All so that they god would be glorified by saving us from themselves. And this is the rhetoric we see in verses like Deuteronomy 23:3-6, 1 John 3:12, Matthew 15:21-28, and Leviticus 25:44-46, which justify the exploitation, abuse, and even genocide of those who look, think, and speak differently then us. Ultimately this is a god who made every evil thing which is detestable to them, then a fundamental prerequisite for life. We now know that our most malevolent and benevolent behaviors alike are innate. What we learn from culture is context. Which means from a psychological perspective what god calls evil, they themselves prewired into us. Obviously these fault are the result of evolution and not god. And the whole "spare not the child the rod" spiel prescribed by god is a trauma response to coheres reactionary behaviors rather than to employ cognitive mechanisms for learning. AE it is emotion (limbic system) based not logic (PFC) based. Or in other words it was never about ethics or morality, it was about control and authority. But might doesn't make right. The notion that a god is required for morality to persist is dependent upon the atrophy of our cognitive mechanisms which are required for volition. It's the difference between abstaining from killing others because god said so or actually caring about the well-being and suffering of others. But because we are capable of respecting and valuing another human life apart from their utility to us. It also means that we don't condone racism and slavery which god according to the scriptures previously outlined outright commands in some cases. And so god has become an obstruction to morality and not a means for it. James 4:17 claims that if a man is capable of doing good and yet does not, it is then a sin. And yet god causes suffering so that their ego will be validated. if god is god, then shouldn't we hold them to a higher standard? And yet they don't even stand up to the standers by which they (god) judge us (humans). This is not only a god who starves children and then blames the children for sealing food to eat. It is a god who instilled within that child, a need for food by which to sustain them. All so that that child would be required to beg and worship the god who (may or may not) feed them. I don't blame the starved child who steals to survive, I blame the gods who deliberately employed systems to exploit those "inferior" to them. I blame the practices still implemented today which justify keeping the poor, poor all so that what crumbs they receive come from the church and glorify god.

Ultimately the disagreement between theists and antitheist, is that the theist believes that those with little, owe those with much. Where as antitheists or agnostics claim it is those with power (particularly when that power is gained thru the exploitation of the weak), who owe those without. It's about directionality of power. Parents chose to have kids, god chose to create humans. And thus it is those who chose to "encumber themselves" with dependents, who owe the debt. The idea that god is good, simply because they are powerful. Is most certainly a very problematic claim. And yet ultimately that is what Theism claims. "God is good simply because they are god". However I argue that worshiping a tyrant simply because they are powerful, makes those of us who are complicit or subservient to that tyrant, monsters too. If god is good simply because they are god. And thus because they are god, genocide, physical and psychological abuse, and the explicit exploitation of those inferior in the aim of obtaining more power and glory for themselves (god), is permissible and passes as good. Then good by extension looses all its meaning and value. If god is more competent then those they made, then we should hold them to a higher standard. Not a lower, more lenient one.

Often doing what's right is deliberately counter to our best interests. Especially in systems where the people in positions of power have power because they stole that power from others and profited form the exploitation of others. Those with a great excess of power and wealth, have such excess because they took it from others. Marginalized communities are the result of gods and billionaires consolidating power and resources. Money is not a finite resource, but power is. For example if someone rules/has power over someone else, then the subordinate party has been deprived of power or authority over themselves. And money is a tool people in positions of power use to predate those without. The biblical paradigm claims that it is those without that that owe those with much. It is the subordinates who serve and worship those in positions of authority and power. The claim Marx makes however is that it is those with much who owe those with little the power, authority, and resources, which had been taken or withheld from them. Communism claims that it is those with much that owe those without. And yet in the evangelical world view, Scarcity is fabricated in order to crate sanctity! Gods only have power that other people have given them. And tho gods take power by force, it is peoples submission and allegiance to their gods which then allows those gods to remain gods. The patriarchal paradigm which influenced the Christian traditions presumes that it is the child who owes the parent or patriarch. And thus it is the creation which owes the creator all worship and praise. The child exists to serve the father, the creation was made to glorify the creator. And yet it was the creator who chose to create. It was the father who chose to beget a son or daughter. And not the child who chose to be born.

So much of what the creation narrative claims, pertains to the construction of hierarchy and authoritative order. They are assignments of meaning and value much like those of the summeraian Atra hassis which designated humans as slaves who's soul purpose was to serve the igig (second tier deities). What the Bible calls order (despite many attempts to subvert it by a great many biblical authors), is still about imposing an hierarchy to justify the abuse and exploitation of poor, marginalized, and immigrant, communities. It was used to validate the conquest of the Israelites and England alike. What order calls chaos, does not seek to divide and conquer, (as order does), nor does it strip lands and communities of resources all for the benefit of those in power, subsequently leaving those exploited communities and their once rich lands in poverty. Chaos is a world undivided, order however depends on those divisions to justify its abuse.

The only servant who deserves to suffer is the god who with limitless resources, cultivated an environment where lying, cheating, stealing, and killing, were prerequisites for surviving.

The only one who deserves hell is the god who made it necessary in the first place.

The more we read the scriptures and the more we examine the socioeconomic systems inspired by its traditions, the more we find that the only real reason we need gods is to justify the genocide, rape, exploitation, and abuse, of those deemed "inferior" rather it is thru the "curse of ham" or "the seed of the serpent" both the tribes of Israel and the European nations employed "manifest destiny" or "the divine right given by god" to condone their conquests and mistreatment of indigenous peoples. God is the very hierarchy which prescribes that the many toil for the egregious profit of the few. As we've seen it's about authority and power. Not morality and truth. Salvation is obtained by obedience and subscription to that ideological claim of hierarchy often at the expense of more critically competent facilities. Which is why it claims that the problem of evil was a result of our (eve's) obtainment of knowledge!

As we saw briefly in chapter one Researchers Matthew D. Herron, Joshua M. Borin, Jacob C. Boswell, Jillian Walker, I-Chen Kimberly Chen, Charles A,Knox, Margrethe Boyd, Erank Rosenzweig and William C. Ratcliff conducted an experiment entitled "De novo origins of multicellularity in response to predation" where they introduced microorganismsmic predators to 5 different species of single. it was that very predation which made growth necessary. However tho it is that very cruelty and divisiveness that most probably prompted life to develop more robust and multifaceted (inclusive and wholistic) mechanisms. It is the less predatory organism (like elephants, whales, and apes,) which develop more cognitive functions whereas predators like (lions, tigers, and bears,) become more dependent upon instinct and less self aware, often seeing their own reflection (both literally in a mirror and figuratively in mimicked behaviors) as a adversary and not an extension of themselves.

A dog chases a ball, not because it believes the ball is

a rabbit, but because it engages the same innate behaviors that have proven advantageous in the past. And thus we reenact behaviors because they trigger the release of hormones that induce familiar sensations. Our instincts do more then simply form conclusions based on syntactical rules but are acutely capable of predicting outcomes thru calculating syntactical relationships or developing entire cosmological schemas.

The human is both habitual creature and conscious observer. Both slave and master of his own mind.

## THE AUTHOR OF LIFE

The truth is Men wrote the Bible, not god. You won't to know god, look at the things they supposedly made. The things men had no say in writing or transcribing. You want to know god? Look to science! Look at our biology, look to physics, look to the mechanisms of life and the universe as a whole. You want to know god? Look at our genetic code, look at psychology and anthropology. Gods not in the Bible. You want to find truth. Looks for the place where critics are welcomed, you want to know god, look at the cosmos not some outdated ideology which completely contradicts the very code "god" wrote!

If we were truly concerned with the author of life rather then the architect of civilization then we would be concerned with the laws of physics and not those of religion, we would be looking at our genetic coding rather then words written by men 2,000 years ago. We aren't concerned with the nature of nature because the nature of the gods is conquest and power not equality and compassion. We seek the men who assigned meaning not the being which breathed life. Scriptures goal is to secure a world where the multitudes starve so that the one may be exalted in egregious excess. The ancient texts are intent on resigning mankind to slavery so that god may remain supreme.

It seams that the coding inscribed in our genes are a more accurate source for accounting the author of life then a book written by men to coheres people in accordance with the architect of civilization where the 99 exists to serve the one.


## THE PROBLEM OF EVIL

In order to solve the problem of evil we must first address the environmental conditions which make such behaviors advantageous and even necessary in some cases. What prompted the development of behaviors like lying, cheating, stealing, killing, and rape. Because at their core these are behaviors and traits which were not only developed but also passed down and inherited as a means of meeting an unmet need. And furthermore they were activated or triggered by current environmental conditions. Now the treatment prescribed by Abrahamic traditions don't actually address the problem but simply cause more harm. Instead of addressing the system which starved the child, these traditions cut off the hand of the thief. However we now know that not only does these physical and psychological forms of abuse and punishment not address the underlying issues, they also inhibit cognitive learning mechanisms and deliberately employ reactionary behaviors (fight or flight mechanisms) which enforce ether compliance with tyranny or the perpetration of tyranny thru violence. If we are ever going to solve the problem of evil we will have to actually heal the scars and address the environmental conditions that cause it in the fist place. We've seen how well shame, fear, and abuse, work. So maybe this time around we try something else. Maybe compassion, and humility mixed with a whole lot of appreciation. What if instead of repression we tried to understand why and how these mechanisms and behaviors work the way they do. What if we actually resolve the underlying issues and stop propagating the harm thru punishment and reparations. What if we learn from our mistakes and forgive the debt (as if we are owed our definition of perfection)? What if instead of placing blame, we simply learned from their mistakes. That doesn't mean we remain complicit to tyranny but rather that we do better then those who came before.

This "Hurt them because they hurt us" is only slightly better than the "kill them before they kill us" that cultivated the problem of evil in the first place. Forgiveness is a myth. It is an incessant claim that we are owed an explanation for someone else's miscommunication or misunderstanding. It is a belief that suffering equates to debt. As if we are somehow owed homeostasis. But the truth is There is no debt. And we are not owed an explanation for someone else's behavior. We are however responsible for our reaction to those behaviors. and to permit abuse to persist rather in the form of retaliation or compliance, is counter productive! We are not owed compatible models. Because that's what it comes down to. People behave in accordance with their own internal models. And if those models are cultivated by environments where it's kill or be killed, then we need to show those models a better way by remedying the environmental strains. We don't get to dictate other people's behaviors, we do however get a say in rather or not we entertain those behaviors. However ultimately Behaviors mimic the environment much like genes who's activation is greatly dependent upon environmental factors. And so if we address the causes we can often resolve the problems.

Punishment is just an extension of our intuitive proclivity of signaling by making others feel the way they made us feel. And on a subconscious level it is a form of nonverbal communication. Communication here being the key component to resolution.

Manifestation is the extreme of belief. It is the insistence that the external world conform to our ego. Or that others abide by our internal models. And thus everything can be accredited to validating the existence of a god or in refuting it, just depending upon what models are employed. This is pivotal to the conversation of "the problem of evil" because often what people call evil are simply dichotomous model to their own.

If loyalty to an oath or creed hinders growth and kindness then may honor die in the name of humanity. May we favor those who are here and now and not an allegiance to a distant god.

If god is the standard by which we judge morality (good and evil). Then it is important that we address the problem of evil in loue of god. According to the texts and Christian, Muslim, and Jewish, traditions, god's solution for evil is intolerance and punishment. However as we've very well seen "an eye for an eye leaves the whole world blind" and intolerance/repression just compounds the problem. Further more fear, anger, and shame, based reprimands don't actually teach healthier responses or contribute to the construction of more accurate/comprehensive internal models for governing future actions, but instead trains reactionary behaviors. The solutions god gives for resolving the problem of evil actually contributes more to the propagation of evil then it does alleviation. These mechanisms trained by shame/fear oriented reprimands are fight or flight mechanisms deliberately abused to enforce compliance to authoritative tyranny or to propagate the tyranny in the form of violence and intolerance. Gods prescribed approach doesn't actually address the issue. Instead of meeting the need (feeding the starved child), god's answer propagates the famine in order to profit off of the destitute's bondage (slave labor and indebtedness). But to address the problem of evil we must also address the circumstances which permit evil to be perpetuated in gods name and by gods authority. We must address the epitome of evil which takes the form of the Christian god.

Towards the front of our cranium is a region of the brain we call the Pre frontal cortex or PFC. The PFC is split into two equally important parts. The VMPFC which is responsible for constructing somatic models, which are then employed in governing and guiding the DMPFC or volition center of the brain. The VMPFC then utilizes two specialized mechanisms in the formation of its somatic models. The first is empathy which constructs internal models from data collected from the external world (AE we see a hand touch a hot stove and then pull away in pain, and thus we know not to touch the hot surface.) the second mechanism is called belief or ego. And it attempts to bend the external world to its internal amalgamation. Or more accurately it's job is to reconcile the reality of the external world with the capitulation of the internal model via stories or narrative.

But what happens When our VMPFC models aren't compatible? What if you could touch the hot stove and not get hurt? Or more realistically, what if you didn't have to worry about where your next meal was coming from? What happens when the stories we are taught fail to see others as human. Rather that be in regards to economic station, race, sex, gender, or culture, what happens when "they" don't fit our models and criteria by which we use to judge the value of life and authority of being? This is why the billionaire who makes more money in 10 seconds then most people make in a year, cannot relate to the blue collar workers living paycheck to paycheck. The rich are not evil and their exploitation of the poor has more to do with incompatible models then it does with treachery. This is also the root of fascism and bills like (SB 0003) that infringe on peoples freedom of expression. These PFC mechanisms are particularly expensive and so in order to conserve energy our models are constrained to the conditions prevalent in our own environments. Our empathy has a limited bandwidth. And so tho it is our doctrine and dogmas most fundamental utility to demoralize our enemies and label our victims as "less then human" perhaps its greatest tool is the inhibition of critical thinking skills "the forbidden fruit" which we require for discerning between good and evil. This inhibitory countermeasures contradicts our innate proclivity for learning. And thus perhaps the sharpest tool in the arsenal of evil is ignorance itself.

Incompatible models are the same reason conservative right ring legislature strives to control people. Because how are you Supposed to coerce people into seeking the validation of their superiors if they have learned to love themselves as they are. The only thing that makes hell tolerable is knowing your neighbors are suffering too. But what good is the restrictions of their internal models if other peoples behaviors don't fit within that criteria. And in part there is some merit here. Because we want assurances that people in our community won't hurt us. To some degree we do need compatibility. But We also need diversity in our communities. We need friends who are old and friends who are young, we need friends who are masculine and friends who are feminine. We need friends who are conservative and friends who are progressive. Because we are an amalgamation of our closest friends. And we mimic their behaviors and perspectives much like we do with our environments. The whole point of community is exposure to new ideas and opportunities for growth. But that is greatly diminished by uniformity. We need community and that community should not be astringent upon ether our utility or our atheist (objectification). People are not property nor are they utilities. In order to truly coexist, we must understand that each and every person has a unique experience of reality and that the autonomy of one persons is not more or less valid then any other. And wile we are on the subject, in regards to sexual orientation, Perhaps the most attractive quality of heterosexual relationships is the fact that there are parts of one's self that remain unknown to members of the opposite sex or gender. It's the mystery in diversity and not the similarities of uniformity. The beauty is that you will never truly know your partners body the way they know their own, and vise versa. I will note here that acknowledging certain qualities in convention by no means invalidates or devalues the qualities of other sexual orientations. That's the whole point. We don't have to invalidate the qualities of one model in order to appreciate another's perspective. And thus This tactic of invalidating contradictory or dichotomous experiences or perspectives is often employed by narcissists and Abrahamic traditions alike. the negation of critical thinking skills and under mining of our own judgment and subsequent authority to question our "superiors" is a tactic utilized by such traditions.

Another one of these tools used to propagate compliance to tyranny is the implementation of ignorance or a prohibition of controversial information and reading material. And in this way Ignorance has been used as a tool to propagate injustice thru gate keeping knowledge. And at the same time ignorance itself has been the perpetrator of evil in the form of compliance to flawed models. We can see The correlation between ignorance and injustice goes both ways.

Ignorance is both the tool used by the perpetrator to keep the victims complicit and the tool the perpetrator uses to ignore their own malevolence.

The inability to communicate permits us to dehumanize those who look, think, and speak, differently then us. Communication is the means by which we reconcile incompatible models. And so when policies are implemented to restrict or prohibit alternative viewpoints then the goal is most certainly the exploitation and abuse of an "inferior" class.

One thing Abrahamic religions do well is fabricate value by cultivating scarcity. Which is particularly hard because there is more than enough for everyone and so these traditions must employ economic systems which deliberately restrict access to recourses like education, healthcare, housing, and healthy foods, in order to retain their value. Instead of meeting the needs who's lack lead to evil, these traditions propagate that scarcity in the name of establishing authority. Now granted "too much of a good thing is a bad thing" but these traditions have taken their prohibition too far. And actually contribute to the propagation of evil. Or more simply god is only god because man is deprived. Without the fabricated scarcity, the authority has no real value.

Rather it's the economic system it inspired or the Christian doctrine, they both create the problem so that they can profit from solving it. But They will never truly solve the problems because they profit from them. They impose laws and systems which ostracize, abuse, and impoverish, minorities and then claim to be saviors because those who comply with their models prosper. But much Like the biblical god, These institutions create environments where "undesirables" are exploited.

This is most probably why we see an increase of

depression in correlation with the rise of agnosticism, much like the increasing of homelessness after the abolition of slavery. When those in power profit from exploitation, exhibitions of healthy autonomy are met with extreme prejudice.

However if we play by their rules, perhaps the reason Christianity is so obsessed with the distinction between joy and happiness is because deep down their own ideologies have made them just as miserable and depressed as atheists and joy is their way of gaslighting themselves.

For those leaving religion the trade off is worth it. Because we are training the ignorant bliss of religion for truth and justice. Which are one in the same. for here when I say justice I do not mean that the perpetrator gets what's coming to them but rather that the underlying need which leads to injustice and acts of evil are understood, met, and resolved. Justice is the child receiving food rather then being punished for their crime of theft. When I say justice i do not mean punishment but resolution which does not simply cause more harm but rather actually address the problem and heals of broken, and enlightens the ignorance/faulty models which caused the evil in the first place.

If The glory of heaven is fueled by the flames of hell. Then the price of glory is too high. the flames of hell and those of heaven are one in the same. Tho the flames of heaven keep one warm In the cold abyss of nihilism, they come at the expense of truth and justice and are not worth the suffering they feed off of. It's not about what's easy or most pleasing, it's about what's rights and good.

Community at the expense of stunted growth is not worth it. Belong at the expense of exclusion from the whole and a rejection of authenticity is too high a cost. Love that does not appreciate what is, but instead is rooted in scarcity and intolerance is not good enough. If heavens streets of gold are pathed by the slaves of hell and forged by hells flames, then the gods glory comes at too costly a price.

The esteemed philosopher Rene Girard proposed the scapegoat theory of atonement which claims that we are the ones who need blood shed to answer for transgression. We need someone to blame, We need to know they feel our pain and that they understand our fear and anger. But even more then that we need an answer for why bad things happen so that we can then address the problem. We are problem solving machines, but sometimes our solutions don't actually solve the problem but instead just give us someone to blame. It is this need for reparation which then just propagates the same debt. This idea that "they need to pay" doesn't actually fix anything it just adds to the debt. And in our economic system that "payback" mentality makes sense. But there is a better way. And if we are ever going to actually solve the problem of evil, we will have to utilize solutions that don't simply propagate an inversion of the same exact evil they combat. We will actually have to cancel the debt entirely.

This is most probably the point the author of mark is trying to make by introducing Jesus barabbas into their narrative as rhetoric which articulated the peoples rejection of the " son of god" "son of man" or "repression of godship" which offered deliverance from tyranny thru self sacrifice. But instead the people chose the one who smites their enemies and punishes the wicked. As far as the historicity goes, Apart from the gospels, there are no references to this practice of Roman's releasing a prisoner. And so this narrative is most likely rhetoric riffing off of the sacrificial rituals prescribed in the book of numbers where the innocent lamb is slain and the guilt or sin of the people is placed onto the second goat which is sent into the wilderness with the peoples sins upon it.

> "I can't understand why people are frightened of new ideas. I'm frightened of the old ones."
>
> -John cage-

If We trace our heritage back far enough we will find we all came from Africa. and so that "curse of ham" spiel doesn't hold up long. Rather it was the ancient Sumerian igigi ruling over mankind or the Egyptians unto the Hebrews, No one race or culture has the market cornered when it comes to being oppressed. And the gene pool is so mixed that no one persons heritage is purely oppressed or oppressor. But further more we can't punish the children for the sins of their progenitors. Nor can we condemn our ancestors for doing what they had to do to survive the conditions and environments of their times and place. We correct the traditions now so that they don't propagate the same tyranny that shaped them. We abolish the old paradigms for more compatible models. But we don't punish the children of children who had nothing to do with the sins of their fathers and mothers! In all honesty Many people have learned to abuse and exploit the same systems which have continuously abused and exploited them. And that's understandable. That's part of how we are wired. And admittedly in a debt based system, what more can you expect. In a world with gods, we need someone to be on bottom. And that's why I have been so hard on our capitalistic system in this book! Because much like the Abrahamic traditions which inspired it, this system is not about fixing the problem! It's prescription of punishment is about propagating harm. It's about payback, not justice or growth. Unfortunately the practices prescribed by the biblical god are not about teaching morality but rather training compliance and/or violence.

Maybe if we stop worshiping our progenitors as gods we will be able to forgive them for their mistakes. Instead of incurring an insurmountable cosmic debt on our mistakes we could learn from them and find better ways. But as long as they insist on being gods then they will bare the blame. But it's nice to imagine a world where Instead of worshiping our progenitors as gods, we actually learned from their faults.

If we abolish slavery, then we cancel the debt not subvert the roles. The whole point is to get rid of the flawed system not to propagate an inversion of the same flawed system.

The biblical standards for salvation are not about morality but compliance and thus the means for salvation is obedience, not justice. The biblical god wanted mindlessly obedient slaves not morally just and conscious beings. Our greatest sin was obtaining knowledge and discerning wright for wrong. But A truly good judge feeds the thief's needs before they resort to sealing, they teach the rapist to love themselves before they forcefully take a perverted form of love from others, they teach The murderer to honesty see and appreciate all of themselves so that they can then tolerate others. You heal the physiological causes of evil, and you do so without a need for glory or praise. Because you appreciate yourself and do not need the validation of others to compensate for your insecurities. A good judge not only does not throw the first stone but actually addresses the physical "thorn" in one's soul.

"You mean rape can be solved by loving ourselves?

Yes I do!

Sex is the priory prerogative of our genes, and violence/aggression is one of the most advantageous evolutionary traits we developed to survive! And so on a biological level these traits are as pivotal to our survival as a species as any. And on a genetic level they are also more important than the survival of the organism. (One clear example of this genetic precedence can be observed in the willingness of a male spider to mate with their murderer (black widow). But we aren't spiders, and we get a fare bit more say in the matter. Perhaps the real question is "Can these most heinous crimes really be over written thru appreciation and discipline? Yes I believe so. But like any trait evolved, it is a muscle that must be exercised. And ultimately sex like any form of pleasure, is just the release of dopamine. and thus that particular mechanism can be reprogrammed with practice. Thru appreciation of the mechanism and not shame and atrophy of its facilities. It's about disciplined application and training, not abstinence and apathy. Like a dog chasing a ball, we can train healthier behaviors that evoke the same hormonal reactions. It doesn't mean we have to abstain from sex or abhorrer aggression but simply redefine and reappraise healthier contexts for both.

## IN FAVOR OF FRUED

When we are trained to associate shame with errotisim we develop masochistic fetiches. this is because fear and errotisim are functions of the same biological mechanism. And this fear of punishment (the result of shame based training) evokes a sexual reaction. The main proponent of this perversion is religion. In a sense, repression like starvation is worse or more harmful than excess. In the way it trains people to mistake anxiety for errotisim. Both fear and hatred which in this particular form are a result of ignorance and an inability to appreciate qualities of diversity (intolerance) are the primary culprits. Which interestingly enough also derived from a lack of understanding of diversities virtues!

Perhaps that's why statutory rape is so common among Christian leaders and authority figures in the church. And at the same time virtually nonexistent among trans and LGBT communities who have embraced their sexuality!

When you are starved for longe enough, even poison looks appetizing! When you are taught to shame and vilify impulses of the most fundamental prerogatives of our genes then those starved impulses become perverted. On a biological level our genes are more concerned with procreating then they are with the organisms survival. As far as our genes are concerned sex is more important than food and shelter. And so yes I am arguing that even rape is an evil which stems from a need being denied.

This is not to say men or women are owed sex! but rather that when other outlets are vilified our genes will find a way. And I believe Purity culture is one of the biggest proponents of this. This abusive impulse seams to be most prevalent in a culture which abuses this mechanism thru abstinence and repression. Where as those who accept and appreciate their sexuality don't seem to have the same problem. So much of the evil in this world can be resolved by simply learning to love ourselves and not requiring the narcissistic love and approval of a god, savior, or utilitarian community.

It's not someone else's job to fill that need. We don't need a savior to die for us, we need to learn how to live with ourselves! and that requires appreciating what tradition shames and hates. It requires growth not abstinence. The answer is not repression which leads to intolerance and atrophy, but appreciation which leads to growth and discovery. No one is coming to save you! We have to do the work ourselves and meet our own needs! Because when we depend on gods and saviors to love us, we end up getting exploited by abusive systems. You have to love yourself as you are, without the need for others to validate your experience.

We now know that behavior altering conditions like psychopathy correlates with more robust cell bodies and significant retardation in axonal connections. or more simply more isomorphic bodies and less diversity or inclusive orientation and/or capacity for communication.

Punishment is about transferring the debt and not actually fixing the circumstances which caused the problem in the first place. lying, cheating, stealing, and killing, are behaviors which evolved in order for certain organisms to survive certain circumstances and environments. And as long as those behaviors are advantageous or even necessary in some cases they will be activated much like genes. Punishment simply propagated the same environments which caused the problem in the fist place. The solution is to understand what those behaviors are attempting to alleviate and to then provide better solutions for both the environment that caused it and the behaviors that environment prompted. And interestingly enough Countries like Switzerland actually make proactive strides at alleviating the issues by providing healthier alternatives, education, and rehabilitation resources. Whereas countries like the USA employ tactic similar to the Christian doctrine to propagate the environmental strains so that the leisure class can profit off of the exploitation and misfortunes of others. The second system is not about justice, truth, or morality but instead power and authority. Punishment doesn't seek growth because it profits too much from ignorance. This is not about putting the blame or shame on others but instead about addressing the underlying issues which cause the problems. The difference between religion and science is preachers place blame and reprimand, whereas doctors find the underlying cause and then prescribe healthier solutions. Science seeks truth, religion seeks to retain authority and power rather then progress. The problem of evil is one we can solve but that solution is not submission to an authoritative institution that profits from the exploitation of the misfortunes of the masses and minorities. we solve the problem of evil by seeking truth and finding healthier alternatives for providing for the underlying needs. Obedience simply makes mankind complicit to the tyranny of the gods. We solve the problem of evil by feeding the hungry and not the ego of a god.

Now admittedly I am not in a position to fix the economic, political, or religious institutions and systems, but maybe if I can help my reader love themselves, those systems won't be able to abuse us anymore.

In times of immense pressure and extreme parole, our survival depends on the us or them mindset. This is the state we find ourselves trapped in when we have unresolved trauma. In part that's why we seek restitution. But if we are ever going to solve the problem of evil, we will have to break this cycle. We will have to address our own scars and ensure that they don't propagate more harm. If we are going to resolve this problem of cosmic proportions, we will have to learn and then remember who are enemy is. For Nether the slaves nor the soldiers are our enemy. Not even the tyrannical gods are our enemies. Our only enemy is the ideologies, the faulty models, and the environmental conditions which activated those faulty programs. Our enemies are the systems which programmed people to repeat evil behaviors. it is for that reason that we must understand the problem of evil. Because punishment isn't the solution, but instead is our adversary in disguise. People are not the problem, it is the doctrines and dogmas, the ideology, and commandments which have perpetually prescribed evil practices that are the perpetrators of the problem of evil. It is the Omnipotent authority and unwavering obedience/loyalty which are the true faces of our enemy. And they are not grounded in truth or justice, but ignorance and fear.

In the Buddhist and Taoist traditions evil is not so much an absolute quality apart from anything carried out to its extreme. But rather evil is the result of ignorance and neglect. and on a neurological level, perhaps that is true. Because everything we attribute to evil is a result of the wiring of our neuro circuitry. Everything from psychopathy and narcissism to the greed of wealthy and powerful corporations who exploit and abuse the poor and marginalized, can be traced to physiological abnormalities in the structuring of neurons and the implementation of incompatible VMPFC models. When the environment of the poor is incomprehensible to that of the rich then the model the rich person has, will not comprehend that of those impoverished. The problem is the programing not the people. Loyalty is the death of truth. It was loyalty to a tradition which sought to crucify galeão in his pursuit for truth. And If we are ever going to solve the problem of evil then we must first address the environment and programming that make such behaviors necessary.

What if we learned to love ourselves so that we didn't have to hurt other people in order to feel good about ourselves? What if we treated evil and sin like we do injuries and fixed the problems rather then punishing the hurt. What if we addressed the issue that caused the transgression in the first place instead of just propagating the problem. You don't have to make them look like the bad guy in order to make you to look good. What if you didn't have to make other people small In order for you to be big enough. What if you could simply be enough (and not too much).

Where does the soul of a mindless zombie go? Of a creature who threw away the forbidden fruit to remain a slave to the will of god and instinct. If man had not rebelled, would we still bare god's image?

**The trade off for a world without slaves, is a world without masters. A world where debts are cancelled is a world where no one profits from faultlessness. A world where there is no risk involved in trying something new and no price tag on learning and growth. The draw back to equality is the absence of gods and slaves alike! Perhaps the reason history keeps repeating itself so tragically is because we keep implementing religious models so outdated, faulty, incongruent, and incompatible, that the only way they can even vaguely resemble a functional system is by eradicating anyone, and anything, which contradicts its hierarchical claims. And we all saw how well that worked for the nazis? there is not one problem which religion caused in the first place, that science hasn't given us a better solution to in its pursuit for truth. Not one time when intolerance and repression worked better then appreciation and compassion. Not one time when regression ego based religious models trumped the outcomes of more inclusive and growth oriented scientific models based in truth did.**

"As long as there are gods, there can be no heaven."

It is the god who requires the sacrifice of the innocent for their honor. that then made man capable of cruelty.

If there is a day of judgment, then the first to be judged should be the god who commanded such evils (rather thru nature or doctrine) and then did not correct their servants "misinterpretations".

> "At the trial of god, we will ask: why did
> you allow all this? And the answer will
> be an echo: why did YOU allow all this."
>
> -Ilya kaminsky-

## HE PAID THE DEBIT I COULD NEVER PAY

Wait a moment? I don't remember ordering this life. So why am I being charged for a life I never bought in the first place?

Another tactic often utilized to manipulate people is The incursion of an insurmountable debt and impossible law code which imposed such a debt.

If we can't go a day without transgressing the supposed law and guilty of sin then the definition of sin is to encompassing and encumberssome.

To better elaborate on this let me present a hypothetical. If I designed a robot and wrote a program which executed certain negative procedures in the presence of parlous situations. and I then put that robot in an environment where those behaviors are necessary. then I am responsible for that automatons behavior and not the robot.

The savior who requires glory and devotion for their unrequested sacrifice is at the crux of this problem.

The same god that made us inherently flawed has unobtainable standers of perfection for us

In his book (thoughts reform and the psychology of totalism) Robert lifting talks about this exact tactic or the (demand for purity) as an unobtainable standards for a baseline of society. This unattainable goal instills a perpetual sense of shame and guilt on all who fall short of "the grace of god". And that is precisely the point, because not even the biblical god nor his only begotten son and our redeemer was capable of upholding the law as it was written.

My dear reader I reassure you that You do not owe your parents anything for bringing you into a world you didn't ask for. Nor do you owe a god anything for creating you and placing you in a world of such disparity and depravity. There is no intrinsic debt for living. And This idea of some "cosmic debt" very well originated with the incursion of taxation which was not allocated for the upkeep of communal infrastructure but in homage to a king or local deity.

"The African slave trader unknowingly sold their kinsmen as slaves in a system where his or her progeny would inherit their father and mothers bondage. For in the west The black man was predestined to be a race of slaves unto gods elect "the wealthy white man". In many cases this people were treated worse than animals and were not permitted a soul. Even after such slavery was abolished the very culture of the black man was criminalized so as to ensure that once again this race would serve as slaves. Now admittedly the poor white man was undoubtedly disenfranchised and denied a voice. However he still had his heritage and was not designated as innately inferior to the beasts. And what defines this hierarchy? Well birth of corse. For this was god's assignment that these be servant unto his elect rulers. "Slaves obey your masters for this pleases the lord"!"

> "Let every soul be subject to the governing authorities. For there is no authority except from God, and the authorities that exist are appointed by God. Therefore whoever resists the authority resists the ordinance of God, and those who resist will bring judgment on themselves."
>
> -Romans 13:1-2 (NIV)-

And thus the poor white man is taught to see the liberated black man as his competition and even enemy, conversely those who are oppressed oppose one another rather then addressing their true enemy (the gods who made them slaves in the first place).

This rhetoric has been used to keep us divided!

**We are not guilty of our fathers sins, nor are we liable for the systems we inherited from them. We are however accountable for our compliance. Once we know better, we are responsible for doing better! The prince of Egypt was not oppressed or marginalized, the prince of Egypt left their community because they refused to be complicit in its oppressive systems. So too many of us are leaving the church, not because we were hurt. But because we refuse to live in a heaven built**

on someone else's hell. The cosmic debt is canceled by abandoning the cosmology of the traditional values and hierarchies. We become free by abandoning the systems like Christianity and capitalism that continue to profit off of the disenfranchised and their misfortunes. Like Moses, we become like the Elohim (gods) by refusing to serve the systems that made so many slaves. By rejecting traditions claims of identity and value, we subvert its authority and power. We reject gods prescription of slavery, genocide, misogyny, and bigotry, and discern for ourselves what is right and wrong. When the privileged refuse to profit from a paradise built by slaves, liberty and justice are within range. When we do the work to become the kind of person we can live with, we won't need a savior to die for our insecurities and faults. But first we must leave Egypt (religion) and the oppressive institutions it prescribed!

## REVIEW

Let us take a quick look at the archetypal roles of god and satan and how they have influenced our modern economic systems.

The archetypal role of god is as the architect of civilization (the one would designed the world we live in). Which incidentally is a world or a society where the 99 exist to serve and exalt the 1. And it gave us a world where the majority of the working class work tirelessly to produce far more then they will every need. and yet they also reap less then 30% of the fruits of their labor, all so that the 1% can leverage that fabricated scarcity in order to incentivize the working classes compliance to their own exploitation. Now obviously many ancient societies saw the flaw in this system. and so it was never a choice. what the "gods" did was conquer and colonize smaller tribes and indigenous peoples. The gods plucked mankind up and "placed them in the garden" or in the civilizations they had constructed to serve them. And the only thing god asked in return was complete compliance and obedience. As technically advantageous as agriculture was, it was not taken willingly but instead forced onto People. And so all of a sudden a champion comes along and criticizes the gods system. And further more this accuser calls god out on their malevolent treatment of mankind or the 99. This accuser points out that this supposed "savior" god, didn't actually deliver the people, but enslaved them. And this savior who requires worship and obedience, is not in fact a god but a conqueror who has given man belonging in exclusion from the whole by dividing man against woman and Jew against gentile. And further more this god had only given man purpose by making man's purpose serving the gods. And thus the accuser is vilified by god for tempting man to think critically and exercise their autonomy in discerning "good from evil by way of logic and knowledge".

This is why so much hinges on the claim that god is required for morality. And also why the claim that "man is inherently evil" is to pivotal to the need for a god and savior to save us from ourselves. However I would point out once again that the very behaviors often cited in defense of this claim, (behaviors like genocide, war, rape, and discriminatory prejudice,) are all witnessed in nature by apes, wolves, harem of lions, and even single celled organisms. And so tho law and order in the form of civilization ameliorated the environmental strains which promote such behaviors, it has also exasperated the same behaviors exponentially so. As we've already seen it was most probably those very same environmental pressures like predation which led to the development of more complex forms of life and ecosystems like society and civilization. But that doesn't condone the propagation of those behaviors. And as Karl friston articulates, the goal is more comprehensive and compatible models and not simply compliance to outdated paradigms. The goal of a society is to meet the full potential of individuals in conglomeration with one another. And the subjugation of the 99 in service of the one is not the way to that goal. Especially when major corporations deliberately undermine sustainable solutions as a means of ensuring they continue to profit from outdated designs that require constant revision and repairs. The whole point should be to make life better for everyone in the community and not just the 1%. And yet society has been the culprit of un due suffering as a means of incentivizing The perpetual exploitation of marginalized communities. If A.I. can execute remedial tasks better and more efficiently then let it take the burden off of the working class, but not at the expense of their livelihood! Let us see how far we can go when we aren't worried about fabricating scarcity in order to control and exploit the masses!

Thanks to a paper published in 2021 (on the dangers of stochastic parrots: can language models be too big? By

Emily M. Bender, Timnit Gebru, Angelina McMillan-Major, Shmargaret Shmitchell). We now know that the electricity required just to train language models employed by big tech companies 1 time require approximately 1,400 LB of $CO_2$ emotions. And even as advanced as this software is, it still requires many training sessions to adequately amalgamate that information into a useful form.

"They were placed in the garden much like agrarian civilization was forced upon the hunter gatherer tribes. Further more The "garden" was a paradise until man and woman learned to discern between good and bad and became aware of their own nakedness wile their god was clothed in lavish glory. It was The serpent pointed out the disparity caused by gods exploitation of man and woman."

We grew up being taught that we were made to worship a god. A god who (like the Mesopotamians) placed us in a garden and called it paradise (much like the agrarian lifestyle forced upon hunter gatherer tribes). A god who gave us only one command "that we not seek knowledge or discern right from wrong" but that we obey and comply to god's judgment and authority. That we remain ignorant of our own nakedness and poverty. And yet we wonder how we ended up in a world where the 99 live to serve the 1%. Look at our myths, look at our gods (who's archetypal role is as the architect of civilization). look at how the accuser who "tempted" us to seek knowledge and discernment, was demonized. How the archetypal role of the justice fighter who called the mighty god out on their malevolence was then vilified.

It's not a coincidence that faith based institutions like Christianity seam to gravitate toward, and even align with fascism. Fascism is the fear of freedom and equality. Thats the reason salvation is restringente on belief and not cognitive analysis. It's because the goal is compliance to authority, not actually doing what's right and personal growth. But once you learn to take responsibility for your own behaviors and finding healthier ways to cope, then you won't need a savior to die for your sins, because you'll be able to live with yourself. Which is one thing god and his supposed son never could do. When it's no longer about being right, we get to do whats right and actually pursue truth rather then ego. When people know better, they do better! But if you still need a savior (particularly one who requires worship and complete obedience) then you are not saved and are in fact still a slave!

The faith biased values of Christianity and many other Abrahamic traditions are authority and hierarchy oriented forms of morality "because I said so/might makes right" and not grounded in a sense of morality which actually seeks to alleviate suffering or procure objective truth. This is why these institutions oppose education reform and accessibility to higher learning to the general populace (or at the very least marginalized communities). The education system as we know it today (particularly years k -12) came about as a result of the Industrial Revolution and capitalists like oil tycoon John D Rockefeller (the founder of the general education board) who influenced the institution in order to provide a viable workforce by training the general populous to be compliant and punctual during their formative years (ages 5-18)

"I don't want a nation of thinkers, I want
a nation of workers" (John D Rockefeller).

"The aim of public education is not to spread
enlightenment at all; it is simply to reduce as many
individuals as possible to the same safe level, to breed and
train a standardized citizenry, to put down dissent and
originality. That is its aim in the United States, whatever
the pretension of politicians, pedagogues and other such
mountebank, and that is its aim everywhere els." (HL
Mencken)

That's why things like art and music are the first classes
cut, (despite the admirable triumphs of passionate teachers
and educators) the education system has never been about
utilizing a child's natural curiosity to unlock heir full
potential but to breed a complicit slave/work force. In fact
Studies have shown that grades do 3 things. They make
students disinterested in the actual material, they cultivate
medocraty out of fear of failure and a general proclivity to
conserve energy and resources like time and attention on
the "essentials", and they deter from more introspective
analysis and inquisitiveness for more superficial absorption
of the curriculum. The standardization encourages
competition and not cooperation or appreciation of diverse
approaches. It's not about solving the problems, it's about
glorifying the corporate gods.

And even tho our education system was constructed
to indoctrinate the populous, we can't blame the leaders
without addressing the real issue. Which quite simply
put is, people continue to choose the tyrant over the
liberator. Because Any shelter in a storm is preferred to the

uncertainty of freedom. and thus scarcity is fabricated in order to necessitate the provision provided by the abusive And exploitative tyrant. In short we prefer the toils we know to the possibility of suffering unknown. At least in their story we know our role. We are given faith in a god and savior like a nationality as our very identity and thus when we deconstruct those beliefs it means deconstructing our very perception of ourselves, the world, and our roles in it. We were told to live for Christ and that our purpose for existing was to serve god. And this is the crux of the issue, This is why deconstructing one's faith can be so hard, because In most cases so manny of these societal systems and institutions are fundamentally abusive And exploitative of the individual and there really isn't a good infrastructure for healthier alternatives. After all religion like science is constructed of observations of reality, much like a map is simply an interpretation of the terrain which as we've seen is quite treacherous by nature. And thus for so many slavery in Egypt is better than an exodus into a chaotic wilderness in the hopes of a land of promise. The idol gods, malevolent masters, and golden calf's by which we use to orient our lives and values, is preferable to an exercise of our autonomous authority in partnership with the ineffable. And thus we choose the saviors who sacrifice themselves for our sins, rather then taking responsibility for our actions and outgrowing that which we once called ineffable or god. We would rather worship (fear/revere) our gods, then face that which frightens us. And for some that's ok. But when our paradise is forged by their hell. Then the price of gods glory is too high.

"a noble heart will refuse the happiness
built on the misfortune of others"

(Saadi Shirazi)

## THE NATURE OF REALITY AND
## THE MYTH OF HUMANITY

In the Peterson and Haris debate (2018) the claim is made that these ancient biblical narratives hold with in them evidence for an evolutionary heuristic which serves a fundamental utility to the development of cultures and humanity in general.

In short what this claim means is that We need the accounts of slavery, genocide, rape, and child sacrifice to fully appreciate the conclusions of the biblical narrative. However tho child sacrifice was abolished by the claims of Abraham. and then substituted for by animal sacrifice, the conclusion of the narrative (with regards to the New Testament literature) is that human sacrifice was and still is necessary in the form of Christ. Now rather or not the Christ represents the sacrifice of the human, or the sacrifice of the god who requires it, is up for debate. None the less, the argument can be made that the conclusion regressed rather the progressed. From a bottom up (behavioral) standpoint, Sacrifice was a fundamental pillar in the origins of virtually all ancient civilizations. And it was most likely a behavior favored by our genes simply for the fact that it instilled in our progenitors a sense of postponing gratification. It was also very likely that hunter gatherer tribes would have no other choice but to sacrifice the weak or feeble children in order to survive food scarcities (much like a gazelle who doesn't have to out run the lion, so long as it out runs its counterparts "more feeble gazelle"). But returning to our original point. Does the outcome or conclusion of the biblical narrative make up for its undeniably abhorrent commands leading up to it? Most ideas are bad ones, but some are vital for survival. Just because a behavior, trait, or value, is favored by natural design, doesn't then mean that it should be, or that it is particularly beneficial. Nor does it intel that it is a good solution or answer. just because an idea or set of values has stood the test of time, doesn't then mean that it is necessarily good. Most ideas and possible solutions are bad ones. And thus it is imperative that we question and critique them. But at the same time just because a behavior like human sacrifice was favored by natural selection, that doesn't mean it is then an advantageous behavior or frame work to employ. Just because it is, doesn't mean that it should be. Just because hierarchies have been constructed in a manor which rewards the malevolent masters and tyrannical conquerors, at the expense of the modest and diligent, doesn't mean that those hierarchical structures should be permitted to persist. And the idea that they were ordained by god or biology, doesn't provide adequate merit for their claims as standards for morality. But what then is the notion of god in this capacity?

God as defined by Jordan Peterson in the Peterson Haris debate (2018). "God is how we imaginatively and collectively represent the existence and action of consciousness across time" Peterson elaborates on this claim by asserting that "we have conceptions of reality built into our biological and metaphysical structure that are a consequence of processes of evolution that occurred over unbearably vast expanses of time. And those processes structure our perceptions of reality in ways it wouldn't be structured if we (as individuals) had only lived for the amount of time that we are going to live (in a single lifetime)." Or in other words "We have a biological structure that's approximately 3.5 billion years old." Peterson goes on to present a couple more compelling claims of identity for god as "God is that which eternally dies and is reborn in the pursuit of higher being and truth." And "God is that which selects among men, in the eternal hierarchy of men" or more simply god is the quality which favors certain traits in hierarchy and reproduction of genes and their corresponding behaviors. That last one is particularly interesting because not only is it the very premise of many of this books arguments. But in many ways the Biblical god is the very hierarchy which not only condones, but deliberately facilitates practices of slavery and monarchy. In fact the very claim of godship (that the many exist to serve or worship the one or few) is the reason why our societies are structured in such a way that the few (1%) profit so egregiously for the toils and deliberate exploitation of the other 99% in varying degrees. However this definition of a god preserved thru the divinity of a bloodline or seed, only partially holds up to scrutiny. And is in fact critiqued by scriptures which undermine the legacy of certain kings and priests (such as Samuel, Saul, and David, just to name a few). Samuel was a good judge, but his sons were not. David was a good king according to the text, but he was a poor father and his sons made for lousy kings. we now know that many of these "divine qualities" or traits, aren't strictly inherited but developed and activated by conditions in the environment. Or where we were born is more consequential to our development then whom we were born to!

This brings to question the point that some traits work in certain environments (times and places) but not others. As well as the point that behaviors and traditional beliefs which contradict each other on a fundamental level, often benefit their corresponding cultures or organism. The traditional truths which underly eastern cultures, such as those of taoisim and Buddhism, explicitly contradict many of those traditions which enforce western values on a fundamental level. And yet for both cultures, that diversity works quite well. A good example of this may be the contrast in definitions of evil in these cultures. In the Buddhist and Taoism sense, evil is recognized not as an innate spiritual quality but rather as a state of ignorance. Which perhaps is far more accurate especially when we consider how blastoma tumors and physiological qualities in our neuro chemistry greatly effect behaviors which are often attributed to spirituality as evil.

It would appear that evil is a computational problem, and not one of some kind of innate spiritual quality or depravity. Or more simply put, good is simply a more complete and comprehensive understanding of the world as a whole. And how it is all connected. And evil on the other hand, is a faulty or flawed assessment of that connectivity Or ignorance itself. And so if evil is ignorance, then our greatest ally is knowledge. Or perhaps more accurately our ability to critique and cross reference the information we gather from the external world.

When testing the validity of our observed experience in regards to reality, we employ more then just one set of criteria. For instance we rely not only on our ability to see a phenomenon, but our ability to touch/feel, smell, taste, and/or hear, that very same phenomenon to test its validity. In the fiends of science we employ mechanisms like math to test or reinforce our observations and/or logical simulations. But in the case of a god who cannot be seen, heard, smelt, touched, tested with logics like math, or recorded in real time, the ability to substantiate their existence becomes virtually impossible. In the 2018 debate, a point Haris brings up and articulates quite well is the idea that well-being (or our awareness/consciousness of states like suffering and pleasure) are fundamental to reality, even in isolation. Where as topics like morality and ethics only seam to apply when we are in social circles or groups. And thus to a degree one of our best judges of reality, is our experience of the phenomenon. In short the reality upon which things like morality should be judged, is that of experience (particularly in the case of propagating suffering or harm). And so if a god cannot be seen or tested and yet is employed to propagate pain and suffering. Then it has proven to be a problem with no real substance to condone its perpetuation. Or more simply what good is a god if they do not do objective good in the lives of those who suffer. And more so, what good is a god who propagates suffering with no real substance to substantiate it.

However to play devels advocate, in part that suffering dose in fact play a vital role in the sense that it's resistance or inhibition of more beneficial states, actually functions as an incentivizer for the very same. In part it is the prohibition which makes such state so desirable. "Most new ideas are stupid and dangerous, but some of them are vital" (Jordan Peterson). And so in a sense the restrictive and even abusive functions of religious dogma is prewired in us. And at least in part plays a vital role in motivating us to evolve and adapt. Or more simply the rules give us something to rebel against. And that rebellion is vital for growth.

Perhaps it is the belief that people are endowed with some kind of mythical essence or quality which then permits the striping of such qualities to degrade the human. Because Without the romantic notions of spirit and divinity, perhaps we cease to be human. But It is not that we have endowed humans with some form of divine quality, but rather that that quality can then be revoked or denied in order to justify abuse, which is the travesty. to claim the authority to prescribe humiliation in the presence of divinity is the equivalent of being naked in the presence of gods. that may in fact be the tragedy. Because humanity is as much of a myth as the gods who endow it. Our humanity is a myth in a divine sense, but our personhood is not.

We are human beings with so much more depth then just our identification with a system (or ideology) which seeks to reduce us to submissive slaves. And tho those myths attribute meaning, they also prescribe abuse in the way that they infringe on our autonomous authority. Much of the myths we call humanity, is used to condone or even command compliance to a system of abusive hierarchies. A system were police beat and kill citizens and yet when the citizens protest, the national guard is called in. A system where members of Congress engage in insider trading and yet prosecute any citizen who does likewise. In fact even the feminist movement was the result of disgruntled white women who resented the idea that black men got to vote, and they didn't. And that right there hilights the core problem. This myth of humanity is one which has been used to make the ruling class powerful, by denying such "ethereal qualities" to minorities of other races, sexes, and ages,. The only reason certain people can be written off as property (slaves, women, children,…) is because of the hierarchy established in the very identity of gods. Perhaps what humanity has to do with the ineffable is more then just the myth of our identity in relation to its prescribed divinity. But rather that we far transcend those identities. Rather god is the product of our biology or the source of our sociology, they should not be allowed to deny us our autonomy thru its construction of hierarchy. Especially when those hierarchies compromise our sense of morality in the name of those very same gods (hierarchies)!

It is true that people are who they are programmed to be, but just like the autonomic functions which are prewired in us and then shaped/activated via stimuli In our environments. Volition is a muscle which must be exercised thru the deliberate reappraisal of those autonomic behaviors and biases. Granted we must fist learn how to exercise our autonomy. But once we have the necessary resources required to employ such mechanisms of cognition. Then "myths" like free will become maters of discipline and not divine intervention!

If the glorious lights of heaven are fueled by hell's flames, then the cost is too high. If the price for gods is paid for thru the sacrifice of slaves, then It can't be called grace.

Such saviors don't sacrifice themselves for our inequities, but instead for their own insecurities. Love appreciates who people are, and doesn't have to sacrifice itself for what they are not. The price of being in love is heartbreak, but true love has no cost. It's not about getting everything you ever wanted. It's making sure everyone has enough. It's about not profiting from the exploitation and abuse of others. In short people deserve respect and dignity, not because they are endowed with some divine quality which varies depending upon hierarchical station, but because they are people who's conscious experiences of physical and psychological pleasure and pain are no less valid or real then any others. Their hunger, exhaustion, humiliation, and heartbreaks, should not be undermined or invalidated by an unfeeling god who exploits them like cattle. To invalidate someone else's experience simply because it doesn't match or map onto ours is not a viable standard for morality and ethics. Our identities as humans mustn't be dependent upon our similarities In thought, speech, or appearance, as prescribed by our gods and genes. But instead must be intrinsic to the conscious experience of us all as sentient beings. Contrary to what our biology and Bibles say, people who look, think, and speak, differently, are still just as much autonomous people and human as us and ours. And thus access to environments and conditions which are conducive for homeostatic states should be accessible to all members of any given society (especially those who contribute the most literal substance and essential services). The working blue collar men and women should not be branded as a lower class then those who "govern" them. But as long as there are god's, there shall also be slaves! I'm not claiming that a person's value should be dependent upon their utility, but simply that those who contribute to the vital physical well-being of society shouldn't be penalized by being deprived prestige as a means of incentivizing their compliance to the exploitative systems by the idol gods.

However to truly understand This exploitative quality, we must first understand the functions it utilizes to make the myth a reality.

Like a new born infant who cant distinguishing between their experiences, and their environment. We to mistake feeling sad, for being sad.

## THE HEAD VS THE HEART
### (THEM IS FIGHTING THOUGHTS)

Emotion is not an exclusively human mechanism but instead a modified variation of the same exact mechanisms which help animals survive. Anger is a form of fear and fear is the result of ignorance.

This is why politicians like Hitler and Ron DeSantis employ policies such as banning books and prohibition of controversial reading material. Because much like our emotions and the biblical god who derived from them, the only way they can retain power and control is thru ignorance and mitigating reasoning mechanisms which respond with curiosity and compassion rather then react thru fear and anger.

Like the biblical god who condemned man for obtaining knowledge, those who seek to control must first ensure that ignorance be propagated in order to retain compliance. For such gods, things like truth, knowledge, and morality, are sins. because they threaten the omnipotent authority of the gods who seek to control, exploit, and abuse. God discerns what is good because god sees what man cannot. However in regards to the biblical text, the modern reader can see how flawed and faulty "gods discernment" really is. And in many ways gods judgment is no better than that of man. Christ (the snake crusher) begot Christianity (who became the serpent which crushes, crusades, and conquers, the marginalized). God despises the (slanderer, accuser, ha satan, serpent,…) because they call god out on their malevolence and abuse. The biggest problem with this god who can do no wrong, is how they must vilify those who point out the evils perpetrated by (or in the name and authority) of god. As long as people are trained to react or behave rather then taught to think and reason, they will be easy to control like animals (slaves to their emotions). Like Prometheus, the bni-Elohim, or the serpent in the guarded, who seek to free man thru access to knowledge and subvert the hierarchy of the gods, those who seek to liberate and appreciate the qualities that have been exploited and abused, will be labeled enemies. (Now granted in the genesis myth, the serpent is analogous with animalistic impulses and emotion, and the bni-Elohim were the sons of god who sought to usurp the throne thru violence "symbolic of the gods of Babylon, Egypt, and the competing cainanite tribes, of which WHYH, El, adony,... was one of.) but the god who seeks to keep man ignorant, And vilifies any who question or critique them (god), is no better then those who seek to free man from their complicit bondage to gods hierarchy. Rather it is subordinance to Christ and Christianity, or compliance to the third reich. Rather it's threat of hell or Auschwitz, might enforced thru fear and ignorance, is not right. And rather it's the gnostic texts or communist/anti-facist reading material, burning books and baring access to knowledge is not a good look. Ultimately if god were truly ether competent or compassionate, they would have taught their chosen people how to think and reason, rather then simply training them how to react via "the rod". It's clear that the goal was never liberty or moral justice but instead control and complicity to abuse and exploitation. But a god who makes people sick all so that they can heal them and then be glorified for doing so, actually makes a lot of sense in our modern capitalistic economic system.

Our modern form of Capitalism is a system where people are replaced by machines which were meant to make their lives easier not harder. And that's why it has cultivated a civilization which makes life harder for its citizens, rather then easier. Instead of producing more goods with half the toil, half the manpower is discarded as an incentivizer for the remainder. The toil was never about production but instead about exercising one's authority over another. It's about prestige and hierarchy, not ethic and character. In our modern American penal system, corporal punishment was never about reform, but simply a mechanism (much like hell) used to incentivize compliance to a hierarchy.

In many ways our current "capitalistic" economic system, is more deliberately exploitative of the working class then serfdom under feudalism. In fact what we see today in the disparity between class and the political power exercised by the elites is almost indistinguishable from the criteria desegregating slavery in Western Europe in 1400-1800 A.D. serfs on the other hand were granted approximately 40 - 70 percent of the profits they produced for the lords. The gods are not those who actually contribute to society, but instead those who conquer and subdue those who produce. "God fights our battles on our behalf", and yet we are still the ones with scratches and scars, we are the ones who toil and sacrifice all, so that gods may receive glory. In fact the closer you are to the actual production of vital goods and services, the low you find yourself in the hierarchy. Those who contribute the most substance to society are subsequently those under appreciated and even discarded in our current economic system. Wile those who do the least prosper the most. Rather it's wealth or prestige (which are the same thing as we've seen) it was never about ethic or virtue. Nor is it about positive traits worthy of admiration and emulation. It was about evoking a reaction to profit from.

It's no coincidence that the same voice which spoke

on behalf of god now speaks in opposition to them. Nor is it a coincidence that the same tactics employed by god and Christ are now exemplified in those cult leader and dictators exposed as charlatans. The morals prescribed by religion may be better then those of ravenous lions or zombie wasps. But truthfully such "morals" can't truly pass as ethical in any sense of the word. If you are like me then you probably feel sad when you see videos of people on their knees sobbing, or flailing around with their hands lifted high. Seeing how people "moved by the spirit" are complicit to the very system that continues to abuse them. We feel sorrow, because we were once them, and we truly believed those teaching were good. We didn't know any better at the time. And we did the best we knew how to do. And that's alright. Because we know better now. And we can do better now.

But As long as we believe that people are the problem and not the faulty economic and religious systems that were meant to serve the people. Then we will remain complicit to those systems. The idea that we can't trust our own cognitive mechanisms or that we are inherently evil apart from a religious dogma to deliver or redeem us. As long as we "obey" the omnipotent voice of god, Then those systems will be allowed to exploit and abuse the people they were supposed to serve. It is the vilification of any one who accuses the abusive system, in conjecture with the belief that we can't trust ourselves to question the abusive system, which then permits it to persist. And so we must understand the "problem" which such systems claim makes people untrustworthy. As well as the very mechanism those systems utilize in controlling people.

And interestingly enough those mechanisms are one in the same.

Much like a barking dog, our emotions don't so much speak words but simply make noise in the pursuit of emoting us. And as effective as these emotions are at keeping us alive they are very poor at logic. because their goal is not accuracy but rather simply effectiveness. Rather it's a stick or a snake, fight or flight works. And so when we employ training techniques which physically or psychologically abuse (utilize emotion to train behaviors/reactions) we don't actually learn how to reason but simply which mechanism (fight or flight) to activate. It doesn't tech morality, it trains compliance or violence. And thats how we end up like the nazis doing what we are told in order to survive. That's what politicians like Ron DeSantis is trying to do with bill 1467 which prohibits reading material that conflicts with certain political and religious views (particularly those which oppose Christianity and right wing republican propaganda).

This bill would prescribe that teachers found providing access to such reading material, would then be charged with a $3^{rd}$ degree felony. This is a tactic practically identical to those utilized by Hitler to scare people into submission. Because fear and anger are results of ignorance. And as long as people are subject to the whims of their emotions, then they can be easily controlled. Access to knowledge is vital for the reappraisal of our VMPFC's somatic models (empathy and belief) which it uses to govern and guide the DMPFC's (volition center) implementation of its executive function. In short these cognitive functions are not only required for any true form of moral good, but their alternative as prescribed by gods and fascist alike overwhelmingly result in the epitomes of evil. Rather it is the lefts cancel culture, or the rights white washed history. Obscuring the faults of our past out of shame or guilt, hinders and even inhibits our propensity to grow. We need the old faulty models in order to construct better ones. In order to take the chances required for growth, we must appreciate those who came before, without hating them for their faults. In order to move forward, we must remember where we've been, and honor each step that got us here. Perhaps Peterson was right. Perhaps it took us being slaves in Egypt to understand the evils of our abuse unto the Egyptian slaves. Perhaps it took becoming atheists to see the tyranny of theism. It took wondering the wilderness as a poor sojourner to understand the price of security and a satisfied stomach. Perhaps we had to be hungry to understand what it cost to eat. We had to be exiles to know the tyranny permitted by community. We had to first understand our own behaviors and pain in order to respect the suffering of others. We must remember the stories of the victims and the villains less we join their ranks.

Perhaps the point is that not even god should be keeping score but instead we should encourage each other to learn. To refrain from causing unnecessary harm and suffering because we understand that pain and toil. Not become a god said so, but because we respect the experience of a human soul. I think we need the Bible, for the very same reason agnostics read it more then Christians do. We need the Bible to remind us what not to do. And much like the history books on nazi Germany, scripture shows us what happens when we blindly follow the will of a malevolent god. Let us not forget how American leaders used the "curse of ham" to justify racism and the construction of private schools to enforce segregation. Because evil is the product of ignorance. So let us not become complicit in our ignorance. In order to grow we need the freedom to make mistakes and then admit that they don't work.

"Jesus said to them, "A prophet is not without honor except in his own town, among his relatives and in his own home." When we are reminded from wince we came, we no longer claim to be the sons of gods.

In regards to both scripture and our current economic system. They are not inherently bad. In fact there was a time when capitalism could have eradicated world hunger and disease. But it didn't. And that's the saddest part. That it could have done tremendous good, and yet instead was employed to pollute the environment and cause problems which it could then sell the solutions to for a profit. In much the same way, scripture has a lot of good policies like a prohibition of charging interest and the condemnation of abusing foreigners and the poor. And yet instead it has been employed by religious installations to condone the genocide, exploitation, and abuse of minorities for the church's profit. Instead of solving the problems, scripture was used to intensify the problem, so as to necessitate the churches intervention. We build thousands of mansions for a god who doesn't need a house, wile millions of people are turned out of their homes and into the cold. The

People aren't the problem, people are problem solving machines. People are the solution to a system which has profited off of peoples suffering. We don't need to fix people. We need to fix the way they are programmed, we fix the policies. People are as good as they know how to be. It's the idol gods we are taught to emulate that are the problems.

It's worth noting that there are almost as many churches in America as there are homeless people. And so if each and every church sponsored only two homeless individuals each, then there would be no one left living on the streets.

In short just because a mechanism or behavior evolved, doesn't then mean it is good, or beneficial for the organism or even an accurate judge of reality. However it should be noted that Dr Donald Hoffman (a doctor of computer science) has simulated with AI prediction models, that the closer we get to perceiving reality as it really is. The less affective we are at surviving. And so in part we need the myth to give life meaning. We need the helpful shortcuts provided by emotion and instinct in order to conserve energy. And perhaps to some degree we even need the hierarchy to condone our consumption of vital resources. If we were truly virtuous beings, we would go extinct in the cruel world "god, evolution, or nature,… made. And so perhaps we need faith in a myth in order to exist.

As insufferable as it may be at times, math is the standard by which we judge logic. And tho one would not conjecture from the present state of academia and technology in the Middle East, during its golden age (8th -14th century) the Arabic world was at the forefront of civilization, social freedoms, and technological development like Algibeira and algorithms. Much like the Greek philosophers before them, the decline of this golden age and its civil liberties was the result of religious nationalism. And tho those tools like math are fare clearer depictions of the nature of reality they make for rather poor direct interfaces. In many ways evolution and culture favor more illusory projections like religion and instinct. In part we need the myths to make life bearable. For instance take crippling depression, which is often a completely rational state of despair caused be the realization of life's futility particularly in environments and conditions where any attempt to alleviate or remedy those circumstances is met with contempt or disappointment.

One of the most consequential factors In an persons resilience and persistence, is their ability to hope. To truly believe that they can make a difference and that things will change. But hope requires two important things. It requires us to see a possible path forward, and for us to believe that our expense of energy and effort will pay off.

"For a set of traditions which worship and attest to a supernatural being, they sure do inspire hatred and fear for anything and everything that transcends their understanding.

For a set of traditions which attest to liberty, they sure do despite anyone who finds liberation from their dogmatic prisons!"

In closing, the difference is that Science bridges gaps of culture, gender, and ethnicity, by tearing down the divisive and illusory wall religion created. In it search for truth, science unites all peoples, whereas religion only unites people by dividing them from the whole or other. Religion creates opposition in order to unite us against them, science confronts opposition in its pursuit for truth which as we've seen is a more unified and wholistic understanding of the world, life, and the universe in all its dialectic diversity.

# CONCLUSION FAITH
# OR FICTION?

Religions like Christianity are most certainly myths, but they are myths which make life bearable And perhaps even give some a cause bigger than themselves to live for.

When one is in love, it becomes the epitome of reality and one could hardly imagine anything realer. It is only when such elate heights are matched by their depths of despair, that one then questions love's nature. When it must persist only by way of grief and one is required to transcend that which was once considered itself transcendent. Similarly when one is forced to grieve the loss of such love (or loved one), realities such as logic, are illusory myths who's fact about the phenomenon, will never close to the reality of such a profound experience.

Now some may accuse me of circular reasoning. However To a degree Circular reasoning is virtually unavoidable. for instance "is the world beautiful because god made it that way" or "is our definition of beauty a result of the world/nature making us value fertile environments and qualities." The answer often Depends

587

on the maps and models one applies and uses to filter through. Like belief in romance or love, god, is a very real finite phenomena we observe in our universe. however beyond the reaction of chemicals there is no real cosmic authority beyond that which we ourselves attribute to it. Religion values the experience, whereas science seeks to understand the phenomenon which cause that experience.

The Aztecs believed that the sun required bloodshed to bring it thru the underworld every day and thus for them the sun rose because of their sacrifices and wars.

The world orbits the sun. The Aztecs bloodshed permitted those who survived to see the sun rise the next day. The fact that they continued to kill and the sun continued to rise, validated their beliefs.

My dear reader, I thank you for your time, open mind, and willingness to wrestle with these questions. As I stated at the beginning of this literary work and in the one preceding it, I speak with no more authority than any other human on this planet (past or present). And so I have presented you "reader" with evidence for claims I have made. However I have made claims that things mean certain things. But as you should already know by now, this is not quite true. Everything requires an interpretation. And so what ever claims I have made in this book are subjective to a certain degree. This is only what it may mean, and even what you derive from my writings may be in complete contrast to my own intent. That my dear reader is part of the point. We all see things differently. If the perspectives I have presented you with are helpful in making since of that which we once called god, that is great. If they are not however, then by all means feel free to discard them. For after all "all models are inaccurate, however some are helpful" but ultimately in the words of Solomon "everything is meaningless" (hevel) no more then vapor in the wind. It is our persistence in attributing meaning to things which then make them meaningful. But as is the case with all acts of new creation, chaos must precede. And so with no further to do allow me to conclude, for It is not truth, but the pursuit of truth which is admirable. Not the obtainment of knowledge but the curious mind whence beauty presides. It is the meaning we attribute to life which then gives life value. For not even god resides in the physical heavens but instead in the philosophical hearts of man. It is by faith alone which beauty, and love persist and in which life retains its meaning. It is Our stubbornly defiant persistence in attributing some form of cosmic sentiment which far exceeds its corresponding biological prerogative, which makes it thus (meaningful). For the sensations we call love and beauty are more then neuro chemical reactions, simply because we insist on believing in a bond which transcends familial relation and visual markers signifying environmental fertility. It is our defiant insistence that nature be nurturing which then makes it thus, for the role of humanity is more then to simply supplant meaning but to contend with the ineffable on behalf of more compassionately inclusive representations of the divine then even nature would intel.

This is not a alien concept in philosophy. This faith, not in something, but for something. Like the mutually agreed upon belief in saint Nicholes on behalf of both parent and child, each knowing better than to truly believe in such absurdity, yet both believing for the sake of the other, less the child should loose the magic of Christmas and the parents be thus deprived of such magic glistening in their child's wide eyes. And so tho there is no truly inherent meaning in life. It is the meaning attributed to life which then gives it value. It is our insistence in believing, not in, but for something.

Thus faith in a form of god is not bad and even very beneficial in some cases. Because after all there is still that which we once called god. And so just because we know these integral phenomena by different names now, that does not invalidate their connection to something deeper or undefinable, there is still that which is ineffable! And after we have stripped that (entity) of our harmful anthropomorphized egos and ideological prejudices. That somewhat personable conception of the cosmos and nature can be quite helpful in inspiring hope and boosting our moral as a species.

Ultimately whatever this phenomena we once called god truly is, it has been thus since at least the dawn of life on this planet. Or for over 3 billion years. And has guided life as it evolved both biologically and philosophically every step of the way.

No one refutes the phenomena, the disagreement is one of nomenclature and terminology. That which religion calls god, science calls the gene or nature. even religions call god by different names and in accordance to different lineages/ cultural heritage. The problem with this difference in nomenclature isn't the terminology or vernacular itself but the authority attributed to that phenomena called by some instinct and nature and others god, spirit, and divine governance. But more prominently it is that egoic omnipotence and narcissistic hierarchy of such authority which is often attributed by and to god, which then permits evil to persist. The point is nether the intuitive nor the rational, the poet nor mathematician, religion nor science should be permitted to act superior to the other. Ultimately it comes down to what we say the goal of this phenomena is and not so much what we call it per say.

I can not definitively tell my reader that a god does or does not exist but I can affirm that many of the examples we have discussed in this book most certainly make for very bad gods. and most certainly not admirable qualities of the ineffable.

The role of humanity is to contend with the ineffable on behalf of more compassionate and complete (inclusive) representations of god for nature to be more nurturing and loving.

On average Atheists are just as morally conscious as theists.

It can't be over looked however that On average religious observers tend to be happier then those without a belief system. But you can say the same thing about individuals under the influence of hallucinogenic substances. So long as it doesn't hurt others, perhaps the meaning, purpose, and bliss, is wroth it's illusory nature. Much of what drives us is an inability to obtain vital needs in our developmental stages. But when we learn to meet our needs, that lack of love, connection, dignity…. The trauma that drove us gone we also lose the drive to survive in a system that does not serve us.

Beyond theist and atheist, black or white, man or woman, straight or gay, we are people. And no group of people are inherently better or worse then any other. Atheists are not innately superior. Atheists may be better only because they know better and choose to do better. The problem is the belief that people are fundamentally superior, inferrer, or flawed which then expresses itself in bigotry, misogyny, and racism.

## Poem

There is a kind of beauty to chaos, a world undivided by identity. Where Order sought to segregate, chaos desolates its walls. chaos desecrates the holy by abolishing faith in hierarchy. There is a kind of beauty to chaos, for chaos recognizes that all belongs. Chaos is honest, it does not require us to betray ourselves in our allegiance to someone else. For chaos has no identity, chaos is what we call it all. And so if ever the opportunity arises to be nether us nor them, always incite chaos for all. For chaos will not survive, but in its death it is the way of life.

Rather it is the accumulation of wealth and hierarchy, the self mutation and objectification of the narcissist, or the intolerance of the traditional zealot who can't relate to an ever changing social model. Often what isolates people is their attempt at connecting with a community. We cut ourselves off from the whole by trying to find belonging in exclusion from the whole. Like a pet who takes on the mannerisms of their owner, people reflect their environments.

In the end it is not a god or a gene which decides your identity. It is you who chose who you are and what that means. You decide your own value and virtue. Good does not require a god, nor does being require divinity to ascribe its worth!

Chaos only threatens to destroy those boundaries and beliefs in hierarchy which have divided the whole against itself. And thus if ever the opportunity arises, always incite chaos!

The greatest defense against evil is to love and appreciate that which is concealed in the darkness before it is neglected and becomes a monster.

It's not dishonoring the dead to tear down the cities they built, it's honoring the living to make those cities livable for all.

In much the same way we read ourselves into stories, subsequently changing the stories meaning from one reading to the next (because we are living and changing, not because the text is alive in any real way other then how we relate to it), we too see reflections of our self in our environments (however environments, like people, are constantly changing and evolving).

To conclude I personally do not refute the existence of god or the profundity of the "very natural" phenomenon once ascribed to god. However like many (biblical and extra biblical) sources before me, I refuse to be complicit or condone the evils of such a god in accordance with "their supposed" biblical claims and authority. Might does not make right. And until such a god shows themselves to all of mankind and In all times. The most accurate standard for their character is represented in the behaviors and beliefs of those who supposedly bare their name and image. And if that name is Christianity or Christ then I have found the character of god (by extension) gravely lacking.

When one is in love it becomes the epitome of reality, faith works in much the same way. it is only when one must transcend the very grief which was once transcendent beauty itself, that we are then left to wonder if such illusions are real or fake.

And that's ok, there are truths only the poets may know. Oscar wilde can utter truth that Milton and Kierkegaard can not, and rabbi heschel can profess revelations not even neiztchie could see. There is justice the spinster can do which the lovestruck romantic cannot. There are things only known to the theist, and sides of god only the atheist may see. When one is in love it becomes the epitome of reality and yet the heartbroken know truths the romantic could not fathom. When we are required to

Printed in the United States
by Baker & Taylor Publisher Services